Monographs in Mathematics
Vol. 86

Masao Nagasawa

Schrödinger Equations and Diffusion Theory

Springer Basel AG

Author
Masao Nagasawa
Institut für Angewandte Mathematik
Universität Zürich
Rämistrasse 74
CH-8001 Zürich

Library of Congress Cataloging-in-Publication Data
Nagasawa, Masao, 1933 Aug. 1 –
Schrödinger equations and diffusion theory / Masao Nagasawa.
  p.   cm. – (Monographs in mathematics ; vol. 86)
Includes bibliographical references and index.
ISBN 978-3-0348-9684-9   ISBN 978-3-0348-8568-3 (eBook)
DOI 10.1007/978-3-0348-8568-3
1. Diffusion processes. 2. Schrödinger equation. I. Title.
II. Series: Monographs in mathematics ; v. 86.
QA274.75.N34 1993
519.2'33—dc20

Deutsche Bibliothek Cataloging-in-Publication Data
Nagasawa, Masao:
Schrödinger equations and diffusion theory / Masao Nagasawa. – Basel ; Boston ; Berlin :
Birkhäuser, 1993
(Monographs in mathematics ; Vol. 86)
ISBN 978-3-0348-9684-9

NE: GT

© 1993 Springer Basel AG
Originally published by Birkhäuser Verlag Basel in 1993
Softcover reprint of the hardcover 1st edition 1993
ISBN 978-3-0348-9684-9

9 8 7 6 5 4 3 2 1

Contents

Chapter III
Duality and Time Reversal of Diffusion Processes

Chapter IV
Equivalence of Diffusion and Schrödinger Equations

Chapter V
Variational Principle

Chapter VI
Diffusion Processes in q-Representation

Chapter VII
Segregation of a Population

Chapter VIII
The Schrödinger Equation can
be a Boltzmann Equation

Chapter IX
Applications of the Statistical Model for Schrödinger Equations

Chapter X
Relative Entropy and Csiszar's Projection

Chapter XI
Large Deviations

Chapter XII
Non-Linearity Induced by the Branching Property

Appendix:

Preface

Since Max Born's "statistical interpretation of wave functions" proposed in 1926, there has been the long-standing problem of "interpretation of Schrödinger equations". However, the emphasis should not be on "interpretation": The right problem to set is

"What is the Schrödinger equation?"

This monograph is devoted to an attempt to give an answer to the problem in terms of diffusion theory, namely, in terms of diffusion processes, and it will be shown that the Schrödinger equation and diffusion equations in duality are equivalent. As a result, Schrödinger's conjecture of 1931 will be solved.

The equivalence of Schrödinger and diffusion equations implies that, roughly speaking, if we add (Brownian) noise and specific additional drift to a "classical particle" then we get the movement of a "quantum particle". It must be emphasized that the quantum particle (= diffusion particle) has its well-defined position but no velocity, and hence it is not at all a "classical particle" any more. Therefore, we must look at it carefully in the context of diffusion theory. Moreover, the theory of diffusion processes for the Schrödinger equation will tell us that we must go further into the theory of systems of (infinitely) many interacting quantum (diffusion) particles.

The contents of the monograph are based on my lectures on diffusion theory (mathematics, not physics) which have been given at various places including Tokyo Institute of Technology, Aarhus University, University of Erlangen, Keio University, University of California at San Diego, and the main part of them at the University of Zürich in the last decade.

This monograph may be regarded as *"An Introduction to the Theory of Diffusion Processes with Applications"*. There will be no difficulty in reading Chapters 2, 3 and 4 for those who have an elementary knowledge of PDE and diffusion processes, and some fundamentals of functional analysis including measure theory. Those who would like to learn in a hurry the equivalence of Schrödinger and diffusion equations and its implications can read Chapter 4 directly, possibly referring to the necessary pages of Chapters 2 and 3. For Chapters 5 and 6, readers are assumed to have slightly more advanced experience in diffusion processes and perhaps some patience, since they contain delicate analysis in connection with the singularity of coefficients of diffusion equations. For Chapters 7 and 8, I have to assume that readers have some elementary knowledge of and intuition for statistical mechanics and more maturity in mathematical experience. Some applications in biology and physics will be given in Chapter 9. Chapters 10 and 11 offer a self-contained exposition on relative entropy and large deviations, needed in Chapters 5 and 8. Non-linearity induced by the branching property will be briefly explained in Chapter 12.

The text is practically self-contained and proofs are given for all theorems except for some proofs in Chapter 7 which are left in the original articles so that the monograph is kept to a reasonable size.

Acknowledgements. I am indebted to my old friends Hiroshi Tanaka, with whom I enjoyed collaboration and discussions on the subject during the last decade, and Hans Föllmer, with whom I jointly organized the "Zürich Seminar on Stochastic Processes" during the decade 1970-1980's. Their important contributions to the subject of the monograph will be seen in many chapters. It is my pleasure to thank those who offered useful comments on the first drafts, in particular, Herbert Amann, Andrew Barbour, Erwin Bolthausen, Nobuyuki Ikeda, Paul-André Meyer, Kohei Uchiyama, Marc Yor, and the audiences at my lectures. Moreover, I would like to thank Robert Aebi who read the drafts at various stages and suggested necessary corrections and improvements. Finally, many thanks to Eiko Nagasawa with whom in the course of discussion I have learned a great deal, which has enabled me to make the exposition clearer; in particular, I have followed her suggestion that it was absolutely necessary to write Section 4.8 on the superposition principle.

Zürich, January 1993   Masao Nagasawa[1]

---

[1] Supported by the Swiss National Foundation (21-29833.90) and European Community (Science Plan, Project "Evolutionary Systems")

Chapter I
Introduction and Motivation

1.1. Quantization

Quantum mechanics was formulated as an abstract algebraic theory (of operator algebras) by Born-Heisenberg-Jordan in 1925-26 based on a successful application to the spectrum of hydrogen atoms.[1] In this theory the so-called "quantization"[2] (commutation relation)

$$(1.1) \qquad pq - qp = \frac{1}{i}\frac{h}{2\pi}I$$

was postulated as a sort of working hypothesis, where h is Planck's constant. This algebraic theory (matrix mechanics) was an astonishing deviation from classical mechanics, which had not given any reasonable way of computing the spectrum of hydrogen atoms. Moreover, if one applies the algebraic theory to a harmonic oscillator, then simple algebraic operations yield immediately, like magic, the energy spectrum of the harmonic oscillator, as one can see in Dirac's textbook.[3] However, it was difficult to extend the applications of the algebraic theory to more complicated systems until the Schrödinger equation appeared in the next year, which enabled further development of the theory. Nonetheless, people were puzzled by the algebraic theory, because it did not give any clear explanation of the fundamental questions: What is a hydrogen atom? If the theory describes the

---

[1] Heisenberg (1925), Born-Jordan (1925), Born-Heisenberg-Jordan (1926). Cf., also e.g. Amano (1943, 48), Van der Waerden (Ed.) (1967)

[2] "Quantization" is a historical naming. Precisely it means "commutation relation" including Dirac's (1958) sophisticated version in terms of Poisson bracket

[3] Cf. Dirac (1958)

movement of an electron in a hydrogen atom, how is it moving? Namely, the operator theory provides the spectrum but no physical picture of either a hydrogen atom or a harmonic oscillator. Should one give up getting it?

Since it was not possible to give any reasonable physical meaning to "quantization" nor to offer a clear physical picture of hydrogen atoms, a common agreement was: "success justifies everything; no further explanation is necessary".[4] However, I must say, if Max Planck would have admitted such a philosophy, he could not have discovered Planck's constant in 1900.[5]

Besides successful applications at the atomic level, "quantization" (commutation relation), together with the Schrödinger equation, was applied to field theory, say, quantum electrodynamics. Another remarkable and successful application of "quantization" in field theory was achieved by Yukawa in 1935. In that year he created a new field of physics through his publication *"On the interaction of elementary particles"*[6] in which he predicted the existence of the π-meson.

Nevertheless, "quantization" has been a constant source of mysticism in quantum mechanics. Is quantization really necessary *as a hypothesis* for quantum mechanics? Is it not just a useful and powerful technical tool? This point will be discussed in Chapter 4.

1.2. Schrödinger Equation

In 1926 Schrödinger invented what he called the *wave equation*

$$(1.2) \qquad i\frac{\partial \psi}{\partial t} + \frac{1}{2}\Delta \psi + i\, b(t,x)\cdot\nabla \psi - V(t,x)\psi = 0,$$

which is nowadays called the Schrödinger equation.[7] It was his great triumph and a shock wave to people working on the subject. Soon afterwards he gave a proof showing that his equation can be formulated in terms of Heisenberg-Born-Jordan's commutation relation.[8] It is clear, however, that this does not explain what the Schrödinger equation is, nor what solutions of the equation mean. Schrödinger considered not only eigenvalues but also wave functions to have physical meaning, and in the case

---

[4] Bohr's "interpretation" with "complementarity" is included in this category

[5] Cf. Planck (1900, 01) and also Einstein (1905, a)

[6] Cf. Yukawa (1935)

[7] Cf. Schrödinger (1926, I, IV)

[8] Cf. Schrödinger (1926)

of hydrogen atoms he thought that wave functions give "distributions of charge".[9] Moreover, he tried to recover a "particle picture" of a free electron from his theory in terms of a wave packet. But this picture had its difficulty, since a wave packet could not keep its shape during time evolution.

Physicists accepted very quickly Born's *statistical interpretation* of wave functions, that

$$(1.3) \qquad | \psi(t,x) |^2 = \psi(t,x)\overline{\psi}(t,x)$$

gives the probability density that something (an electron) is found at (t, x), since Born's interpretation explained experimental facts well, especially in scattering theory (cf. Born (1926)). In formula (1.3), $\psi(t,x)$ denotes a solution of the Schrödinger equation (1.2) and $\overline{\psi}(t,x)$ is its complex conjugate, which satisfies

$$(1.4) \qquad - i \frac{\partial \overline{\psi}}{\partial t} + \frac{1}{2} \Delta \overline{\psi} - i\, b(t,x) \cdot \nabla \overline{\psi} - V(t,x)\overline{\psi} = 0.[10]$$

Schrödinger himself kept his own idea of wave functions, and then tried to solve this contradictory situation, considering the interrelation between quantum mechanics and Brownian motions, as will be explained.

In contrast to its successful applications in various fields in physics and chemistry, quantum mechanics with Born's statistical interpretation did not give satisfactory answers to fundamental questions such as: What is an electron? "Wave" or "particle"? Should one think of it as something being "wave *and* particle" at the same time? Practically speaking people were forced to give up having concrete physical pictures and were told: "Don't ask such questions. *Pictures* are not important in physics. This is your new experience with quantum mechanics. Because quantum mechanics is a revolutionary theory, you must accept it as it is".[11]

If quantum mechanics is really such, one cannot stop having suspicions that something important is missing in the theory, or that the theory is formulated in terms of the wrong language. In fact, it is well known that M. Planck and A. Einstein[12] kept their distance from the philosophy or attitude of the school of Copenhagen toward quantum mechanics.

---

[9] Cf. Jammer (1974), Doring (1987) for various attempt of interpreting the Schrödinger equation

[10] For an interpretation by Eddington (1932) of this conjugate equation see Section 8.4

[11] For such a comment cf., e.g. Dirac (1958)

[12] Cf. e.g. Einstein-Podolsky-Rosen (1935)

1.3. Quantum Mechanics and Diffusion Processes

The Schrödinger equation is nothing else but a "wave equation" and hence there can be logically *no particle notion* with it. Schrödinger proposed the equation as such and kept himself consistent in this respect. This fact must be emphasized.

The well-known inconsistency with the wave-notion was brought (or created) by Born with his "statistical interpretation" of wave functions, through which he *artificially attached a particle-notion* to wave functions, namely, a flow

$$(1.5) \qquad \mu_t(x) = \overline{\psi_t}(x)\,\psi_t(x), \quad (t,x) \in [a,b] \times \mathbf{R}^d \,,$$

where $-\infty < a < b < \infty$, was defined and *interpreted* as a probability density that a "particle" (say, an electron in a hydrogen atom) is found at $(t,x) \in [a,b] \times \mathbf{R}^d$.

If one accepts Born's "particle" notion attached to solutions of Schrödinger equations, one should establish a "particle theory" which justifies his claim. But this was not done, because this point was not clearly recognized by most physicists. People regarded Born-Heisenberg-Jordan's theory as a "particle-theory", since its formulation was based on classical Hamiltonians of a particle(s). However this algebraic theory contains no built-in structure to conclude that $\overline{\psi_t}\psi_t$ is a probability density of a particle. Namely, both Heisenberg-Born-Jordan's and Schrödinger's theories cannot justify nor conclude Born's statistical interpretation. It is thereby quite understandable that the conflict between Born-Heisenberg-Jordan's "particle-theory", Schrödinger's "wave-theory", and Born's "statistical-interpretation" has remained unsolved. Consequently, Born's idea has just always been an *interpretation*, not integrated into quantum mechanics as a firm (mathematical) structure, and a source of conceptual confusion on quantum mechanics.

It was clear that a "particle" which should be attached to a solution of the Schrödinger equation could not be a Newtonian particle for which well-defined position and velocity can be assigned at the same time. Hence, one must find, as an alternative, something not Newtonian. Non-Newtonian "particles" known at the time were Brownian motions which have position but *no velocity*. Therefore, it was a promising attempt to take Brownian motions as candidates and to investigate their relations to quantum mechanics in order to recover a reasonable "particle picture" for it. Schrödinger

(1931) considered this possibility seriously and formulated Brownian motions in a symmetric form with respect to time reversal, because Schrödinger equations are time reversible; the formula

$$\mu_t(x) = \overline{\psi_t}(x)\psi_t(x) = \hat{\phi}_t(x)\phi_t(x)$$

was established, where $\hat{\phi}_t(x)$ and $\phi_t(x)$ are real-valued functions determined in terms of time reversal of a Brownian motion. However, he could not deduce the Schrödinger equation from his Brownian (particle) theory, and closed his article with no clear perspective.

For a flow $\mu_t(x)$ of probability densities given in (1.5), it is always possible to construct a (singular)[13] space-time diffusion process $\{(t, X_t), Q\}$. This will be shown in Chapters 5 and 6.

To establish a "particle theory" for quantum mechanics in terms of diffusion processes, we must show that it recovers the Schrödinger equation.[14] This will be done in Chapter 4, in which we will show that the Schrödinger equation is equivalent to a pair of diffusion equations in duality.

Therefore, Born's "statistical interpretation" on $\psi\overline{\psi}$ is *no longer an interpretation*, but it will turn out to be a *formula* $\psi_t\overline{\psi}_t = Q \circ X_t^{-1}$ in our particle (diffusion) theory based on the equivalence between Schrödinger and diffusion equations (cf. Chapters 3 and 4).

In connection with the relation between quantum mechanics and probability theory, a deeply rooted problem appears in the following simple question related to the superposition of states in quantum mechanics. Dirac (1930) writes:

"*the superposition that occurs in quantum mechanics is of an essentially different nature from that occurring in the classical theory*".

According to the superposition principle in quantum theory, if one adds two wave functions ψ_1 and ψ_2,

---

[13] "Singularity" will play an essential role, since the distribution density μ_t has zeros in general. Cf. Chapters 5, 6, 7, and 8

[14] I will call this Schrödinger's conjecture

(1.6) $\psi_1 + \psi_2,$

then one gets another wave function.   We ignore normalization here for simplicity.

The superposition (1.6) combined with (1.5) yields

(1.7) $|\psi_1 + \psi_2|^2 = |\psi_1|^2 + |\psi_2|^2 + 2\mathcal{R}e(\psi_1 \overline{\psi_2}).$

The real part $\mathcal{R}e(\psi_1\overline{\psi_2})$ of cross terms represents the "interference" of the wave functions ψ_1 and ψ_2. It has been claimed that the interference is a typical "wave effect" with no reasonable explanation for it to be found since 1926 in either probability theory or stochastic processes. However, it will be shown in Chapter 4 that the "*interference of diffusion processes*" can be defined exactly like that of wave functions, which means that the particle theory in terms of diffusion processes is consistent with the "wave picture" in terms of Schrödinger equations.

1.4.   Equivalence of Schrödinger and Diffusion Equations

Let us begin with the formal resemblance between a system of Schrödinger equations (1.2) and (1.4) in duality and a system of diffusion equations

(1.8) $\dfrac{\partial u}{\partial t} + \dfrac{1}{2}\Delta u + \boldsymbol{b}(t, x)\cdot\nabla u + c(t,x)u = 0,$

(1.9) $-\dfrac{\partial \widehat{u}}{\partial t} + \dfrac{1}{2}\Delta \widehat{u} - \boldsymbol{b}(t, x)\cdot\nabla\widehat{u} + c(t,x)\widehat{u} = 0.$

These two equations are in duality with respect to the volume element dx under a gauge condition div $\boldsymbol{b}(t,x) = 0$.[15] The formal similarity is too close to be overlooked. As a matter of fact Schrödinger (1931) himself took this as a starting point and gave an interesting formulation of Brownian motions, which could have been a good choice for quantum mechanics. This will be discussed in Chapters 3 and 4.

---

[15] This condition is harmless but technically necessary, and will be assumed in this monograph

"Resemblance" simply means if one removes the "i" from the Schrödinger equations (1.2) and (1.4) and replaces $-V$ by c, one gets diffusion equations (1.8) and (1.9), respectively. Mathematically this resemblance is superficial and does not make much sense, since it is clear that diffusion equations and wave equations have completely different analytical characters, and one cannot transfer a property of diffusion equations to wave equations, and vice versa. A fundamental difference is this: solutions of a wave equation induce a group of unitary operators, while those of a diffusion equation induce a semi-group of non-negative operators. They are analytically completely different mathematical objects. Nevertheless can one expect Schrödinger and diffusion equations to be equivalent?

The existence of a simple transformation will be shown between solutions of the pair of Schrödinger equations (1.2) and (1.4) and the pair of diffusion equations (1.8) and (1.9) in Chapter 4, in which we shall find a typical non-linear dependence between the "potential function V" in the Schrödinger equations and the "creation and killing c" in the diffusion equations. This non-linear dependence disappears when we treat stationary states. To analyze the "equivalence", we must look at duality more closely.

1.5. Time Reversal and Duality

The transition of the diffusion process $\{X_t, Q\}$ which has the probability density $\mu_t = \overline{\psi}_t \psi_t$, $t \in [a, b]$, cannot be governed by the diffusion equation (1.8) or (1.9). This is clear because the diffusion equations (1.8) and (1.9) contain a term of "creation and killing c", and hence they *do not give a probability* density.[16] It will be shown that the transition probability density of the diffusion process $\{X_t, Q\}$ evolves according to a pair of diffusion equations of a different kind

$$(1.10) \qquad \frac{\partial q}{\partial t} + \frac{1}{2}\Delta q + \{b(t,x) + a(t,x)\}\cdot\nabla q = 0,$$

$$(1.11) \qquad -\frac{\partial \widehat{q}}{\partial t} + \frac{1}{2}\Delta \widehat{q} + \{-b(t,x) + \widehat{a}(t,x)\}\cdot\nabla \widehat{q} = 0,$$

with additional drift terms $a(t,x)$ and $\widehat{a}(t,x)$, respectively. The creation and killing term with $c(t,x)$ in the diffusion equations (1.8) and (1.9) has disappeared and has been replaced by drift terms with $a(t,x)$ and $\widehat{a}(t,x)$ in equations (1.10) and (1.11). Equations (1.10) and (1.11) are in duality with

---

[16] As a matter of fact this claim will turn out to be wrong in a sense. Cf. Chapters 3 and 4

respect to the probability distribution $\mu_t dx$, and hence the additional drift coefficients $a(t, x)$ and $\widehat{a}(t, x)$ satisfy a "duality relation"

$$(1.12) \qquad a(t, x) + \widehat{a}(t, x) = \mathrm{grad}\{\log \mu_t(x)\}.$$

This will be shown in Chapter 3. Formula (1.12) implies that the additional drift coefficients $a(t, x)$ and $\widehat{a}(t, x)$ have a strong singularity at the zero set of the distribution density $\mu_t = \overline{\psi_t}\psi_t$. Therefore, the diffusion process $\{X_t, Q\}$ turns out to be singular and must be analyzed more carefully.

The additional drift coefficients $a(t, x)$ and $\widehat{a}(t, x)$ in the equations (1.10) and (1.11) are determined not by $\mu_t = \overline{\psi_t}\psi_t$, but directly by wave functions and specify the movement of corresponding diffusion processes. Namely, assuming that a wave function is given in the form of an exponential function as $\psi(t, x) = e^{R(t, x) + iS(t, x)}$,[17] we can write the drift coefficients explicitly as

$$a(t, x) = \mathrm{grad}\{R(t, x) + S(t, x)\},$$

(1.13)

$$\widehat{a}(t, x) = \mathrm{grad}\{R(t, x) - S(t, x)\}.$$

The existence of diffusion processes with such singular drift is not at all evident. The existence problem will be discussed in Chapters 5 and 6.

First of all we will clarify the relation between the pair (1.8) and (1.9) of diffusion equations with creation and killing and the pair (1.10) and (1.11) with additional drift. This analysis will reveal the relation between a pair $\{\psi_t, \overline{\psi_t}\}$ of wave functions and the probability density $\mu_t = \overline{\psi_t}\psi_t$. In Chapter 3 it will be explained that the duality relation of drift (1.12) or the duality of diffusion equations (1.10) and (1.11) is an analytic expression of the "time reversal" of diffusion processes.

Let us adopt the diffusion picture and assume that an electron moves as a Brownian particle with additional drift. Consider an electron not confined in an atom moving with high energy, which means that the drift coefficient of the Brownian particle is relatively large compared to the diffusion coefficient, and hence that one can neglect Brownian noise. Such a model explains well why we can observe paths of electrons (drift of the diffusion process) in a bubble chamber or in pictures on film layers like the paths of Newtonian particles.

---

[17] It should be remarked that, unlike the wave function $\psi(t, x)$, the functions $R(t, x)$ and $S(t, x)$ have a severe singularity at the zero set of $\psi(t, x)$, as formula (1.13) shows

However, is it possible to have the same kind of picture for an electron in a hydrogen atom? Does it move like a Brownian particle also in a hydrogen atom ? What is the drift coefficient of the Brownian motion in a hydrogen atom ? How can it be confined in a hydrogen atom? These points will be discussed in later chapters.

1.6. QED and Quantum Field Theory

Tomonaga succeeded in expressing the Schrödinger equation for wave fields in a relativistically invariant form in 1946.[18] Soon afterwards, he applied his relativistically invariant theory to quantum electrodynamics. The theory of quantum electrodynamics achieved great success through the work of Tomonaga, Feynman, Dyson, and Schwinger, from 1948-53.[19] However, this did not overcome the difficulty of self-interaction. Instead, it rather revealed how deeply rooted the trouble is. Since then, tremendous effort has been invested in establishing a consistent quantum field theory, with success less than expected. After learning from this bitter experience, a group of specialists has gradually moved into simple stochastic models such as Markov chains (fields) on a lattice and their limit theorems.

On the other hand, in the interim, so-called "gauge theory" has achieved remarkable progress. However, one should notice that gauge theory is not quantum field theory but a classical field theory.

Let us look further back at the early 1920's. It is well-known that Schrödinger was strongly influenced by Boltzmann. In fact, he closed his inaugural lecture in 1922 at the University of Zürich by saying:[20]

"····· ich muß mich auf die folgende grundsätzliche Bemerkung beschränken, die zugleich unser Ergebnis kurz zusammenfaßt:
    Die von Exner aufgestellte Behauptung geht darin : es ist sehr wohl möglich, daß die Naturgesetze samt und sonders statistischen Charakter haben. Das hinter dem statistischen Gesetz heute noch ganz allgemein mit Selbstverständlichkeit postulierte absolute Naturgesetz *geht über die Erfahrung hinaus*. Eine derartige doppelte Begründung der Gesetzmäßigkeit in der Natur ist an sich unwahrscheinlich. *Die Beweislast obliegt den Verfechtern, nicht den Zweiflern an der*

---

[18] Cf. Tomonaga (1946), also Schwinger (Ed.) (1958)

[19] Cf. Schwinger (Ed.) (1958)

[20] Schrödinger "Was ist ein Naturgesetz?" Dec. 22, 1922, at the University of Zürich

absoluten Kausalität. Denn daran zu zweifeln ist heute bei weitem das *Natürlichere.*

Diesen Beweis nun zu erbringen, erscheint die Elektrodynamik des Atoms aus dem Grund ungeeignet, weil sie nach allgemeinem Urteil selbst noch an schweren inneren Widersprüchen krankt, die vielfach als logische empfunden werden. Ich halte es für wahrscheinlicher, daß die Befreiung von dem eingewurzelten Vorurteil der absoluten Kausalität uns bei der Überwindung der Schwierigkeiten helfen, als daß, umgekehrt, die Theorie des Atoms das Kausalitätsdogma dennoch als - sozusagen zufällig - richtig erweisen wird."

We have learned a lot since 1922, but do we have a better perspective? As we shall see, there is a good reason for assuming that Schrödinger firmly kept his thought which was deeply rooted in statistical mechanics, when and after he published his papers on wave mechanics in 1926. It would be, therefore, not meaningless to reconsider the Schrödinger equation in the context of statistical mechanics under the light of the nowadays well-developed theory of stochastic processes.

1.7. What is the Schrödinger Equation ?

Since Schrödinger and diffusion equations are equivalent, the conventional "interpretation" of quantum mechanics which has been kept in the last 70 years needs careful reinvestigation. Therefore, it might be reasonable to go back to the starting point, start afresh, and ask what the Schrödinger equation is. According to the abstract formulation of quantum mechanics by Born-Heisenberg-Jordan-Dirac in terms of operator algebra, the element of a Hilbert space (wave function) plays a small role. We need it only at the last stage to compute probability densities or correlations. The physical importance of "wave functions" has been more or less put aside in this abstract theory.[21]

In contrast, Schrödinger always considered that "wave functions must have a physical meaning". One of his attempts is in Schrödinger (1931) as already mentioned. Following his idea, we will establish a theory which unifies Born-Heisenberg-Jordan's algebraic theory, Schrödinger's wave theory, and diffusion theory, based on the equivalence of the Schrödinger and diffusion equations in duality in Chapter 4. Then, we shall immediately find that we must proceed further, because the single-electron-picture does

---

[21] Cf. Dirac (1930, 58), von Neumann (1932)

not work in excited states of hydrogen atoms. Namely, we must assume the existence of infinitely many (virtual) electrons, because the diffusion process with the distribution density $\mu_t = \overline{\psi_t}\psi_t$ cannot cross over the zero set of the density. One possibility to overcome these difficulties is to interpret the function $\mu_t = \overline{\psi_t}\psi_t$ as a spatial *statistical* distribution density of (infinitely) many interacting diffusion particles. This will be discussed in Chapters 7 and 8.

To justify the statistical interpretation mentioned above, we will construct a mathematical model in which solutions of a Schrödinger equation describe the spatial statistical distributions of interacting diffusion processes, as the Boltzmann equation does for a gas. As an example, we consider a system of virtual photons (or electrons) in an excited state of a hydrogen atom. Applying a diffusion approximation, we regard the system of virtual photons as a system of interacting diffusion processes. In other words, we find Schrödinger equations are equations in statistical mechanics.

As a general theory, the statistical mechanics for Schrödinger equations is not confined to conventional quantum mechanics. If we can find an appropriate specific diffusion coefficient (it is Planck's constant in quantum mechanics), this statistical theory can be applied to various fields with different orders of magnitude. This will be explained in Chapters 7 and 8. Taking it for granted that the statistical theory for Schrödinger equations is established, some applications to biology and physics will be discussed in Chapter 9.

1.8. Mathematical Contents

The text of the monograph is deductive, partly because the theme is motivated by problems in physics and biology. However, contents of the monograph are purely mathematical, although they might provide new aspects and implications to some problems in applications.

For mathematicians who like an inductive (axiomatic) presentation of mathematical contents, I would recommend reading the text backwards from Chapter 9 to Chapter 3 (once more) as a theory of systems of interacting diffusion processes.

The mathematical contents of the monograph are:

(1)   Theory of (singular) diffusion processes;

(2)   Transformations of diffusion processes;

(3)   Renormalization of Kac's semi-groups;

(4)   Feller's one-dimensional diffusion processes;

(5)   Diffusion equations and Schrödinger equations;

(6)   Time reversal and duality of diffusion processes;

(7)   Kolmogoroff's and Schrödinger's representations of a diffusion process;

(8)   Equivalence of diffusion and Schrödinger equations;

(9)   Variational principle of diffusion processes;

(10) An entropy characterization of the Markov property;

(11) Skorokhod's problem with singular drift;

(12) A limit theorem of systems of interacting diffusion processes;

(13) Schrödinger equations with severely singular potentials;

(14) The propagation of chaos of diffusion processes;

(15) Relative entropy and Csiszar's projection;

(16) Large deviations of diffusion processes;

(17) Systems of interacting diffusion processes;

(18) Statistical mechanics for Schrödinger equations;

(19) Non-linearity induced by the branching property;

(20) Revival theorem and construction of branching processes.

Chapter II

Diffusion Processes and their Transformations

This chapter is devoted to a brief exposition on the theory of diffusion processes in order to fix definitions and notations which will be necessary in the following chapters. Moreover, transformations of diffusion processes by means of multiplicative functionals, a renormalization of Kac's semi-groups, and Feller's one-dimensional diffusion processes will be explained.

2.1. Time-Homogeneous Diffusion Processes

Let us denote by A an elliptic differential operator

$$(2.1) \qquad A = \frac{1}{2}\Delta + b(x)\cdot\nabla.$$

In this monograph Δ denotes the Laplace-Beltrami operator[1]

$$(2.2) \qquad \Delta = \frac{1}{\sqrt{\sigma_2(x)}} \frac{\partial}{\partial x^i} \left(\sqrt{\sigma_2(x)} \, (\sigma^T\sigma(x))^{ij} \frac{\partial}{\partial x^j} \right),$$

unless otherwise stated, where $\sigma^T\sigma(x)$ is a positive definite diffusion matrix, $\sigma_2(x) = |(\sigma^T\sigma(x))_{ij}|$, and $b(x)$ denotes a drift coefficient. Moreover, let us assume the existence of a unique fundamental solution $p_t(x,y)$, for $t > 0$ and $x, y \in \mathbf{R}^d$, of the diffusion equation

$$(2.3) \qquad \frac{\partial p}{\partial t}(t,x) = A p(t,x).$$

---

[1] This is for convenience to discuss duality in the following chapters, but it is not absolutely necessary

It is well known that there exists such a fundamental solution $p_t(x,y)$ of the diffusion equation (2.3) if $\partial^2(\sigma^T\sigma(x))^{ij}/\partial x^k \partial x^h$ and $\partial b^i(x)/\partial x^k$ are locally uniformly Hölder continuous.[2]

In terms of the fundamental solution $p_t(x,y)$ we define a transition probability $P_t(x,B)$ through

$$P_t(x,B) = \int p_t(x,y)1_B(y)dy, \quad for \ \ t > 0,$$

$$= \delta_x(B), \qquad\qquad for \ \ t = 0,$$

where $\delta_x(B)$ denotes a point measure at x, and $p_t(x,y)$ will be called the transition probability density. In this chapter we assume that a transition probability density $p_t(x,y)$ is given. The transition probability satisfies the Chapman-Kolmogoroff equation

$$P_{s+t}(x,B) = \int P_s(x,dy)P_t(y,B), \quad for \ \ s, t \geq 0,$$

and the normality condition

$$P_t(x, \mathbf{R}^d) = 1.$$

For a given initial distribution μ and a transition probability $P_t(x,B)$, following Kolmogoroff (1931, 33), we can construct a probability measure on the space $\Omega = C([0,\infty), \mathbf{R}^d)$ of $\mathbf{R}^d$-valued continuous functions by means of finite dimensional distributions:

(2.4) $P_\mu[f(X_0, X_{t_1}, \ldots, X_{t_{n-1}}, X_{t_n})]$

$$= \int \mu(dx_0)P_{t_1}(x_0, dx_1)P_{t_2 - t_1}(x_1, dx_2) \cdots$$

$$\cdots P_{t_n - t_{n-1}}(x_{n-1}, dx_n) f(x_0, \ldots, x_n),$$

in terms of transition probabilities $P_t(x, dy)$, where $0 = t_0 < t_1 < \cdots < t_n$, and f is any bounded measurable function on the product space $(\mathbf{R}^d)^{n+1}$.[3]

---

[2] Cf. e.g., S. Ito (1957), Friedman (1964), Dynkin (1965)

[3] In general the diffusion coefficient may degenerate, in which case there is no transition density. This happens when space-time processes will be considered

For $\omega \in \Omega$ we denote

$$(2.5) \qquad\qquad X_t(\omega) = X(t, \omega) = \omega(t).$$

We call $\{X_t, P_\mu\}$ a diffusion process with the initial distribution μ and the transition probability density $p_t(x, y)$. The space $\Omega = C([0, \infty), \mathbf{R}^d)$ is called the "path space" and $\omega(t)$ a (sample) path of the diffusion process.

The probability measure P_μ is defined first on the product space $(\mathbf{R}^d)^{[0, \infty)}$ through the formula (2.4) and then Kolmogoroff's continuity theorem is applied, in order to restrict it on $\Omega = C([0, \infty), \mathbf{R}^d)$:

Kolmogoroff's continuity Theorem. *Let* $\{X_t, P\}$ *be a stochastic process. If there are positive constants* α, β, *and* c *such that*

$$P[|X_t - X_s|^\alpha] \leq c|t - s|^{1 + \beta}, \quad for \ any \quad s, t \in [a, b],$$

then there exists a continuous modification of the process X_t, $t \in [a, b]$.[4]

By $P[f]$ we denote the expectation of f with respect to a measure P (we avoid the notation $E[f]$ for the expectation), and by $P[f | \mathcal{F}]$ the conditional expectation of f under a condition of a σ-field $\mathcal{F}$.

Since the fundamental solution satisfies

$$p_t(x, y) \leq \kappa t^{-d/2} e^{-\lambda |x - y|^2/t}$$

in any finite time interval with positive constants κ and λ,[5] we have

$$P_\mu[|X_t - X_s|^4] \leq \kappa \int x^4 (t - s)^{-d/2} e^{-\lambda |x|^2/(t-s)} dx$$

$$= c|t - s|^2$$

with a positive constant c.

Therefore, the probability measure P_μ can be defined on the space $\Omega = C([0, \infty), \mathbf{R}^d)$ by Kolmogoroff's continuity theorem.

---

[4] Cf., e.g. Bauer (1981, 90, 91), Dynkin (1965), or any other standard textbooks on Markov and diffusion processes

[5] Cf., e.g. Friedman (1964), Dynkin (1965)

Formula (2.4) yields immediately the *Markov property* of the diffusion process $\{X_t, P_\mu\}$ in the following practical form: with $P_x = P_{\delta_x}$

$$(2.6) \qquad\qquad P_x[\,Gf(X_{t+s})\,] = P_x[\,G\,P_{X_t}[f(X_s)]\,],$$

for a bounded measurable function f on $\mathbf{R}^d$ and $G = \prod\limits_{k=1}^{n} f_k(X_{t_{k-1}})$, where f_k are bounded measurable functions on $\mathbf{R}^d$, $t_{k-1} \le t$, and $k \le n$.

With the help of the monotone class lemma it is easy to see that the Markov property (2.6) holds for any bounded $\mathcal{F}_t$ measurable function G on Ω, where $\mathcal{F}_t$ denotes the standard σ-field generated by $\{X_r : \forall\, r \le t\}$.

The Markov property (2.6) is often written in a more general form in terms of the conditional expectation

$$(2.6') \qquad\qquad P_\mu[F \circ \theta_t \,|\, \mathcal{F}_t] = P_{X_t}[F], \qquad P_\mu\text{- a.e.},$$

where F is any bounded measurable function on Ω, and θ_t is the shift operator defined by $\theta_t \omega(s) = \omega(t + s)$ for $\omega \in \Omega$. This can be shown easily from (2.6) with the help of the monotone class lemma.

Formula (2.6) of the Markov property implies immediately the semi-group property

$$P_{s+t}f = P_s P_t f, \quad s, t \ge 0,$$

of the system of non-negative operators $\{P_t : t \ge 0\}$ defined through

$$P_t f(x) = P_x[\,f(X_t)\,].$$

The semi-group $\{P_t : t \ge 0\}$ is defined on the space of bounded measurable functions. In many cases it can be considered as a semi-group on the space of bounded continuous functions. The semi-group description of diffusion processes will be needed in Chapter 3, when duality of diffusion equations will be discussed in connection with time reversal of diffusion processes.

If we are given a normalized semi-group $\{P_t : t \ge 0\}$ of non-negative linear operators on the space of, say, bounded continuous functions, we can get a transition probability $P_t(x, B)$ with the help of the Riesz-Markov theorem (cf. e.g., Yosida (1965)). In terms of the transition probability a Markov process can be constructed. The method by means of semi-groups is a well known powerful analytic tool and often adopted to construct Markov processes. However, this analytic semi-group method is really not well-suitable to handle singular diffusion processes which will be treated in later chapters, and hence we will not use it in this monograph to construct (singular) diffusion processes.

The diffusion process such as $\{X_t, P_\mu\}$ explained above will be a basic (unperturbed) stochastic motion, which is our starting point but not our main interest. There exist always such diffusion processes under mild regularity conditions on diffusion and drift coefficients as explained (see also Section 2.4). Therefore, it is harmless to assume the existence of the basic diffusion process, relying on standard textbooks on diffusion equations and diffusion processes and we will do so. Then, we shall apply "perturbation" to the basic diffusion processes. Severely perturbed diffusion processes will be our main concern in this monograph. Since the perturbation which will be treated is necessarily singular, as will be seen, the existence of perturbed diffusion processes is no longer evident and cannot be found in standard textbooks. Chapters 5 and 6 will be devoted to the existence problem of such singular diffusion processes.

2.2. Time-Inhomogeneous Diffusion Processes [6]

If time dependent diffusion and drift coefficients $(\sigma^T \sigma)^{ij}(t,x)$ and $b(t,x)$, respectively, are given, we consider a time-inhomogeneous diffusion equation

$$(2.7) \qquad \tilde{L}p(t,x) = 0, \quad (t,x) \in (a,b) \times \mathbf{R}^d,$$

where $-\infty < a < b < \infty$, and $\tilde{L}$ is a time-dependent parabolic differential operator

$$(2.8) \qquad \tilde{L} = \frac{\partial}{\partial t} + \frac{1}{2}(\sigma^T \sigma(t,x))^{ij} \frac{\partial}{\partial x^i} \frac{\partial}{\partial x^j} + \tilde{b}(t,x)^i \frac{\partial}{\partial x^i},$$

or as a special case

$$(2.8') \qquad L = \frac{\partial}{\partial t} + \frac{1}{2}\Delta + b(t,x) \cdot \nabla,$$

where Δ denotes the Laplace-Beltrami operator.

---

[6] Cf., e.g. Chapter 1 of Gihman-Skorohod (1975), Vol. II

If $\partial(\sigma^T\sigma(t,x))^{ij}/\partial t$, $\partial^2(\sigma^T\sigma(t,x))^{ij}/\partial x^k\partial x^h$, and $\partial\tilde{b}^i(t,x)/\partial x^k$ are locally uniformly Hölder continuous, then there exists a unique fundamental solution $p(s,x;t,y)$ to the parabolic differential equation (2.7).[7] We consider a time-inhomogeneous diffusion process $\{X_t, t \in [a,b], P_\mu\}$ characterized by the fundamental solution $p(s,x;t,y)$ in terms of the finite dimensional distributions:

$$(2.9)\quad P_\mu[f(X_a, X_{t_1}, \dots, X_{t_{n-1}}, X_b)]$$

$$= \int \mu(dx_0)p(a,x_0;t_1,x_1)dx_1 p(t_1,x_1;t_2,x_2)dx_2 \cdots$$

$$\cdots p(t_{n-1},x_{n-1};b,x_n)dx_n f(x_0,x_1,\dots,x_n),$$

where $a < t_1 < \cdots < t_{n-1} < b$, $f(x_0, x_1, \dots, x_n)$ is any bounded measurable function on the product space $(\mathbf{R}^d)^{n+1}$, and P_μ is a probability measure on the path space $\Omega = C([a,b], \mathbf{R}^d)$. Since the fundamental solution satisfies

$$p(s,x;t,y) \leq \kappa(t-s)^{-d/2}e^{-\lambda|x-y|^2/(t-s)}$$

in any finite time interval with positive constants κ and λ,[8] we have

$$P_\mu[|X_t - X_s|^4] \leq c|t-s|^2, \quad for \ any \quad s,t \in [a,b],$$

with a positive constant c, and hence with the help of Kolmogoroff's continuity theorem, one can define P_μ on the path space $\Omega = C([a,b], \mathbf{R}^d)$ through formula (2.9).[9]

It is clear by the definition that a time-homogeneous diffusion process is a special case of time-inhomogeneous diffusion processes. However, when we treat a time-inhomogeneous diffusion process $\{X_t, t \in [a,b], P_\mu\}$, we employ a standard trick: Instead of the process X_t on $\mathbf{R}^d$ we consider a *space-time diffusion process* (t, X_t) on an enlarged (space-time) state space $[a,b]\times\mathbf{R}^d$.

It is well known that space-time diffusion processes are *time-homogeneous* with transition probabilities

---

[7] Cf., e.g. S. Ito (1957), Friedman (1964)

[8] Cf., e.g. Friedman (1964), Dynkin (1965)

[9] We can define a diffusion process in terms of solutions of a stochastic differential equation. See the following sections

(2.10) $$\widetilde{P}_r((s,x), d(t,y)) = p(s,x;t,y)\delta_{s+r}(dt)dy,$$

where $\delta_r(dt)$ denotes the point measure at r, and hence we can apply the theory of time-homogeneous processes to space-time processes.

Let us denote by $\mathcal{F}_s^t$ the standard σ-field generated by the random variables $\{X_r: s \leq \forall r \leq t\}$. With the help of the monotone class lemma, we define $P_{(s,x)}[F]$ for any bounded $\mathcal{F}_s^b$-measurable function F through (2.9), replacing μ by δ_x, and a by s, and requiring $s < t_1 < \cdots < t_{n-1} < b$.

The *time inhomogeneous Markov property* :

(2.11) $$P_\mu[F \mid \mathcal{F}_a^s] = P_{(s,X_s)}[F], \quad P_\mu\text{- a.e.,}$$

follows from (2.9) with the help of the monotone class lemma, where F is any bounded (or non-negative) $\mathcal{F}_s^b$-measurable function on Ω. Formula (2.11) can be regarded as a space-time version of the time-homogeneous Markov property (2.6) or (2.6'). We apply the Markov property (2.11) often in the following practical form

$$P_{(r,x)}[Gf(t,X_t)] = P_{(r,x)}[GP_{(s,X_s)}[f(t,X_t)]],$$

where $a \leq r \leq s \leq t \leq b$, f is a bounded measurable function on $[a,b] \times \mathbf{R}^d$, and G a bounded $\mathcal{F}_r^s$-measurable function on Ω.

The semi-group of a space-time process can be defined with

$$P_{t-s}f((s,x)) = P_{(s,x)}[f(t,X_t)], \quad if \quad t-s \geq 0,$$

for any bounded measurable functions f on $[a,b] \times \mathbf{R}^d$. The semi-group property follows immediately from the Markov property (2.11).

2.3. Brownian Motions

The most typical and fundamental time-homogeneous diffusion processes are d-dimensional **Brownian motions**,[10] which we can define, applying (2.4), with the Brownian transition probability density

---

[10] Cf. Einstein (1905, b)

$$(2.12) \qquad p_t(x, y) = (2\pi t)^{-d/2} \exp(-\frac{\|x - y\|^2}{2t}),$$

which is the fundamental solution of the diffusion equation (2.3) with $\sigma = \delta_{ij}$ and $b \equiv 0$.

An explicit construction of one-dimensional Brownian motion is due to P. Lévy :[11] let $\{g_0, g_{k, 2^{-n}} : odd\ k < 2^{-n}\}$ be a Gaussian family of random variables on a probability space $\{W, \mathbf{F}, P\}$, and by $\{f_0, f_{k, 2^{-n}} : odd\ k < 2^{-n}\}$ denote the family of Schauder functions

$$f_{k, 2^{-n}}(t) = \int_0^t h_{k, 2^{-n}}(s)ds,$$

where the system of functions $\{ h_{k, 2^{-n}}(s) \}$ is defined by

$$h_{k, 2^{-n}}(s) = 2^{(n-1)/2}, \quad (k-1)2^{-n} \le s < k2^{-n}$$

$$= -2^{(n-1)/2}, \quad k2^{-n} \le s < (k+1)2^{-n}$$

$$= 0, \quad \text{otherwise.}$$

The family of Schauder functions gives a collection of little tents, though superposition of which with independent Gaussian random coefficients g's, we define

$$(2.13) \qquad B(t) = g_0 f_0(t) + \sum_{n=1}^{\infty} \sum_{odd\ k < 2^n}^{\infty} g_{k, 2^{-n}} f_{k, 2^{-n}}(t).$$

It is not difficult to show[12] that this converges uniformly in $t \in [0, 1]$, and has the distribution density

$$P[B(t) \in dx] = (2\pi t)^{-1/2} \exp(-\frac{|x|^2}{2t})dx.$$

Therefore, $B(t)$ is a one-dimensional Brownian motion starting from the origin. It is routine to extend the time parameter from $[0, 1]$ to $[0, \infty)$ and for arbitrary starting points.

---

[11] N. Wiener adopted another basis system and constructed Brownian motions (Wiener measure), cf. Itô-Nisio (1968)

[12] Cf. McKean (1969)

A d-dimensional Brownian motion can be obtained by

$$\boldsymbol{B}(t) = (B_1(t), \dots, B_d(t)),$$

where $B_i(t)$'s are independent copies of $B(t)$. This is a probabilists' favorite construction of Brownian motions, since the continuity of the process is evident by definition; formula (2.13) shows clearly that Brownian motions are sums of independent random increments, and finally the method does not depend too heavily on analysis.

There is another method of constructing a Brownian motion as a limit of Markov chains with an appropriate scaling. This method plays an important role in diffusion approximations, but we will not discuss it in this monograph.

Remark. For Brownian motions there are many excellent textbooks and monographs which have different characters. One can refer to *e.g.* Revuz-Yor (1991) and the references given there.

2.4. Stochastic Differential Equations

If diffusion and drift coefficients are smooth or merely Lipschitz continuous, we apply Itô's theory of stochastic differential equations.[13] When we handle the parabolic differential operator L with the Laplace-Beltrami operator Δ given at (2.8'), we rewrite it as

$$(2.14) \qquad L = \frac{\partial}{\partial t} + \frac{1}{2}(\sigma^T \sigma(x))^{ij}\frac{\partial}{\partial x^i}\frac{\partial}{\partial x^j} + \tilde{\boldsymbol{b}}(t,x)^i\frac{\partial}{\partial x^i},$$

as appears in Itô's formula, where

$$\tilde{\boldsymbol{b}}(t,x) = \boldsymbol{b}(t,x) + \boldsymbol{b}_\sigma(x),$$

$$\boldsymbol{b}_\sigma(x)^i = \frac{1}{2}\frac{1}{\sqrt{\sigma_2(x)}}\frac{\partial}{\partial x^j}(\sigma^T\sigma(x)^{ij}\sqrt{\sigma_2(x)}).$$

By $\{B_t\}$ we denote a d-dimensional Brownian motion defined on a probability space $\{W, \mathcal{F}, P\}$. Then we consider the stochastic differential equation (SDE)

---

[13] For stochastic differential equations cf., e.g., McKean (1969), Ikeda-Watanabe (1981, 89), Chung-Williams (1983, 88)

$$(2.15) \qquad X_t = X_a + \int_a^t \sigma(r, X_r) dB_r + \int_a^t \tilde{b}(r, X_r) dr.$$

Let us assume, for simplicity, that the initial value X_a is square integrable and independent of the Brownian motion B_r. If $\sigma(t, x)$ and $\tilde{b}(t, x)$ are Lipschitz continuous and satisfy a growth condition

$$\| \sigma(t, x) \|^2 + \| \tilde{b}(t, x) \|^2 \le c_1 (1 + \| x \|^2),$$

then there exists a pathwise unique solution X_t which is square integrable.

In terms of the solution we can define a diffusion process $\{X_t, P_\mu\}$ corresponding to the parabolic differential operator L given in (2.14) or more generally $\tilde{L}$ given in (2.8). Solutions of the SDE (2.15) can be obtained through the standard successive approximation as follows: To demonstrate the standard way we consider (2.15) simply in one-dimension and assume $\sigma(t, x)$ and $\tilde{b}(t, x)$ are bounded in the following.

By the assumption $\sigma(t, x)$ and $\tilde{b}(t, x)$ are Lipschitz continuous:

$$| \sigma(t, x) - \sigma(t, y) |, \ | \tilde{b}(t, x) - \tilde{b}(t, y) | \le c_0 | x - y |.$$

A solution can be obtained through successive approximation. We set

$$X_t^{(0)} = X_a,$$

$$X_t^{(n)} = X_a + \int_a^t \sigma(r, X_r^{(n-1)}) dB_r + \int_a^t \tilde{b}(r, X_r^{(n-1)}) dr, \ for \ n \ge 1.$$

First of all it is clear that

$$P[| X_t^{(1)} - X_t^{(0)} |^2] \le c, \ for \ t \in [a, b].$$

For arbitrary $n > 1$, we have

$$(2.16) \qquad P[| X_t^{(n)} - X_t^{(n-1)} |^2] \le c K^{n-1} \frac{t^{n-1}}{(n-1)!},$$

which can be verified through induction as follows:

$$P[|X_t^{(n+1)} - X_t^{(n)}|^2] \leq 2P[|\int_a^t \{\sigma(s, X_s^{(n)}) - \sigma(s, X_s^{(n-1)})\}dB_s|^2]$$

$$+ 2P[|\int_a^t \{\widetilde{b}(s, X_s^{(n)}) - \widetilde{b}(s, X_s^{(n-1)})\}ds|^2],$$

where we have applied an inequality $(a + b)^2 \leq 2(a^2 + b^2)$. The first integral on the right-hand side is dominated by

$$2(c_0)^2 P[\int_a^t |X_s^{(n)} - X_s^{(n-1)}|^2 ds],$$

because of the Lipschitz continuity of $\sigma(t, x)$, and the second one by

$$2P[\int_a^t |\widetilde{b}(s, X_s^{(n)}) - \widetilde{b}(s, X_s^{(n-1)})|^2 ds \int_a^t ds]$$

$$\leq 2(c_0)^2(b - a)P[\int_a^t |X_s^{(n)} - X_s^{(n-1)}|^2 ds],$$

where we have applied Schwarz's inequality and the Lipschitz continuity of $\widetilde{b}(t, x)$. Thus we have shown, with a constant $K > 0$,

$$P[|X_t^{(n+1)} - X_t^{(n)}|^2] \leq KP[\int_a^t |X_s^{(n)} - X_s^{(n-1)}|^2 ds].$$

A substitution of (2.16) on the right hand side verifies (2.16) for any n.

Therefore, we have

$$P[\sup_{t \in [a,b]} |X_t^{(n)} - X_t^{(n-1)}|^2] \leq const. \frac{(K(b - a))^{n-1}}{(n - 1)!},$$

and hence

$$P[\sup_{t \in [a,b]} |X_t^{(n)} - X_t^{(n-1)}|^2 > \frac{1}{2^{n-1}}] \leq const. \frac{(K(b - a))^{n-1}}{(n - 1)!}.$$

With the help of the Borel-Cantelli lemma, $X_t^{(n)}$ converges uniformly in $t \in [a, b]$, P-*a.e.* and also in L^2. Moreover,

$$X_t = \lim_{n \to \infty} X_t^{(n)}$$

satisfies the SDE (2.15).

The uniqueness of solutions follows immediately from

Gronwall's Lemma. *Let $A(t)$ be a non-negative integrable function on $[a, b]$ satisfying*

$$A(t) \leq \kappa \int_a^t A(s)ds + C(t), \quad \kappa > 0,$$

where $C(t)$ is also integrable. Then

$$A(t) \leq \kappa \int_a^t e^{\kappa(t - s)}C(s)ds + C(t).$$

In particular if $C(t)$ is non-negative and non-decreasing, then

$$A(t) \leq e^{\kappa(t - a)}C(t).$$

Proof is a good exercise (apply iteration).[14]

Now let X_t^1 and X_t^2 be solutions of equation (2.15) and set

$$Z_t = |X_t^1 - X_t^2|^2.$$

Then, $P[Z_t] < \infty$ and

$$P[Z_t] \leq K \int_a^t P[Z_s]ds.$$

Therefore, by Gronwall's lemma $P[Z_t] = 0$, for $t \in [a, b]$, which proves the uniqueness of solutions.

Applying Itô's formula, we will show that the diffusion process $\{X_t, P\}$ is determined by the parabolic differential operator $\tilde{L}$:

---

[14] Cf. e.g., Gikhman-Skorokhod (1969), p. 393, or Revuz-Yor (1991), p. 499

Let us denote the i-th component of X_t by X_t^i. Then, the random variable $Y_t = f(t, X_t^1, \dots, X_t^d)$, for any bounded $f \in C^2([a,b] \times \mathbf{R}^d)$, satisfies

$$(2.17) \qquad dY_t = \partial_t f \, dt + \sum_{i=1}^d (\partial_i f) \cdot dX_t^i + \frac{1}{2} \sum_{i,j=1}^d (\partial_i \partial_j f) \cdot dX_t^i dX_t^j,$$

which is called *Itô's formula*, where $\partial_t = \partial/\partial t$, $\partial_i = \partial/\partial x^i$,

$$dX_t^i = \sigma_j^i \, dB_t^j + \tilde{b}^i \, dt,$$

$$dB_t^i dB_t^j = \delta^{ij} dt, \quad and \quad dt \, dB_t^i = 0.$$

Therefore, we get

$$(2.18) \quad f(t, X_t) - f(s, X_s) - \int_s^t \tilde{L} f(r, X_r) dr = \sum_{i,j=1}^d \int_s^t (\sigma_j^i \, \partial_i f)(r, X_r) dB_r^j,$$

the right-hand side of which is a martingale as a sum of Itô's stochastic integrals with respect to Brownian motions, the expectation of which vanishes consequently. Let $t = b$, $s = a$, and let f be any C^∞-function with a compact support in $(a,b) \times \mathbf{R}^d$, and take the expectation of both sides of (2.18). Then, except for the third term on the left-hand side, all terms vanish, and hence

$$\int_a^b P[\tilde{L} f(t, X_t)] dt = 0.$$

Therefore, denoting by $\mu_t(x)$, the probability density of the diffusion process X_t, we have

$$(2.19) \qquad \int_{(a,b) \times \mathbf{R}^d} \mu_t(x) \tilde{L} f(t, x) dt dx = 0,$$

namely, $\mu_t(x)$ is a weak solution of

$$(2.20) \qquad\qquad\qquad \tilde{L}^\circ \mu = 0,$$

where $\tilde{L}^\circ$ is the formal adjoint of $\tilde{L}$. When L of (2.14), which is the same as (2.8') with the Laplace-Beltrami operator Δ, is handled, replace $\tilde{L}^\circ$ by

(2.21) $$L^\circ g = -\frac{\partial g}{\partial t} + \frac{1}{2}\Delta g - \frac{1}{\sqrt{\sigma_2}}\nabla\{\sqrt{\sigma_2}\,b(t,x)g\}.$$

Let us denote by X the mapping from W to $\Omega = C([a,b], \mathbf{R}^d)$

$$X : w \rightarrow X_t(w),$$

and define a probability measure P_μ on $\Omega = C([a,b], \mathbf{R}^d)$ by

$$P_\mu = P \circ X^{-1}.$$

In this way we get a diffusion process $\{X_t, P_\mu\}$ defined on the space of continuous paths $\Omega = C([a,b], \mathbf{R}^d)$.

Maruyama (1954) proved that the probability measure P_μ is absolutely continuous with respect to the probability measure of a d-dimensional Brownian motion defined on the space $\Omega = C([a,b], \mathbf{R}^d)$, if the diffusion coefficient is non-degenerate, [15] and hence the transition probability of the diffusion process $\{X_t, P_\mu\}$ has a density function $p(s,x;t,y)$. This will be called Maruyama's absolute continuity theorem.

Assume the diffusion coefficient is non-degenerate. Then, the equation (2.19) or (2.20) implies that the transition probability density $p(s,x;t,y)$ of the diffusion process X_t satisfies

(2.22) $\widetilde{L}p = 0,$

weakly, as we have wanted to show.

Remark. In later chapters, diffusion process with singular drift will be treated. Truncating and approximating drift coefficients, we can apply the SDE method even in such singular cases. However, since this approximation procedure makes things complicated, we will not do it. Instead we will construct singular diffusion process in other ways, cf. Chapters 5 and 6. In some cases the truncation method is dangerous and must be handled with great care; see examples in Section 7.9.

Remark. Based on formula (2.18) a method of the so-called martingale problem was developed by Stroock-Varadhan (1970), which allows diffusion

---

[15] This is known as Girsanov's theorem (1960), being not aware of Maruyama's paper. See the next section

and drift coefficients merely to be continuous. This martingale method combines SDE and semi-group methods and generalizes both. For this we refer to Stroock-Varadhan (1970).

Remark. For a detailed treatment of stochastic differential equations one can refer to *e.g.* McKean (1969), Liptser-Shiryayev (1977) and Ikeda-Watanabe (1981, 89).

2.5. Transformation by a Multiplicative Functional

In the theory of diffusion processes a transformation of probability measures in terms of multiplicative functionals is a strong tool in perturbing a given diffusion process into a new one.

Let $\{(t, X_t), P_{(s,x)}, (s,x) \in [a,b] \times R^d\}$ be a space-time (unperturbed) basic diffusion process.[16]

A functional $M_s^t(\omega), a \le s \le t \le b,$ is a *multiplicative functional*, if it is $\mathcal{F}_s^t$ -measurable and satisfies the *multiplicativity*

$$(2.23) \qquad M_r^t(\omega) = M_r^s(\omega)M_s^t(\omega), \quad for \ r \le s \le t.[17]$$

For simplicity we assume $M_s^t(\omega)$ is continuous in t for fixed s .

If a multiplicative functional satisfies in addition the *normality condition*

$$(2.24) \qquad\qquad P_{(s,x)}[M_s^t] = 1,$$

then it will be called a normal multiplicative functional, or martingale multiplicative functional, since it turns out to be a martingale; in fact

$$P_{(s,x)}[M_s^t \mid \mathcal{F}_s^r] = P_{(s,x)}[M_s^r M_r^t \mid \mathcal{F}_s^r] = M_s^r P_{(s,x)}[M_r^t \mid \mathcal{F}_s^r]$$

$$= M_s^r P_{(r,X_r)}[M_r^t]$$

$$= M_s^r, \qquad P_{(s,x)} - a.e.,$$

---

[16] We consider a family of probability measures $P_{(s,x)}$ with arbitrary starting points (s,x)

[17] In general the equality may have an exceptional set of measure zero, cf Meyer (1962, 63), Blumenthal-Getoor (1968), Sharpe (1988)

for $s < r < t$, where we have applied at the third and fourth equalities the Markov property (2.11) and normality condition (2.24), respectively.

Let us define a system of new probability measures $Q_{(s,x)}$ by

$$(2.25) \qquad Q_{(s,x)}[F] = P_{(s,x)}[M_s^b F],$$

for any bounded $\mathcal{F}_s^b$-measurable function F, namely, M_s^b is a density of the probability measure $Q_{(s,x)}$ with respect to the (unperturbed) probability measure $P_{(s,x)}$. We will denote $Q_{(s,x)} = M_s^b P_{(s,x)}$ (instead of the standard notation $dQ_{(s,x)} = M_s^b dP_{(s,x)}$).

In this way we can obtain a new space-time diffusion process $\{(t,X_t), Q_{(s,x)}, (s,x) \in [a,b]\times\mathbf{R}^d\}$. This is the so-called *transformation of a space-time diffusion process* $\{(t,X_t), P_{(s,x)}, (s,x) \in [a,b]\times\mathbf{R}^d\}$ *by a multiplicative functional* M_s^t. The Markov property of the transformed process is immediate: For any bounded $\mathcal{F}_s^r$-measurable G on Ω and bounded measurable f on $[a,b]\times\mathbf{R}^d$,

$$(2.26) \qquad Q_{(s,x)}[G f(t,X_t)] = P_{(s,x)}[M_s^b G f(t,X_t)]$$

$$= P_{(s,x)}[M_s^r G P_{(s,x)}[M_r^b f(t,X_t) \mid \mathcal{F}_s^r]]$$

$$= P_{(s,x)}[M_s^r G P_{(r,X_r)}[M_r^b f(t,X_t)]]$$

$$= Q_{(s,x)}[G Q_{(r,X_r)}[f(t,X_t)]].$$

The transformed space-time process $\{(t,X_t), Q_{(s,x)}, (s,x) \in [a,b]\times\mathbf{R}^d\}$ also inherits the strong Markov property of the unperturbed space-time diffusion process $\{(t,X_t), P_{(s,x)}, (s,x) \in [a,b]\times\mathbf{R}^d\}$, since one can apply the same manipulation as in (2.26) for random stopping (or optional) times.[18]

Thus we have shown

Theorem 2.1. *Let* $\{(t,X_t), P_{(s,x)}, (s,x) \in [a,b]\times\mathbf{R}^d\}$ *be a space-time diffusion process. If a continuous multiplicative functional* $M_s^t(\omega)$ *satisfies normality condition (2.24), then the system of transformed probability measures* $Q_{(s,x)} = M_s^b P_{(s,x)}$ *defines a new (perturbed) space-time diffusion process* $\{(t,X_t), Q_{(s,x)}, (s,x) \in [a,b]\times\mathbf{R}^d\}$.

[18] Cf. Meyer (1962, 63), Blumenthal-Getoor (1968), Sharpe (1988)

A typical example of multiplicative functionals is Kac's multiplicative functional

$$m_s^t = \exp(\int_s^t c(r, X_r)dr),$$

where $c(r, x)$ is a bounded measurable function. The multiplicativity (2.23) of the functional is immediate. However, the Kac multiplicative functional does not satisfy normality condition (2.24), and hence Theorem 2.1 cannot be applied. To overcome the difficulty, we will renormalize Kac's functionals in Section 2.7. The renormalization of Kac's multiplicative functionals will play an important role in this monograph (see Chapters 5 and 6).

Another well-known multiplicative functional for a d-dimensional Brownian motion B_t is the *Maruyama-Girsanov density*

$$(2.27) \qquad M_s^t = \exp(\int_s^t b(r, B_r) \cdot dB_r - \frac{1}{2}\int_s^t \|b(r, B_r)\|^2 dr),$$

(cf. Maruyama (1954), Girsanov (1960). See also Liptser-Shiryayev (1977), Ikeda-Watanabe (1981, 89)). If the vector function $b(t, x)$ is bounded measurable, then it is easy to see, applying Itô's formula, that the M_s^t in (2.27) is well-defined and satisfies normality condition (2.24).

The boundedness assumption on the drift coefficient $b(t, x)$ is not necessary, but for simplicity. A well-known sufficient condition for the normality is Novikov's condition[19]

$$(2.28) \qquad P_{(s, x)}[\exp(\frac{1}{2}\int_s^b \|b(r, B_r)\|^2 dr)] < \infty.$$

This condition allows the drift vector $b(t, x)$ to be singular to some extent.[20]

We will prove that the transformation in terms of the Maruyama-Girsanov density (2.27) induces the drift term with $b(t, x)$. Therefore, it is often called "drift transformation" or the "Maruyama-Girsanov transformation". Let us consider the case of one-dimensional Brownian motion for simplicity, and assume normality condition (2.24).

---

[19] Cf., e.g. Liptser-Shiryayev (1977), Ikeda-Watanabe (1981, 89), Revuz-Yor (1991)
[20] For interesting examples see Stummer (1990)

For a proof we apply formulae of Itô's stochastic calculus:

(2.29) $$d(X_tY_t) = (dX_t)Y_t + X_t(dY_t) + (dX_t)(dY_t),$$

(2.30) $$(dB_t)^2 = dt \quad and \quad (dt)(dB_t) = 0.$$

We apply formula (2.29) to

$$X_t = f(t, B_t) \quad and \quad Y_t = M_s^t = e^\alpha,$$

$$\alpha = \int_s^t b(r, B_r)dB_r - \frac{1}{2}\int_s^t b(r, B_r)^2 dr,$$

where f is any C^∞-function of compact support in $(s, b) \times \mathbf{R}$.

Since

$$dX_t = \frac{\partial f}{\partial t}dt + \frac{\partial f}{\partial x}dB_t + \frac{1}{2}\frac{\partial^2 f}{\partial x^2}dt, \quad and \quad dY_t = e^\alpha b(t, B_t)dB_t,$$

we get

(2.31) $$d(X_tY_t) = Lf(t, B_t)e^\alpha dt + \{ f(t, B_t)b(t, B_t) + \frac{\partial f}{\partial x}(t, B_t) \}e^\alpha dB_t ,$$

because of Itô's formula (2.17) with (2.30), where

$$L = \frac{\partial}{\partial t} + \frac{1}{2}\frac{\partial^2}{\partial x^2} + b(t, x)\frac{\partial}{\partial x} .$$

Therefore, taking the expectation of both sides of (2.31) and integrating over $[s, b]$, since f is of compact support in $(s, b) \times \mathbf{R}$, we get

$$\int_s^b P_{(s,x)}[Lf(t, B_t)M_s^t]dt = 0.$$

If we denote by $\mu_t(x)$ the probability density of B_t, *with respect to the transformed probability measure* $Q_{(s, x)} = M_a^b P_{(s, x)}$, then

(2.32) $$\int_{(s,b)\times\mathbf{R}} \mu_t(x)Lf(t, x)dtdx = 0,$$

namely, the *diffusion process under the transformed probability measure* $Q_{(s,x)} = M_a^b P_{(s,x)}$ *has an additional drift term* $b(t,x)$.

To extend the above arguments to higher dimensions is more or less routine.[21]

2.6. Feynman-Kac Formula

Let $\{(t,X_t), P_{(s,x)}, (s,x) \in [a,b] \times \mathbf{R}^d\}$ be the diffusion process determined by the parabolic differential operator $\tilde{L}$ given in (2.8) and let $c(t,x)$ be a measurable function. Then, the Feynman-Kac formula, which will be given in (2.35), represents the solution $u(s,x)$ of the diffusion equation

$$(2.33) \quad \{\frac{\partial}{\partial s} + \frac{1}{2}(\sigma^T\sigma(s,x))^{ij}\frac{\partial}{\partial x^i}\frac{\partial}{\partial x^j} + \tilde{b}(s,x)^i\frac{\partial}{\partial x^i} + c(s,x)\}u = 0,$$

with

$$u(t,x) = f(t,x),$$

in terms of the diffusion process. In (2.33), $s \in (a,t)$ and $t \in (a,b]$ is arbitrary but fixed. We assume in this section that $c(s,x)$ and $u(s,x)$ are bounded.

If the function $c(s,x)$ is bounded, there is no problem showing the Feynman-Kac formula. There are various ways of treating the formula: purely analytically as perturbation, in terms of semi-group theory, or using Itô's formula. However, if $c(s,x)$ is unbounded (or singular), there are several points which must be carefully treated, and the advantage or disadvantage of the three methods mentioned above will come out. This will be discussed later on.

Let us assume the diffusion process X_r is given as a solution of the stochastic differential equation

$$dX_r = \sigma(r,X_r)\cdot dB_r + \tilde{b}(r,X_r)dr,$$

where B_r is a d-dimensional Brownian motion.[22] Then, applying formulae (2.29) and (2.30) of Itô's stochastic calculus to

---

[21] Cf., e.g. Liptser-Shiryayev (1977), Ikeda-Watanabe (1981, 89)

[22] If $\sigma(t,x)$ and $\tilde{b}(t,x)$ are bounded and Lipschitz continuous, then solutions exist

$$Z_r = p(r, X_r), \quad Y_r = e^\alpha, \quad \alpha = \int_s^r c(u, X_u)du, \quad for \ \ r \in [s, t],$$

we have, after a routine manipulation using Itô's stochastic calculus,

$$(2.34) \qquad u(s, X_s) - f(t, X_t) \exp(\int_s^t c(r, X_r)dr) = a \ martingale,$$

because of (2.33), where the expectation of the right-hand side vanishes.

Therefore, taking the expectation of both sides of (2.34), we get the *Feynman-Kac formula*

$$(2.35) \quad u(s, x) = P_{(s, x)}\left[\exp(\int_s^t c(r, X_r)dr)f(t, X_t)\right], \quad for \ \ s \in (a, t).$$

Conversely, if we define a function $u(s, x)$ by (2.35), it is easy to see that $u(s, x)$ satisfies an integral equation

$$(2.36) \quad u(s, x) = P_{t-s}f(s, x) + \int_s^t P_{(s, x)}[c(r, X_r)u(r, X_r)]dr, \quad for \ \ s \in (a, t),$$

where P_t denotes the semi-group of the unperturbed space-time diffusion process $\{(t, X_t), P_{(s, x)}, (s, x) \in [a, b] \times \mathbf{R}^d\}$. In fact, expanding Kac's multiplicative functional in the right-hand side of (2.35) as

$$\exp(\int_s^t c(r, X_r)dr) = 1 + \sum_{k=1}^\infty \frac{1}{k!}(\int_s^t c(r, X_r)dr)^k$$

$$= 1 + \int_s^t c(r, X_r)dr \sum_{k=1}^\infty \frac{1}{(k-1)!}(\int_r^t c(u, X_u)du)^{k-1};$$

taking the expectation and applying the Markov property, we have

$$P_{(s, x)}[\int_s^t c(r, X_r)dr \sum_{k=1}^\infty \frac{1}{(k-1)!}(\int_r^t c(\tau, X_\tau)d\tau)^{k-1}f(t, X_t)]$$

$$= \int_s^t P_{(s, x)}[c(r, X_r)u(r, X_r)]dr,$$

and hence the right-hand side of (2.36). Therefore, the $u(s, x)$ defined at (2.35) satisfies equation (2.33) weakly.

If the function $c(t, x)$ is not bounded, we shall see that the method of applying Itô's formula explained above is not always best for the Feynman-Kac formula. This point will be discussed in Chapter 6.

Now let us adopt an analytic method: in this case we consider the integral equation (2.36) instead of the diffusion equation (2.33). A solution can be constructed as follows: Define successively

$$u^{(0)}(s, x) = P_{t-s}f(s, x) = P_{(s,x)}[f(t, X_t)],$$

$$u^{(k)}(s, x) = \int_s^t P_{(s,x)}[c(r, X_r)u^{(k-1)}(r, X_r)]dr, \text{ for } k \geq 1.$$

Then it is easy to see that

$$u(s, x) = \sum_{k=0}^{\infty} u^{(k)}(s, x)$$

converges, is bounded, and satisfies equation (2.36). Moreover we can show easily by induction

$$u^{(k)}(s, x) = P_{(s,x)}[\frac{1}{k!}(\int_s^t c(r, X_r)dr)^k f(t, X_t)].$$

Therefore, the solution $u(s, x)$ has the expression of (2.35). The uniqueness of solutions of the integral equation (2.36) is easy to show, if $c(t, x)$ is bounded. For uniqueness, see Lemma 6.1 and Section 6.5, in which the case of singular $c(t, x)$ will be treated.

2.7. Kac's Semi-Group and its Renormalization

Let $\{X_t, P_{(s,x)}; (s, x) \in [a, b] \times \mathbf{R}^d\}$ be a basic (unperturbed) space-time diffusion process and let $c(t, x)$ be a bounded measurable function.[23]

Kac's multiplicative functional is defined by

---

[23] The boundedness assumption is just for simplicity, and will be removed in later sections

(2.37)
$$m_s^t = \exp(\int_s^t c(r, X_r)dr),$$

which does not satisfy normality condition (2.24).

In terms of Kac's multiplicative functional, we can define a new semi-group P_{t-s}^c by

(2.38)
$$P_{t-s}^c f((s,x)) = P_{(s,x)}[m_s^t f(t, X_t)],$$

which is called *Kac's semi-group*.

The semi-group property of $\{P_{t-s}^c\}$ follows immediately from the multiplicativity of Kac's functional and the Markov property of the basic unperturbed process $\{X_t, P_{(s,x)}; (s,x) \in [a,b] \times \mathbf{R}^d\}$:

$$P_{r-s}^c P_{t-r}^c f((s,x)) = P_{(s,x)}[m_s^r P_{(r, X_r)}[m_r^t f(t, X_t)]]$$

$$= P_{(s,x)}[m_s^r m_r^t f(t, X_t)]$$

$$= P_{t-s}^c f((s,x)).$$

for $a \le s \le r \le t \le b$.

If the function $c(t,x)$ is non-positive, then $P_{t-s}^c 1((s,x)) \le 1$, and hence we can construct a space-time diffusion process *with killing* on an extended probability space such that its semi-group coincides with the semi-group $\{P_{t-s}^c\}$ defined by (2.38). This is well known.[24] However, if the function $c(t,x)$ takes both positive and negative values, one cannot construct a diffusion process which has the semi-group $\{P_{t-s}^c\}$.

In fact, when we define a probability measure applying the formula (2.9) of finite dimensional distributions, we need a normalized, i.e., transition *probability*. The well-known problem with Kac's semi-group $\{P_{t-s}^c\}$ is that it is not normalized, i.e., it happens to be $P_{t-s}^c 1 \ge 1$ ($P_{t-s}^c 1 \le 1$ causes no trouble, as remarked above). Therefore, one cannot apply formula (2.9) to the transition function defined through the semi-group $\{P_{t-s}^c\}$. Probabilistically the positive part of the function $c(t,x)$ represents the existence of "creation of particles", which is problematic. We will encounter

---

[24] Cf., e.g. Dynkin (1965), Blumenthal-Getoor (1968). However, the killed processes will not be employed in this monograph; instead, we will apply "renormalization"

this problem later on in Chapters 3, 5 and 6.[25]

On the other hand, we can "comfortably" define measures $P^c_{(s,x)}$ with creation and killing by

$$(2.39) \qquad\qquad P^c_{(s,x)}[F] = P_{(s,x)}[m^b_s F],$$

where F is any bounded $\mathcal{F}^b_s$-measurable function on Ω. However, the system of measures $\{P^c_{(s,x)}, (s,x) \in [a,b] \times \mathbf{R}^d\}$ does not define a Markov process or a semi-group. Nonetheless, through the **renormalization** of the measure with creation and killing $P^c_{(s,x)}$ we can get a Markov process as follows.

Let us define

$$(2.40) \qquad\qquad \xi(s,x) = P^c_{(s,x)}[1] = P_{(s,x)}[m^b_s].$$

Since $c(t,x)$ is bounded, it is clear that

$$(2.41) \qquad\qquad 0 < \xi(s,x) < \infty.$$

With the function $\xi(s,x)$ we define a system of **renormalized measures** $\{\overline{P}_{(s,x)}\}$ of $\{P^c_{(s,x)}\}$ by

$$(2.42) \qquad\qquad \overline{P}_{(s,x)}[F] = \frac{1}{\xi(s,x)} P^c_{(s,x)}[F].$$

Then, the renormalized measures define a new space-time diffusion process $\{(t,X_t), \overline{P}_{(s,x)} : (s,x) \in [a,b] \times \mathbf{R}^d\}$, which will be called the **renormalized process**. Its semi-group $\overline{P}_{t-s} f$ as a space-time process is given by

$$(2.43) \qquad \overline{P}_{t-s} f((s,x)) = \frac{1}{\xi(s,x)} P_{(s,x)}\big[\, e^{\int_s^t c(r,X_r)\,dr} f(t,X_t)\xi(t,X_t)\big],$$

which is the ξ-*transformation* of Kac's semi-group P^c_{t-s} defined in (2.38), namely,

$$\overline{P}_{t-s} f((s,x)) = \frac{1}{\xi} P^c_{t-s} (f\xi)(s,x).$$

---

[25] There are various ways to handle "creation of particles", introducing additional structures and interpreting induced semi-groups, cf. Nagasawa (1969), Mitro (1979). See Chapter 12 on branching processes

Formula (2.43) can be shown easily applying the Markov property of the basic unperturbed process:

$$\overline{P}_{t-s} f((s,x)) = \frac{1}{\xi(s,x)} P_{(s,x)}[m_s^t f(t, X_t) m_t^b]$$

$$= \frac{1}{\xi(s,x)} P_{(s,x)}[m_s^t f(t, X_t) P_{(t, X_t)}[m_t^b]]$$

$$= \frac{1}{\xi(s,x)} P_{(s,x)}[m_s^t f(t, X_t) \xi(t, X_t)],$$

where $P_{(t,x)}[m_t^b] = \xi(t,x)$ is substituted, and hence we have (2.43).[26]

Thus we have shown[27]

Theorem 2.2. *Let* $\overline{P}_{(s,x)}$ *be the renormalization of* $P_{(s,x)}^c$ *defined in (2.42) and let* $\{(t, X_t), \overline{P}_{(s,x)} : (s,x) \in [a,b] \times \mathbf{R}^d\}$ *be the renormalized process. Then, its semi-group* $\overline{P}_{t-s}$ *is the* ξ-*transformation of Kac's semi-group* P_{t-s}^c *defined by (2.38), namely,*

(2.44) $$\overline{P}_{t-s} f((s,x)) = \frac{1}{\xi} P_{t-s}^c (f\xi)(s,x).$$

We can formulate Theorem 2.2 as a corollary of Theorem 2.1 applied to the *renormalization of Kac's functional* m_s^t defined by

(2.45) $$n_s^t = \frac{1}{\xi(s, X_s)} m_s^t \xi(t, X_t).$$

In fact we have

Theorem 2.3. *The renormalized Kac functional* n_s^t *defined in (2.45) satisfies normality condition (2.24).*

Proof. Because of definition (2.40) and of the multiplicativity of m_s^t, we have

---

[26] Therefore, the renormalized process is a conditional space-time diffusion process in terms of the survival condition $\xi(s,x) = P_{(s,x)}^c[1]$

[27] The case of unbounded or singular $c(t,x)$ can be handled similarly under an integrability condition, see Chapters 5 and 6

$$P_{(s,x)}[n_s^c] = \frac{1}{\xi(s,x)} P_{(s,x)}[m_s^t \xi(t,X_t)] = \frac{1}{\xi(s,x)} P_{(s,x)}[m_s^t P_{(t,X_t)}[m_t^b]]$$

$$= \frac{1}{\xi(s,x)} P_{(s,x)}[m_s^t m_t^b] = \frac{1}{\xi(s,x)} P_{(s,x)}[m_s^b]$$

$$= \frac{1}{\xi(s,x)} P_{(s,x)}^c[1] = 1,$$

completing the proof.

Therefore, when one treats Kac's semi-group, it is better to consider the renormalized process $\{(t, X_t), \overline{P}_{(s,x)}: (s, x) \in [a, b] \times \mathbf{R}^d\}$, from which one can always recover Kac's semi-group. Namely, one uses the renormalized process (it is a conservative diffusion process !) in computation, and when one needs Kac's semi-group, one applies formula (2.44) the other way round

(2.46) $$P_{t-s}^c(f) = \xi \overline{P}_{t-s}(f\frac{1}{\xi}).$$

The crucial fact is this: $\overline{P}_{t-s}$ is the semi-group of a *diffusion process* but Kac's one P_{t-s}^c is not.

Remark. The renormalized process will play an important role in Chapter 5, in which we consider the case of creation and killing $c(t, x)$ with singularity. If $c(t, x)$ is singular, some additional conditions will be needed to guarantee property (2.41) of the function $\xi(s, x)$ defined in (2.40).

2.8. Time Change

In this and the following sections, we consider time-homogeneous diffusion processes. If we observe a diffusion process $\{X_t, P_x\}$ with a defective clock, then the movement of the process looks slower or faster even though it stays on the same path. This is the so-called "time change".

Let $c(x)$ be a positive continuous function and set

$$\tau(t, \omega) = \int_0^t c(X_s(\omega))ds,$$

and the *time-change function*

$$\tau^{-1}(s, \omega) = \sup \{t: \tau(t, \omega) \le s\}.$$

We define a new process by

(2.47)
$$Y_t = X_{\tau^{-1}(t)}, \quad t < \zeta,$$

$$= \Delta, \qquad t \ge \zeta,$$

where $\zeta(\omega) = \tau(\infty, \omega)$ and Δ is an extra point.

Lemma 2.1. (Nagasawa-Sato (1963))   *Let G_λ and G_λ^Y, $\lambda > 0$, be the resolvent operators of $\{X_t, P_x\}$ and $\{Y_t, \zeta, P_x\}$, respectively. Then, they satisfy*

(2.48)
$$G_\lambda^Y f = G_\lambda(cf) - \lambda G_\lambda\{(c - 1)G_\lambda^Y f\},$$

(2.49)
$$G_\lambda f = G_\lambda^Y(fc^{-1}) - \lambda G_\lambda^Y\{(c^{-1} - 1)G_\lambda f\}.[28]$$

Proof.   We set $f(\Delta) = 0$.   By the definition of resolvent operators of semi-groups

(2.50) $$G_\lambda^Y f(x) = P_x[\int_0^\infty f(Y_t)e^{-\lambda t} dt] = P_x[\int_0^\infty f(X_s)e^{-\lambda \tau(s)} d\tau(s)].$$

Therefore, we have

$$\lambda G_\lambda\{(c - 1)G_\lambda^Y f\} = \lambda P_x[\int_0^\infty \{c(X_t) - 1)G_\lambda^Y f(X_t)e^{-\lambda t} dt]$$

$$= \lambda \int_0^\infty dt P_x[e^{-\lambda t}(c(X_t) - 1)P_{X_t}[\int_0^\infty d\tau(r)f(X_r)e^{-\lambda \tau(r)}]].$$

Because of the Markov property and $\tau(t + r, \omega) = \tau(t, \omega) + \tau(r, \theta_t\omega)$, where θ_t is the shift operator, $X_r(\theta_t\omega) = X_{t+r}(\omega)$,

$$= \lambda \int_0^\infty dt P_x[e^{-\lambda t}(c(X_t) - 1)e^{\lambda \tau(t)}\int_t^\infty d\tau(s) f(X_s)e^{-\lambda \tau(s)}]$$

---

[28] For a general form of the formulae, cf. Theorem 2.1 in Nagasawa-Sato (1963)

$$= P_x[\int_0^\infty d\tau(s)e^{-\lambda\,\tau(s)}f(X_s)\int_0^s dt(\lambda c(X_t) - \lambda)e^{-\lambda t+\lambda\,\tau(t)}]$$

$$= P_x[\int_0^\infty d\tau(s)e^{-\lambda\,\tau(s)}f(X_s)(e^{-\lambda s+\lambda\,\tau(s)} - 1)]$$

$$= P_x[\int_0^\infty d\tau(s)e^{-\lambda s}f(X_s)] - P_x[\int_0^\infty d\tau(s)e^{-\lambda\,\tau(s)}f(X_s)]$$

$$= G_\lambda(cf) - G_\lambda^Y(f),$$

which proves formula (2.48). Formula (2.49) can be shown in the same way.

Then we have

Theorem 2.4. *The time changed process* $\{Y_t, \zeta, P_x\}$ *is a diffusion process with the generator*

(2.51) $$A^Y = \frac{1}{c}A,$$

where A *and* A^Y *denote the generators of the diffusion process* $\{X_t, P_x\}$ *and* $\{Y_t, \zeta, P_x\}$ *defined respectively through*

$$\lambda - A = G_\lambda^{-1},$$

$$\lambda - A^Y = (G_\lambda^Y)^{-1}.$$

Proof. Apply $(\lambda - A)$ to both sides of formula (2.48). Then

$$(\lambda - A)G_\lambda^Y f = (\lambda - A)G_\lambda\{cf - \lambda(c - 1)G_\lambda^Y f\}$$

$$= cf - \lambda(c - 1)G_\lambda^Y f,$$

since $(\lambda - A) = G_\lambda^{-1}$. Therefore,

$$\lambda c G_\lambda^Y f - A G_\lambda^Y f = cf,$$

from which follows

$$(\lambda - \frac{1}{c}A)\, G_\lambda^Y f = f,$$

which implies (2.51).

2.9. Dirichlet Problem

Let $\{X_t, P_x\}$ be a diffusion process on $\mathbf{R}^d$ determined by an elliptic differential operator A given in (2.1). Let D be a compact connected domain in $\mathbf{R}^d$ with a smooth boundary ∂D, and let T be the first hitting time to the boundary ∂D. Moreover, let $g(x)$ be a continuous function on the boundary ∂D. Then

(2.52) $$u(x) = P_x[e^{-\lambda T} g(X_T)]$$

solves the Dirichlet problem

(2.53)
$$(\lambda - A)u(x) = 0, \quad in \quad D,$$

$$u(x) = g(x), \qquad on \quad \partial D.$$

This assertion is treated in standard textbooks on Markov processes and potentials under more general problem setting,[29] but we shall need no such generality in this book.

Let $U = U_\varepsilon$ be the first leaving time from an ε-neighbourhood of a point $x \in D$. Then

(2.54) $$P_x[e^{-\lambda U} u(X_U)] = P_x[e^{-\lambda U} P_{X_U}[e^{-\lambda T} g(X_T)]]$$

$$= P_x[e^{-\lambda (U(\omega) + T(\theta_U \omega))} g(X_{U(\omega) + T(\theta_U \omega)}(\omega))]$$

$$= P_x[e^{-\lambda T} g(X_T)]$$

$$= u(x),$$

where $U(\omega) + T(\theta_U \omega) = T(\omega)$ and the strong Markov property have been applied.

---

[29] Cf. Dynkin (1965), Blumenthal-Getoor (1968), Port-Stone (1978), Doob (1984), ...

Let $f(x)$ be a bounded continuous function on $\mathbf{R}^d$. Then

$$(2.55) \quad P_x[\int_0^U dt e^{-\lambda t} f(X_t)] = P_x[\int_0^\infty dt e^{-\lambda t} f(X_t)] - P_x[\int_U^\infty dt e^{-\lambda t} f(X_t)]$$

$$= G_\lambda f(x) - P_x[e^{-\lambda U} P_{X_U}[\int_0^\infty dt e^{-\lambda t} f(X_t)]]$$

$$= G_\lambda f(x) - P_x[e^{-\lambda U} G_\lambda f(X_U)].$$

Since $(\lambda - A)u = f$ holds for $u(x) = G_\lambda f(x)$, formula (2.55) yields

$$(2.56) \quad \frac{1}{P_x[U]} \{P_x[e^{-\lambda U} u(X_U)] - u(x)\} = \frac{1}{P_x[U]} P_x[\int_0^U dt e^{-\lambda t} (A - \lambda)u(X_t)],$$

for any u in the domain of the generator A.

The generator A is in the sense of the one in Theorem 2.4, which coincides with Dynkin's one in our case, cf. Dynkin (1965). Since various generators are defined for a semi-group depending on purposes, when we speak of "the generator" of a semi-group, we must be aware of its domain of definition. For detail see books mentioned at footnote 29.

Since the function $u(x)$ defined at (2.52) is λ-harmonic, as is shown in (2.54), if we define "λ-harmonic measure" by

$$H_U(x, B) = P_x[e^{-\lambda U} 1_B(X_U)],$$

then we have

$$u(x) = \int_{\partial U} u(\xi) H_U(x, d\xi),$$

where ∂U denotes the boundary of the ε-neighbourhood, and hence it is differentiable.

Letting ε tend to zero, formula (2.56), which is called Dynkin's formula, yields the first equation of the Dirichlet problem (2.53) for the function $u(x)$ defined in (2.52). Therefore, the second equality being clear, the $u(x)$ solves the Dirichlet problem.

2.10. Feller's One-Dimensional Diffusion Processes

Let us consider a second order differential operator

$$(2.57) \qquad A = \tfrac{1}{2} a(x) \frac{d^2}{dx^2} + b(x) \frac{d}{dx}, \quad a(x) > 0,$$

in an open interval (α, β), and define

$$(2.58) \qquad W(x) = \int_c^x dy \, \frac{2b(y)}{a(y)},$$

where $c \in (\alpha, \beta)$ is arbitrary but fixed. Then, with the function $W(x)$, the operator A can be represented in a divergence form

$$(2.59) \qquad A = \tfrac{1}{2} a(x) e^{-W(x)} \frac{d}{dx} (e^{W(x)} \frac{d}{dx}).$$

As an example let us consider

$$A = \tfrac{1}{2} \frac{d^2}{dx^2} + (\frac{d-1}{2} \frac{1}{x} - x) \frac{d}{dx}, \quad d \geq 2,$$

in $(0, \infty)$. Then, $W(x) = (d - 1) \log x - x^2$ and hence

$$A = \tfrac{1}{2} x^{1-d} e^{x^2} \frac{d}{dx} (x^{d-1} e^{-x^2} \frac{d}{dx}).$$

This case will be treated in Section 7.9 as an example of diffusion processes of Schrödinger equations with singular potentials.

In general, with a given continuous function $W(x)$, we define Feller's *canonical scale* $S(x)$ by

$$(2.60) \qquad S(x) = \int_c^x dy \, e^{-W(y)},$$

where $c \in (\alpha, \beta)$ is arbitrary but fixed, and Feller's *speed measure M*, with a positive continuous function $a(x)$, by

$$(2.61) \qquad \frac{dM}{dx} = \frac{1}{a(x)} e^{W(x)}.$$

In terms of the canonical scale S and the speed measure M, we define Feller's *canonical operator* $\mathcal{A}$ by

(2.62) $$\mathcal{A} = \frac{1}{2} \frac{d}{dM} \frac{d^+}{dS}$$

on a subset

$$\mathcal{D}(\mathcal{A}) = \{f : f \in C([\alpha, \beta]), \ d^+f \ll dS,$$

$$d(\frac{d^+f}{dS}) \ll dM, and \ \frac{d}{dM}(\frac{d^+f}{dS}) \in C([\alpha, \beta]\},[30]$$

where d^+f/dS denotes the Radon-Nikodym derivative of the signed measure induced by f with respect to the measure induced by S. If f is differentiable, then $d^+f/dS = df/dS$.

The diffusion process determined by Feller's canonical operator $\mathcal{A}$ with an appropriate boundary condition is called *Feller's one-dimensional diffusion process*.[31] Feller's diffusion process was constructed by Feller with the help of Hille-Yosida's semi-group theory; and by Itô-McKean (1965) using the transformation theory of diffusion processes; their method will be explained in the following.

First we construct a diffusion process determined by

(2.63) $$\frac{1}{2} \frac{d}{dS} \frac{d^+}{dS}.$$

Assume that $S(x)$ is defined on $\mathcal{D}_S = [\alpha, \beta]$, and let $\mathcal{R}_S = $ *the range of* S.

Let $\{B_t, P_x\}$ be a one-dimensional Brownian motion, and set

(2.64) $$\zeta(\omega) = \inf \{t : B_t(\omega) \notin \mathcal{R}_S\}.$$

Then we define a diffusion process on the transformed state space $\mathcal{D}_S$ by

(2.65) $$Y_t = \begin{cases} S^{-1}(B_t), & for \ t < \zeta, \\ \Delta, & for \ t \geq \zeta, \end{cases}$$

---

[30] The interval may be half-open or open

[31] Interesting phenomena of Feller's diffusion process are discussed in Brox (1986) when $W(x)$ is a Brownian path. Cf. also Tanaka (1987), Kawazu-Tamura-Tanaka (1992)

$$Q_x = P_{S(x)}, \qquad for \quad x \in \mathbf{D}_S,$$

where Δ denotes an extra point.

Lemma 2.2. *The diffusion process* $\{Y_t, t < \zeta, Q_x, x \in \mathbf{D}_S\}$ *defined in* (2.65) *is determined by the second order differential operator*

$$\frac{1}{2}\frac{d}{dS}\frac{d^+}{dS}.$$

Proof. The (strong) Markov property of the transformed process is easy to show and left as an exercise. For $f \in C^2(\mathbf{D}_S)$, $f(\Delta) = 0$,

$$\lim_{h \downarrow 0} \frac{1}{h}\{Q_x[f(Y_h)] - Q_x[f(Y_0)]\}$$

$$= \lim_{h \downarrow 0} \frac{1}{h}\{P_{S(x)}[f(S^{-1}(B_h))] - P_{S(x)}[f(S^{-1}(B_0))]\}$$

$$= \frac{1}{2}\frac{d}{dy}\frac{d}{dy}f(S^{-1}(y)), \quad \text{where set } y = S(x),$$

$$= \frac{1}{2}\frac{d}{dS(x)}\frac{d}{dS(x)}f(x),$$

which completes the proof.

Let us consider a simple example:

$$\mathcal{A} = \frac{1}{2}x^2\frac{d}{dx}(x^2\frac{d}{dx}),$$

with

$$S(x) = -\frac{1}{x} + 1,$$

where $\mathbf{D}_S = (0, \infty)$ and $\mathcal{R}_S = (-\infty, 1)$. Then

$$T_0(Y) = \inf\{t: Y_t = 0\} = \inf\{t: B_t = -\infty\} = T_{-\infty}(B) = \infty,$$

$$T_\infty(Y) = \inf\{t: Y_t = \infty\} = \inf\{t: B_t = 1\} = T_1(B) < \infty.$$

Therefore, the origin $\{0\}$ is an inaccessible point of the diffusion process Y_t, while $\{\infty\}$ is accessible. Since $S^{-1}(y) = (1 - y)^{-1}$,

$$Y_t = \frac{1}{1 - B_t} \quad for \quad t < \zeta,$$

$$Q_x = P_{1 - \frac{1}{x}} \quad for \quad x \in \boldsymbol{D}_S = (0, \infty).$$

Now returning to the starting point, we consider the diffusion process determined by Feller's canonical operator

$$\mathcal{A} = \frac{1}{2} \frac{d}{dM} \frac{d^+}{dS}$$

$$= \frac{1}{2} a(x) e^{-2W(x)} \frac{d}{dS} \frac{d^+}{dS},$$

the expression of which suggests an application of time change. Define Kac's additive functional

(2.66)
$$\tau(t) = \int_0^t dr \, \frac{e^{2W(Y_r)}}{a(Y_r)},$$

with which we apply "time-change" to the diffusion process $\{Y_t, t < \zeta, Q_x, x \in \boldsymbol{D}_S\}$ in Lemma 2.2.

Then we get

Theorem 2.5. (Itô-McKean (1965)) *Feller's canonical diffusion process* $\{Z_t, t < \tau(\zeta), Q_x, x \in \boldsymbol{D}_S\}$[32] *determined by*

$$\mathcal{A} = \frac{1}{2} \frac{d}{dM} \frac{d^+}{dS}$$

$$= \frac{1}{2} a(x) e^{-W(x)} \frac{d}{dx} (e^{W(x)} \frac{d^+}{dx})$$

is given through time change of the diffusion process Y_t *in Lemma* 2.2, *namely,*

(2.67)
$$Z_t = Y(\tau^{-1}(t)) = S^{-1}(B(\tau^{-1}(t))),$$

$$Q_x = P_{S(x)}, \quad for \quad x \in \boldsymbol{D}_S,$$

---

[32] ζ is defined at (2.59)

where τ is defined at (2.66) and, with $c \in (\alpha, \beta)$,

$$S(x) = \int_c^x dy \, e^{-W(y)}.$$

2.11. Feller's Test[33]

Let $\{X_t, t < \tau(\zeta), Q_x, x \in \boldsymbol{D}_S = (\alpha, \beta)\}$[34] be Feller's canonical diffusion process. We assume that the process is regular in (α, β), namely,

$$P_x[T_y < \infty] > 0, \quad for \quad \forall \, x, y \in (\alpha, \beta),$$

where T_y is the first hitting time

(2.68) $T_y = \inf \{t > 0 : X_t = y\}.$

A classification of boundary points was given by Feller (1957) in terms of the canonical scale S and the speed measure M. Let us formulate it for the left boundary point $\{\alpha\}$. Denote

$$S(\alpha, x] = S(x) - \lim_{y \downarrow \alpha} S(y),$$

(2.69)

$$M(\alpha, x] = M((\alpha, x]).$$

Then, the boundary point $\{\alpha\}$ is classified as follows (*Feller's Test*):

$\{\alpha\}$ is if	**Regular**	**Exit**	**Entrance**	**Natural**
$S(\alpha, x]$	$< \infty$	$< \infty$	$= \infty$	$= \infty$
and				
$M(\alpha, x]$	$< \infty$	$= \infty$	$< \infty$	$= \infty$

---

[33] At the first reading this section may be skipped until Section 7.9
[34] We denote Feller's canonical diffusion process again by X_t instead of Z_t

Theorem 2.6. (Feller (1957)) *If the left boundary point $\{\alpha\}$ is*

(i) *regular, then $\{\alpha\}$ is accessible from (α, β), and (α, β) is accessible from $\{\alpha\}$,*

(ii) *exit, then $\{\alpha\}$ is accessible from (α, β), but (α, β) is inaccessible from $\{\alpha\}$,*

(iii) *entrance, then $\{\alpha\}$ is inaccessible from (α, β), but (α, β) is accessible from $\{\alpha\}$,*

(iv) *natural, then $\{\alpha\}$ is inaccessible from (α, β), and (α, β) is inaccessible from $\{\alpha\}$.*

Proof. Let us define for $\lambda > 0$ and $y \in (\alpha, \beta)$

(2.70)
$$P_\alpha[e^{-\lambda T_y}] = \lim_{x \downarrow \alpha} P_x[e^{-\lambda T_y}],$$
$$P_y[e^{-\lambda T_\alpha}] = \lim_{x \downarrow \alpha} P_y[e^{-\lambda T_x}].$$

It is clear that $\{\alpha\}$ is accessible from $y \in (\alpha, \beta)$ if $P_y[e^{-\lambda T_\alpha}] > 0$, while inaccessible if $P_y[e^{-\lambda T_\alpha}] = 0$; and (α, β) is accessible from $\{\alpha\}$, if $P_\alpha[e^{-\lambda T_y}] > 0$, but inaccessible if $P_\alpha[e^{-\lambda T_y}] = 0$. Therefore, our proof is reduced to the evaluation of

(2.71) $$u^y(x) = P_x[e^{-\lambda T_y}] \quad or \quad w^y(x) = \frac{1}{P_y[e^{-\lambda T_x}]}.$$

Let $y \in (\alpha, \beta)$ be fixed. We have shown in Section 2.9 that

(2.72) $$u(x) = P_x[e^{-\lambda T_y} g(X_{T_y})]$$

solves the Dirichlet problem

(2.73)
$$(\lambda - \mathcal{A})u = 0, \quad in \quad (\alpha, \beta) \setminus \{y\},$$
$$u(y) = g(y),$$

in one-dimension. Therefore, $u(x) = u^y(x) = P_x[e^{-\lambda T_y}]$, for a fixed y, satisfies (2.73) with $g(y) = 1$, namely it is λ-harmonic in $(\alpha, \beta) \setminus \{y\}$.

Now let us define

(2.74)

$$\sigma_S\,[c,y] = \int_c^y S[c,z]M(dz),$$

$$\sigma_M\,[c,y] = \int_c^y M[c,z]dS(z),$$

and

(2.75)

$$\sigma_S(\alpha,y] = \lim_{c\downarrow\alpha}\sigma_S[c,y],$$

$$\sigma_M(\alpha,y] = \lim_{c\downarrow\alpha}\sigma_M[c,y].$$

Then, it is routine to check that the boundary point $\{\alpha\}$ can be classified also in terms of $\sigma_S(\alpha,y]$ and $\sigma_M(\alpha,y]$ as follows (*Feller's Test*):

$\{\alpha\}$ is if	**Regular**	**Exit**	**Entrance**	**Natural**
$\sigma_S(\alpha,y]$	$< \infty$	$< \infty$	$= \infty$	$= \infty$
and				
$\sigma_M(\alpha,y]$	$< \infty$	$= \infty$	$< \infty$	$= \infty$

Lemma 2.3.

(2.76) $P_y[\,e^{-\lambda T_\alpha}\,] > 0,\ \text{for } y \in (\alpha,\beta) \iff \sigma_S(\alpha,\cdot\,] < \infty,$

(2.77) $P_y[\,e^{-\lambda T_\alpha}\,] = 0,\ \text{for } y \in (\alpha,\beta) \iff \sigma_S(\alpha,\cdot\,] = \infty;$

namely the left boundary point $\{\alpha\}$ is accessible, if it is regular or exit, while inaccessible if it is entrance or natural.

Proof. Assume $\sigma_S(\alpha,\cdot\,] < \infty$. Then $S(\alpha,\cdot\,] < \infty$. Let $w(x)$ be a non-negative decreasing solution of

(2.78) $(\lambda - \mathcal{A})w = 0,$ in $(\alpha, y),$ $with$ $w(y) = 1,$

and hence

$$\frac{1}{2}\frac{d}{dM}\frac{dw}{dS} = \lambda w, \; in \; (\alpha, y).$$

Integrating twice with respect to dM and then dS, we have

$$\frac{dw}{dS}(y)S[x, y] - \{w(y) - w(x)\} = 2\lambda \int_x^y dS(\xi)\int_\xi^y w(z)dM(z),$$

which yields, through the substitution $w(y) = 1$,

(2.79) $w(x) = 1 + (-\dfrac{dw}{dS}(y))S[x, y] + 2\lambda \displaystyle\int_x^y dS(\xi)\int_\xi^y w(z)dM(z).$

Since $w(x) \geq w(z)$ for $x \geq z$ and

(2.80) $\displaystyle\int_x^y dS(\xi)\int_\xi^y dM(z) = \int_x^y dS(\xi)M[\xi, y] = \int_x^y S[x, \xi]dM(\xi) = \sigma_S[x, y]$

by partial integration, we have

$$w(x) \leq 1 + (-\frac{dw}{dS}(y))S[x, y] + 2\lambda w(x)\sigma_S[x, y].$$

Because of the assumption $\sigma_S(\alpha, \cdot] < \infty$, there exists y_0 such that

$$2\lambda \sigma_S[x, y_0] < \frac{1}{2},$$

and hence

$$\frac{1}{2}w(x) \leq 1 + (-\frac{dw}{dS}(y_0))S[x, y_0],$$

which yields

$$\lim_{x \downarrow \alpha} w(x) \leq 2\{1 + (-\frac{dw}{dS}(y_0))S(\alpha, y_0]\} < \infty.$$

Let $\alpha < c < x < y$. Then, by the strong Markov property, we have

(2.81) $$P_y[e^{-\lambda T_c}] = P_y[e^{-\lambda T_x}] P_x[e^{-\lambda T_c}].$$

Therefore, for a fixed y,

(2.82) $$w(x) = w^y(x) = \frac{1}{P_y[e^{-\lambda T_x}]} = \frac{P_x[e^{-\lambda T_c}]}{P_y[e^{-\lambda T_c}]} = \text{const } u(x),$$

where $u(x) = P_x[e^{-\lambda T_c}]$ for a fixed c, and hence the $w(x)$ is monotone decreasing and satisfies $w(y) = 1$ and

$$(\lambda - \mathcal{A})w = 0, \quad in \quad (\alpha, y),$$

since we can let $c \downarrow \alpha$, it consequently satisfies (2.78). Therefore, we have

$$\lim_{x \downarrow \alpha} w^y(x) = \frac{1}{P_y[e^{-\lambda T_\alpha}]} < \infty,$$

and hence

$$P_y[e^{-\lambda T_\alpha}] > 0.$$

Conversely, assume $\sigma_S(\alpha, \cdot \,] = \infty$. Since $w(z) \geq w(y) = 1$ for $z \leq y$, formula (2.79) yields

$$w(x) \geq 1 + (-\frac{dw}{dS}(y)) S[x, y] + 2\lambda \int_x^y dS(\xi) \int_\xi^y dM(z),$$

and hence because of (2.80) we have

$$\lim_{x \downarrow \alpha} w(x) \geq 1 + (-\frac{dw}{dS}(y)) S(\alpha, y] + 2\lambda \sigma_S(\alpha, y] = \infty.$$

Since $w(x) = w^y(x) = \frac{1}{P_y[e^{-\lambda T_x}]}$ is a non-negative decreasing solution of (2.78), we have

$$\lim_{x \downarrow \alpha} w^y(x) = \lim_{x \downarrow \alpha} \frac{1}{P_y[e^{-\lambda T_x}]} = \infty,$$

and hence

$$P_y[e^{-\lambda T\alpha}] = 0,$$

which completes the proof of the lemma.

Lemma 2.4. (i) *If the left boundary point $\{\alpha\}$ is regular or entrance, then*

$$P_\alpha[e^{-\lambda T_x}] > 0, \quad for \quad x \in (\alpha, \beta);$$

namely (α, β) is accessible from $\{\alpha\}$.

(ii) *If $\{\alpha\}$ is exit or natural, then*

$$P_\alpha[e^{-\lambda T_x}] = 0, \quad for \quad x \in (\alpha, \beta);$$

namely (α, β) is inaccessible from $\{\alpha\}$.

Proof. Assume that the left boundary point $\{\alpha\}$ is regular or entrance. Let $u(x)$ be an increasing non-negative solution of the Dirichlet problem

(2.83) $(\lambda - \mathcal{A})u = 0 \quad in \quad (\alpha, y), \quad with \quad u(y) = 1,$

and hence

$$\frac{1}{2}\frac{d}{dM}\frac{du}{dS} = \lambda u,$$

from which we have

(2.84) $$\frac{du}{dS}(x) - \frac{du}{dS}(c) = 2\lambda \int_c^x u(z)M(dz).$$

Since $u(x)$ is increasing, $du/dS \geq 0$. On the other hand

$$\frac{d}{dx}\left(\frac{du}{dS}\right) = 2m\mathcal{A}u = 2m\lambda u > 0,$$

where $m = e^W/a$, and hence du/dS is strictly increasing in the interval (α, y). Consequently, we have

(2.85) $$1 \geq u(y) - u(\alpha) = \int_\alpha^y dS\frac{du}{dS} > \frac{du}{dS}(\alpha)S(\alpha, y],$$

which yields

(2.86) $\frac{du}{dS}(\alpha) S(\alpha, y] \leq \gamma \{u(y) - u(\alpha)\}, \quad with \quad \gamma < 1.$

Therefore, (2.84) together with (2.86), implies

$$u(y) - u(\alpha) = 2\lambda \int_{\alpha}^{y} dS(\xi) \int_{\alpha}^{\xi} u(z) M(dz) + \gamma \{u(y) - u(\alpha)\}.$$

Since $u(z) \leq u(y)$ for $z \leq y$, we have

$(1 - \gamma)\{u(y) - u(\alpha)\}$

$$\leq 2\lambda u(y) \int_{\alpha}^{y} dS(\xi) \int_{\alpha}^{\xi} M(dz) = 2\lambda u(y) \sigma_M(\alpha, y] < \infty,$$

where $\sigma_M(\alpha, y] < \infty$, because of the assumption that $\{\alpha\}$ is regular or entrance. Since $\sigma_M(\alpha, y] \downarrow 0$ (as $y \downarrow \alpha$), there exist $y_0 \in (\alpha, y)$ such that $2\lambda\sigma_M(\alpha, y_0] < 1 - \gamma$, and hence we have

$$u(y_0) - u(\alpha) < u(y_0),$$

namely,
$$u(\alpha) > 0.$$

Therefore, since the function

$$u(x) = u^y(x) = P_x[e^{-\lambda T_y}]$$

is an increasing non-negative solution of the Dirichlet problem (2.83), we have

$$u^y(\alpha) = P_\alpha[e^{-\lambda T_y}] > 0,$$

which proves the first assertion of the lemma.

Let us prove the second assertion. Assume $\sigma_M(\alpha, \cdot] = \infty$. Integrating both sides of (2.84) with respect to dS, we have

$$u(x) - u(c) - \frac{du}{dS}(c)\,S[c,x] = 2\lambda \int_c^x dS(\xi) \int_c^\xi u(z)\,M(dz).$$

Since $u(c) \le u(z) \le u(y) = 1$,

$$1 \ge u(x) - u(c) \ge \frac{du}{dS}(c)\,S[c,x] + 2\lambda u(c) \int_c^x dS(\xi)\,M[c,\xi],$$

and hence

$$1 \ge 2\lambda \lim_{c \downarrow \alpha} u(c)\,\sigma_M(\alpha,x],$$

which implies $\lim_{c \downarrow \alpha} u(c) = 0$, since $\sigma_M(\alpha,x] = \infty$. Applying this to $u(x) = u^y(x) = P_\alpha[e^{-\lambda T_y}]$, we have

$$\lim_{c \downarrow \alpha} u(c) = P_\alpha[e^{-\lambda T_y}] = 0,$$

which proves the second assertion of the lemma.

Lemma 2.3 and Lemma 2.4 complete the proof of Theorem 2.6.

As an example let us consider

(2.87)
$$A = \frac{1}{2}\frac{d^2}{dx^2} + \varepsilon\frac{1}{x}\frac{d}{dx},$$

in $(0, \infty)$, which includes the case of the radial part of the d-dimensional Brownian motion, i.e., d-Bessel process with $\varepsilon = (d - 1)/2$. Then,

(2.88)
$$W(x) = 2\varepsilon \log x,$$

and hence

(2.89)
$$M(c,x] = \int_c^x dy\,e^{W(y)} = \int_c^x dy\,y^{2\varepsilon}$$

$$= \begin{cases} \dfrac{1}{1 + 2\varepsilon}(x^{1+2\varepsilon} - c^{1+2\varepsilon}), & \text{for } \varepsilon \ne -1/2, \\ \log x - \log c, & \text{for } \varepsilon = -1/2. \end{cases}$$

Therefore

$$(2.90) \qquad M(0,x] \begin{cases} < \infty, & if \ \varepsilon > -1/2, \\ = \infty, & if \ \varepsilon \le -1/2. \end{cases}$$

On the other hand

$$(2.91) \qquad S[c,x] = \int_c^x dy \, e^{-W(y)} = \int_c^x dy \, y^{-2\varepsilon}$$

$$= \begin{cases} \dfrac{1}{1-2\varepsilon}(x^{1-2\varepsilon} - c^{1-2\varepsilon}), & for \ \varepsilon \ne 1/2, \\ \log x - \log c, & for \ \varepsilon = 1/2, \end{cases}$$

and hence

$$(2.92) \qquad S(0,x] \begin{cases} < \infty, & if \ \varepsilon < 1/2, \\ = \infty, & if \ \varepsilon \ge 1/2. \end{cases}$$

Consequently, the origin $\{0\}$ is

$$(2.93) \qquad \begin{array}{ll} "exit", & if \ \varepsilon \le -1/2, \\ "regular", & if \ -1/2 < \varepsilon < 1/2, \\ "entrance", & if \ 1/2 \le \varepsilon. \end{array}$$

Another example which will be considered in Chapter 7 is

$$(2.94) \qquad A = \frac{1}{2}\frac{d^2}{dx^2} + (\varepsilon\frac{1}{x} - x)\frac{d}{dx}, \quad in \ (0,\infty).$$

In this case

$$(2.95) \qquad W(x) = 2\varepsilon \log x - x^2.$$

Since the term $-x^2$ vanishes near the origin, it does not contribute the divergence or convergence of the integrals $M(0,x]$ and $S(0,x]$, and hence (2.93) also holds for the diffusion process determined by (2.94).

Chapter III
Duality and Time Reversal of Diffusion Processes

Duality and time reversal of time-homogeneous diffusion processes will be discussed in the first and second sections. Theorems in these sections will be applied to time-inhomogeneous diffusion processes in the third and fourth sections. Moreover, two different representations of a diffusion process will be established. They will play a crucial role in connection with quantum mechanics in Chapter 4.

3.1. Kolmogoroff's Duality

Motivated by Schrödinger (1931), Kolmogoroff proved the existence of a system of absolute probabilities (an entrance law) of a given transition matrix and characterized time reversal of Markov chains in terms of duality of transition matrices with respect to an entrance law. In his paper published in 1936, he begins his article with

"Die nachfolgenden Betrachtungen scheinen mir, trotz ihrer Einfachheit, neu und nicht ohne Interesse für gewisse physikalische Anwendungen zu sein, insbesondere für die Analyse der Umkehrbarkeit der statistischen Naturgesetze, welche Herr Schrödinger[1] im Falle eines speziellen Beispiels durchgeführt hat."

During the next year, generalizing Schrödinger (1931) and the result mentioned above, Kolmogoroff (1937) published a diffusion version of time reversal, in which he characterized the time reversal of a diffusion process

---
[1] Cf. Schrödinger (1931)

in terms of a duality relation of the drift coefficients of the diffusion process and its time reversal. Especially in time symmetric cases he gave his celebrated formula establishing a relation between the drift coefficient $a(x)$ and positive invariant distribution density $\mu(x) > 0$:

$$(3.1) \qquad a(x) = \frac{1}{2} \sigma^T \sigma(x) \nabla \log \mu(x),$$

where the diffusion process is determined by an elliptic differential operator

$$A = \frac{1}{2} \Delta + a(x) \cdot \nabla,$$

where, Δ denotes the Laplace-Beltrami operator. We assume $\sigma(x)$ and $a(x)$ are bounded and sufficiently smooth.

Kolmogoroff remarks that formula (3.1) identifies the invariant density $\mu(x)$ with the Gibbs distribution of a potential $U(x)$, namely

$$\mu = e^U,$$

when $a(x)$ has a potential function $U(x)$,

$$a = \frac{1}{2} \sigma^T \sigma \nabla U.$$

This fact will find an important implication in Chapter 8.

It will be said that a time-homogeneous diffusion process $\{X_t, P_\mu\}$ is *determined by an elliptic differential operator* A, when the process is defined in terms of the fundamental solution $p_t(x, y)$ of the diffusion equation

$$\frac{\partial p}{\partial t}(t, x) = A p(t, x), \quad t > 0,$$

as discussed in Chapter 2, or $\{X_t, P_\mu\}$ is called the diffusion process with the *initial distribution* μ and the *transition probability density* $p_t(x, y)$.

Generalizing the results in Kolmogoroff (1937), let us consider reflecting diffusion processes on $\overline{D} = D \cup \partial D$, where D is a domain in $\mathbf{R}^d$ with a piece-wise smooth boundary ∂D.

A *reflecting diffusion process* on $\overline{D}$ can be constructed by means of finite dimensional distributions (2.4) in terms of the fundamental solution $p_t(x, y)$ of a diffusion equation with the reflecting boundary condition:

$$\frac{\partial p}{\partial t}(t,x) = A p(t,x), \quad in \ D, \ t > 0,$$

$$\frac{\partial p}{\partial n}(t,x) = 0, \qquad on \ \partial D, \ t > 0,$$

where $\partial/\partial n$ denotes the inner normal derivative on the boundary ∂D. If $\partial^2(\sigma^T\sigma(x))^{ij}/\partial x^k\partial x^h$ and $\partial b^i(x)/\partial x^k$ are locally uniformly Hölder continuous and the boundary ∂D is piece-wise C^3, there exists a unique fundamental solution $p_t(x,y)$ for the diffusion equation with the reflecting boundary condition.[2] If the time symmetry is not assumed, we obtain a pair of reflecting diffusion processes in duality, which will be shown to be time reversal of each other in the next section.

Using stochastic differential equations with local times, one can construct a reflecting diffusion process on $\overline{D} = D \cup \partial D$, that does not rely on the existence of the fundamental solution.[3] This will be explained in Chapter 7 when we shall apply it to a system of interacting diffusion processes.

Let μ be a non-negative measurable function. Then, $\mu(x)\sqrt{\sigma_2(x)}dx$ is an *invariant measure* for the reflecting diffusion process, if

$$(3.2) \qquad \int_{\overline{D}} P_t f(x)\, \mu(x)\sqrt{\sigma_2(x)}\, dx = \int_{\overline{D}} f(x)\, \mu(x)\sqrt{\sigma_2(x)}\, dx,$$

for $\forall f \in C(\overline{D})$, where P_t denotes the semi-group defined by

$$P_t f(x) = \int p_t(x,y)\sqrt{\sigma_2(y)}\, dy\, f(y)$$

with the transition probability density $p_t(x,y)$, which is the fundamental solution of the diffusion equation with the reflecting boundary condition.

From the definition, it immediately follows that $\mu(x)$ is an invariant distribution density if and only if

$$<\mu, Af> = 0, \ \ for \ \ \forall f \in C^2(\overline{D}) \cap \{f: \frac{\partial f}{\partial n} = 0, on \ \partial D\},$$

---

[2] Cf. S. Ito (1957)

[3] Cf., e.g. Tanaka (1979), Lions-Sznitman (1984), Ikeda-Watanabe (1981, 89),

where

$$<f,g> = \int_D f(x)g(x)\sqrt{\sigma_2(x)}\,dx.$$

Therefore, with the help of the Gauss-Green formula it is easy to see that $\mu \in C^2(\overline{D})$ is an invariant density, if and only if

$$A^\circ \mu = 0, \quad in \ D,$$

$$\frac{1}{2}\frac{\partial \mu}{\partial n} - a_n\, \mu = 0, \quad on \ \partial D,$$

where n denotes the inner normal vector on ∂D, $a_n = a \cdot n$, and A° is the formal adjoint of A,

$$A^\circ \mu = \frac{1}{2}\Delta \mu - \frac{1}{\sqrt{\sigma_2}}\nabla \cdot (\sqrt{\sigma_2}\,a\mu).$$

The formal adjoint A° of A is defined with respect to the bilinear form $<f,g>$ defined above in terms of the volume element $\sqrt{\sigma_2(x)}\,dx$.

Besides the formal adjoint, when we discuss *duality* of a pair of reflecting diffusion processes on $\overline{D}$, we adopt another bilinear form

$$<f,g>_\mu = \int_{\overline{D}} f(x)g(x)\mu(x)\sqrt{\sigma_2(x)}\,dx,$$

where μ satisfies the Fokker-Planck equation $A^\circ \mu = 0$ in D with the boundary condition $\partial \mu/\partial n - 2a_n\, \mu = 0$ on ∂D.

A pair of reflecting diffusion processes, in other words their semi-groups P_t and $\hat{P}_t$, are *in duality* with respect to an invariant density μ, if

(3.3) $$<g,P_t f>_\mu = <\hat{P}_t g, f>_\mu,$$

for $\forall f, g \in C(\overline{D})$. This is equivalent to the equation (3.8) below in terms of a pair of elliptic operators A and $\hat{A}$ with the reflecting boundary condition.

It should be remarked that the measure $\mu(x)\sqrt{\sigma_2(x)}dx$ has an intrinsic meaning as an invariant measure, but the volume element $\sqrt{\sigma_2(x)}dx$ has no such probabilistic meaning.

Theorem 3.1. (Nagasawa (1961)) *Let $\mu \in C^2(\overline{D})$ be positive in D. Then, $\mu(x)\sqrt{\sigma_2(x)}dx$ is an invariant measure for a pair of reflecting diffusion processes with drift coefficients $a(x)$ and $\hat{a}(x)$, respectively, and they are in duality with respect to the measure $\mu(x)\sqrt{\sigma_2(x)}dx$, if and only if the drift coefficients satisfy*

(3.4) $$a + \hat{a} = \sigma^T \sigma \nabla(\log \mu), \quad in \ \overline{D},$$

(3.5) $$\frac{1}{\sqrt{\sigma_2}} \nabla \{ \sqrt{\sigma_2}(a - \hat{a})\mu \} = 0, \quad in \ \overline{D},$$

$$(a_n - \hat{a}_n)\mu = 0, \quad on \ \partial D.$$

Moreover, if the difference of the drift coefficients is given by a potential function S, then

(3.6)
$$a + \hat{a} = 2\sigma^T \sigma \nabla R,$$

$$a - \hat{a} = 2\sigma^T \sigma \nabla S,$$

where $\log \mu = 2R$.

Let us call (3.4) (or (3.6)) with (3.5) simply the **duality relation** of drift coefficients (cf. (3.8) and (3.12) below).

Proof. We consider a pair of reflecting diffusion processes on $\overline{D}$ determined by $\{A, \partial/\partial n\}$ and $\{\hat{A}, \partial/\partial n\}$, where

$$Af = \tfrac{1}{2}\Delta f + a \cdot \nabla f,$$

$$\hat{A} g = \tfrac{1}{2}\Delta g + \hat{a} \cdot \nabla g,$$

and apply a duality formula between A and $\hat{A}$, which can be verified easily using the Gauss-Green formula; namely,[4] for $f, g, \mu \in C^2(\overline{D})$,

---

[4] Cf. Nagasawa (1961)

(3.7)
$$\int_D \{(Af)g - f(\widehat{A}g)\}\mu \sqrt{\sigma_2}\,dx$$

$$= -\frac{1}{2}\int_{\partial D} \{\frac{\partial f}{\partial n} g - f\frac{\partial g}{\partial n}\}\mu \sqrt{\sigma_2}\,d\tilde{x}$$

$$+ \int_D fg\,(A^\circ\mu)\sqrt{\sigma_2}\,dx$$

$$+ \int_{\partial D} fg\,\{\frac{1}{2}\frac{\partial \mu}{\partial n} - a_n\,\mu\}\sqrt{\sigma_2}\,d\tilde{x}$$

$$- \int_D f\{a + \widehat{a} - \sigma^T\sigma\nabla\log\mu\}\cdot\nabla g\,\mu \sqrt{\sigma_2}\,dx,$$

where dx (resp. $d\tilde{x}$) denotes the Lebesgue measure in D (resp. on the boundary ∂D), $a_n = a\cdot n$, n the inner normal vector on ∂D.

Let us first assume that the two diffusion processes are in duality with respect to the invariant measure $\mu(x)\sqrt{\sigma_2(x)}\,dx$, namely

(3.8)
$$<Af, g>_\mu = <f, \widehat{A}g>_\mu$$

holds for $\forall f, g \in C^2(\overline{D}) \cap \{f : \dfrac{\partial f}{\partial n} = 0 \text{ on } \partial D\}$.

Then, on the right-hand side of the formula in (3.7), all integrals except for the last one vanish, since

$$A^\circ\mu = 0, \ \ in\ D, \ \ and \ \ \frac{1}{2}\frac{\partial \mu}{\partial n} - a_n\,\mu = 0, \ \ \frac{\partial f}{\partial n} = 0, \ \ \frac{\partial g}{\partial n} = 0, \ \ on\ \partial D.$$

Therefore, formula (3.7) combined with (3.8) implies

$$\int_D f\{a + \widehat{a} - \sigma^T\sigma\nabla\log\mu\}\cdot\nabla g\,\mu \sqrt{\sigma_2}\,dx = 0\,,$$

for $\forall f, g \in C^2(\overline{D}) \cap \{f : \dfrac{\partial f}{\partial n} = 0 \text{ on } \partial D\}$, and hence

$$a + \widehat{a} - \sigma^T \sigma \nabla \log \mu = 0,$$

which is (3.4).

The function $\mu(x)$ is an invariant distribution density of both processes if and only if

(3.9)
$$\frac{1}{2}\Delta\mu - \frac{1}{\sqrt{\sigma_2}}\nabla\cdot(\sqrt{\sigma_2}\,a\,\mu) = 0, \ in \ D, \quad \frac{1}{2}\frac{\partial\mu}{\partial n} - a_n\,\mu = 0, \ on \ \partial D,$$

$$\frac{1}{2}\Delta\mu - \frac{1}{\sqrt{\sigma_2}}\nabla\cdot(\sqrt{\sigma_2}\,\widehat{a}\,\mu) = 0, \ in \ D, \quad \frac{1}{2}\frac{\partial\mu}{\partial n} - \widehat{a}_n\,\mu = 0, \ on \ \partial D,$$

hold at the same time. Therefore, subtracting the second line from the first, we get (3.5).

Conversely, the formula in (3.4), namely

$$(a + \widehat{a})\mu = \sigma^T\sigma\nabla\mu$$

implies

$$\frac{1}{\sqrt{\sigma_2}}\nabla\cdot(\sqrt{\sigma_2}\,(a + \widehat{a})\,\mu) = \frac{1}{\sqrt{\sigma_2}}\nabla\cdot(\sqrt{\sigma_2}\,\sigma^T\sigma\nabla\mu),$$

which, combined with (3.5), yields (3.9), and hence $\mu(x)\sqrt{\sigma_2(x)}\,dx$ is an invariant measure of the pair of reflecting diffusion processes. Moreover, because of (3.4) the right-hand side of (3.7) vanishes; and hence we have (3.8), namely, the reflecting diffusion processes are in duality. This completes the proof.

If the distribution density $\mu(x)$ is positive and bounded on $\overline{D}$, there aren't many interesting mathematical problems in this context. However, if the density $\mu(x)$ vanishes or tends to infinity on a subset of the boundary ∂D, then we meet immediately a difficult but interesting problem of the same kind as was analyzed thoroughly and completely by Feller in one dimension in the 1950's (see Section 2.10, but we will not rely on Feller's results except in Section 7.9), because the drift coefficient a (resp. $\widehat{a}$) diverges on the set $\{x: \mu(x) = 0 \ or \ \mu(x) = \infty\}$ as is clear from the formula (3.4).[5] In higher dimensions, this problem of Feller's is not yet solved except in some special cases. As a matter of fact, concerning the zero set $\{x: \mu(x) = 0\}$, even though we will allow μ to vanish on a subset of ∂D in this monograph, we shall see the zero sets of invariant distribution densities are of a special kind, namely *inaccessible* in typical cases, so that we can avoid complicated mathematical problems which would be encountered

---

[5] See examples in Section 7.9, and also in appendix of Nagasawa-Tanaka (1985)

otherwise.

Typical examples are provided by solutions of Schrödinger equations. Let us consider an excited state of the one-dimensional harmonic oscillator

$$-\frac{1}{2}\frac{d\varphi}{dx^2} + \frac{1}{2}x^2\,\varphi = E\varphi,$$

say, the first exited state

$$\varphi(x) = \beta^{-1/2}x\,e^{-x^2/2}.$$

The distribution density $\mu(x) = \varphi^2(x)$ with $\varphi \in C^2(\mathbf{R}^1)$ has zeros, and the zero set of μ is inaccessible according to the Feller test (cf., Section 2.10, but this will be shown in higher dimensions independent of Feller's test, cf. also Section 7.9); namely, by the zero set of $\mu(x) = \varphi^2(x)$ ergodic decomposition occurs. Therefore, it is clear that a single diffusion particle *cannot* realize (reproduce) the distribution density $\mu(x)$ but just a part of it. In order to recover the whole $\mu(x)$ we need many diffusion particles which are segregated by the zero set of the given $\mu(x)$. Such ergodic decomposition can occur even in the ground states[6] if an additional singular potential such as $V_1(X) = \gamma^2/2x^{2(1+\gamma)}$ is added, since it induces singular repulsive drift $a_o(x) = \gamma/|x|^{1+\gamma}$ (see Theorem 3.1 and Chapter 7) in the neighbourhood of the origin, and hence it becomes inaccessible.

The relation between diffusion processes and Schrödinger equations was first analyzed by Schrödinger (1931), who established a special kind of symmetric representation (in time reversal) of Brownian motions, which will be discussed later on in this chapter. He looked for an interrelation between diffusion processes and Schrödinger equations. Because of the parallelism between quantum mechanics and the diffusion theory in his formulation, he was convinced that diffusion theory must provide a "better understanding of quantum mechanics", but failed to give definitive conclusion to it. Therefore, let us formulate this as

Schrödinger's Conjecture:

Quantum Mechanics must be a Diffusion Theory.

Secondly, the segregation (*ergodic decomposition*) by the zero set of a flow of distribution densities (in other words the zero set of a solution of the Schrödinger equation, say, for a hydrogen atom) suggests to us that we need

---

[6] Cf. Faris-Simon (1975), see also Ezawa-Klauder-Shepp (1975), Glimm-Jaffe (1987)

many (perhaps infinitely many) particles for the Schrödinger equation. In order to handle a system of infinitely many interacting diffusion particles, we must establish a statistical mechanics for solutions of Schrödinger equations. Therefore, there must be

Statistical Mechanics *for* Schrödinger Equations.

These two problems are the main motivation of this monograph.

A solution to Schrödinger's conjecture will be given in Chapter 4, while the second problem of establishing statistical mechanics for Schrödinger equations will be discussed in Chapter 7 with simple one-dimensional models; general cases in higher dimensions will be treated in Chapter 8.

3.2. Time Reversal of Diffusion Processes

We have established a duality relation of a pair of reflecting diffusion processes (semi-groups) with respect to an invariant measure in the preceding section, especially in terms of their drift coefficients $a(x)$ and $\hat{a}(x)$. In this section we will clarify the probabilistic meaning of duality of diffusion processes. As a matter of fact duality is an analytic counterpart of time reversal of diffusion processes, as will be seen.

Let $\{X_t, P_\mu\}$ be a diffusion process with a transition probability density $p_t(x, y)$ and an initial distribution density $\mu(y)$ with respect to the volume element dy (we omit $\sqrt{\sigma_2}$ for brevity). Moreover, we assume that $\mu(y)$ is an invariant density, and consider its time reversal.

Time reversal of a diffusion process $\{X_t, P_\mu\}$ is defined by

(3.10) $$\widehat{X}_t = X_{L-t}, \quad for \ t \in [0, L],$$

where L is a positive constant.

The time-reversal $\{\widehat{X}_t, P_\mu\}$ is also a diffusion process. To show this we compute its finite dimensional distributions:

For $0 = t_0 < t_1 < \cdots < t_n = L$,

$$P_\mu[f(\widehat{X}_{t_0}, \ldots, \widehat{X}_{t_n})]$$

$$= P_\mu[f(X_{L-t_0}, \ldots, X_{L-t_n})]$$

$$= \int \mu(x_n)dx_n p_{t_n-t_{n-1}}(x_n, x_{n-1})dx_{n-1} p_{t_{n-1}-t_{n-2}}(x_{n-1}, x_{n-2})dx_{n-2} \cdots$$

$$\cdots p_{t_2-t_1}(x_2, x_1)dx_1 p_{t_1}(x_1, x_0)dx_0 f(x_0, \ldots, x_n)$$

$$= \int \mu(x_0)dx_0 \frac{1}{\mu(x_0)} p_{t_1}(x_1, x_0)\mu(x_1)dx_1 \frac{1}{\mu(x_1)} p_{t_2-t_1}(x_2, x_1)\mu(x_2)dx_2 \cdots$$

$$\cdots p_{t_{n-1}-t_{n-2}}(x_{n-1}, x_{n-2})\mu(x_{n-1})dx_{n-1} \frac{1}{\mu(x_{n-1})} p_{t_n-t_{n-1}}(x_n, x_{n-1})\mu(x_n)dx_n f(x_0, \ldots, x_n),$$

where we require the zero set $\{x: \mu(x) = 0\}$ of μ to be Lebesgue measure zero.

Define the adjoint transition probability by

$$\widehat{P}_t(x, dy) = \widehat{p}_t(x, y)dy,$$

with *the time reversed transition probability density*

(3.11)
$$\widehat{p}_t(x, y) = \frac{1}{\mu(x)} p_t(y, x) \mu(y).$$

Then we have

$$P_\mu[f(\widehat{X}_{t_0}, \ldots, \widehat{X}_{t_n})] = \int \mu(x_0)dx_0 \widehat{P}_{t_1}(x_0, dx_1)\widehat{P}_{t_2-t_1}(x_0, dx_1) \cdots$$

$$\cdots \widehat{P}_{t_{n-1}-t_{n-2}}(x_{n-2}, dx_{n-1})\widehat{P}_{t_n-t_{n-1}}(x_{n-1}, dx_n)f(x_0, \ldots, x_n),$$

which proves that the time reversed process $\{\widehat{X}_t, P_\mu\}$ is a diffusion process with the adjoint transition probability $\widehat{P}_t(x, dy)$ and with the standard σ-field $\widehat{\mathcal{F}}_t$ generated by by $\{\widehat{X}_r: \forall\, r \le t\}$.[7]

It is clear that the semi-groups P_t and $\widehat{P}_t$ of $\{X_t, P_\mu\}$ and $\{\widehat{X}_t, P_\mu\}$, respectively, are in duality with respect to the invariant measure $\mu\, dx$.

---

[7] We have shown the claim assuming transition densities, but it is not necessary, cf. Nagasawa (1964, 1970/71)

In fact

$$< g, P_t f >_\mu = \int dx \mu(x) g(x) \int p_t(x, y) f(y) dy$$

$$= \int dx \mu(x) g(x) \int p_t(x, y) \frac{1}{\mu(y)} f(y) \mu(y) dy$$

$$= \int dy \mu(y) f(y) \int \hat{p}_t(y, x) g(x) dx$$

$$= < \hat{P}_t g, f >_\mu .$$

Thus we have shown

Theorem 3.2. *Let* P_t *and* $\hat{P}_t$ *be the semi-groups of a diffusion process* $\{X_t, P_\mu\}$ *and of its time reversal* $\{\hat{X}_t, P_\mu\}$, *respectively, where the initial distribution is an invariant measure* $\mu(x)dx$. *Then,* **duality** *holds for the semi-groups with respect to the invariant measure* μdx ; *namely,*

(3.12) $< g, P_t f >_\mu = < \hat{P}_t g, f >_\mu ,$

for any bounded measurable f *and* g.

If we choose an initial distribution $\mu(x)dx$ to not be an invariant (excessive) measure, then $\hat{p}_t$ defined in (3.11) is not a transition probability density anymore. Therefore, for the time reversed process to be a *time homogeneous* diffusion process in the case of an arbitrary initial distribution, we must reverse time in (3.10) not from a constant time L but from the last exit times $L(\omega)$ of subsets, more generally, from *L-times* (cf. Nagasawa (1964)), later called *co-optional* times to emphasize "duality" with optional times.

$L(\omega)$, $\omega \in \Omega$, is a **co-optional time (the last occurrence time of an event)**, if

(3.13) $\{s < L - t\} = \{s < L \circ \theta_t\}, \quad t, s \geq 0,$

where θ_t is the shift operator: $\theta_t \omega(s) = \omega(t + s)$. The property (3.13) is equivalent to

(3.14) $L \circ \theta_t = (L - t)^+, \; for \; t \geq 0.$

Time reversal from a co-optional time $L(\omega)$ of a diffusion process $\{X_t, P_\mu\}$ is defined by

(3.15) $Y_t(\omega) = X_{L(\omega) - t}(\omega).$

Then we have

Theorem 3.3. (Nagasawa (1964)) *Let* $\{X_t, P_\mu\}$ *be a diffusion process with a transition semi-group* P_t. *Then, its time reversal* $\{Y_t, P_\mu\}$ *from a co-optional time* $L(\omega)$ *is a diffusion process with the semi-group* $\widehat{P}_t$ *which is in duality with* P_t :

(3.16) $< g, P_t f >_m = < \widehat{P}_t g, f >_m$

with respect to a measure m *defined by*

(3.17) $m(B) = P_\mu[\int_0^\infty dt 1_B(X_t)],$

which is an excessive measure [8] *of the diffusion process* $\{X_t, P_\mu\}$.

The semi-group $\widehat{P}_t$ of the time-reversed process *does not* depend on co-optional times, but does depend *on the initial distribution* μ through the excessive measure m defined by (3.17). Therefore, we must fix an initial distribution, when we discuss time reversal. This fact should be kept in mind.

Proof. Let us prove the Markov property of the time reversed process $Y_t(\omega) = X_{L(\omega) - t}(\omega)$, namely

(3.18) $P_\mu[f(Y_t)g(Y_{t+s})F] = P_\mu[f(Y_t)\widehat{P}_s g(Y_t)F],$

where $F = \prod_{j=1}^n f_j(Y_{t_j}), \; 0 < t_1 \leq \cdots \leq t_n = r \leq t,$ and $f, g, f_j \in C_K.$[9]
Equation (3.18) implies

---

[8] m is excessive, if $\int m P_t f \leq \int m f$, for any non-negative bounded measurable f

[9] C_K denotes the space of continuous function of compact support

$$P_\mu[g(Y_{t+s}) \mid \widehat{\mathcal{F}_t}] = \widehat{P}_s g(Y_t), \quad P_\mu\text{-}a.e.,$$

where $\widehat{\mathcal{F}_t}$ stands for the standard σ-field generated by $\{Y_r : \forall r \leq t\}$. Denote the left- and right-hand sides of (3.18) by $A(s,t)$ and $B(s,t)$, respectively. Then, for (3.18) it is sufficient to verify

$$(3.19) \qquad \int_r^\infty e^{-(\alpha+\beta)t}dt \int_0^\infty e^{-\beta s}dsA(s,t) = \int_r^\infty e^{-(\alpha+\beta)t}dt \int_0^\infty e^{-\beta s}dsB(s,t),$$

for $\forall \alpha, \beta > 0$, since $A(s,t)$ and $B(s,t)$ are continuous in s and t, and hence uniquely determined by their Laplace transforms. To show (3.19) we prepare

Lemma 3.1. (Nagasawa (1964)) *Let $\alpha > 0$ and $\gamma \geq 0$. Then*

$$(3.20) \qquad \int_r^\infty e^{-\alpha t}dt P_\mu[e^{-\gamma L}f(Y_t)\,F] = \int m_\gamma(dx)\,f(x)P_x[e^{-(\alpha+\gamma)L}F],^{10}$$

where $F = \prod_{j=1}^n f_j(Y_{t_j}),\ 0 < t_1 \leq \cdots \leq t_n = r \leq t$ *and*

$$m_\gamma(B) = P_\mu[\int_0^\infty e^{-\gamma s}1_B(X_s)ds].$$

Proof. The left-hand side of (3.20) is equal to

$$(3.21) \qquad P_\mu[\int_r^\infty dt\,e^{-\alpha t-\gamma L}f(X_{L-t})\,\widetilde{F}1_{\{t<L\}}]$$

$$= P_\mu[\int_0^\infty e^{-\alpha(L-u)-\gamma L}du\,f(X_u)\,\widetilde{F}1_{\{u<L-r\}}],$$

with $\widetilde{F} = \prod_{j=1}^n f_j(X_{L-t_j})$. Since L is a co-optional time,

$$\{u < L-r\} = \{r < L\circ\theta_u\} \quad and \quad L = u + L\circ\theta_u,$$

and hence, by the Markov property of $\{X_t, P_x\}$, the right-hand side of (3.21) is equal to

---

[10] $P_x = P_\mu$ with $\mu = \delta_x$

$$= P_\mu\left[\int_0^\infty e^{-\gamma u}\, du\, P_{X_u}[e^{-(\alpha+\gamma)L} f(X_u)\, \tilde{F} 1_{\{t<L\}}]\right]$$

$$= \int m_\gamma(dx) f(x) P_x[e^{-(\alpha+\gamma)L}\, \tilde{F} 1_{\{t<L\}}],$$

which is the right-hand side of (3.20). This completes the proof of the lemma.

Now applying Lemma 3.1 twice to the left-hand side of (3.19), we get, with $m(B) = m_0(B)$,

$$\int m(dx) g(x) G_\beta h(x),$$

where $h(x) = f(x) P_x[e^{-(\alpha+\beta)L} F]$ and G_β denotes the resolvent of the semi-group P_t :

$$G_\beta f = \int_0^\infty e^{-\beta t} P_t f\, dt.$$

Let $\widehat{G}_\beta$ be the resolvent of the semi-group $\widehat{P}_t$. Then, the duality relation (3.16) of the pair of semi-groups yields

$$\int m(dx) g(x) G_\beta h(x) = \int m(dx) h(x) \widehat{G}_\beta g(x),$$

from the right-hand side of which we can recover the right-hand side of (3.19), applying again Lemma 3.1 but in the opposite direction. This completes the proof of Theorem 3.3.

Up to now we have treated duality and time reversal of time-homogeneous diffusion processes. However, the results we have obtained are not restrictive, but can be applied to time-inhomogeneous diffusion processes. As we have remarked in the preceding chapter, we will employ a trick for this: Instead of a time-inhomogeneous process X_t, we consider the so-called space-time process (t, X_t).

3.3. Duality of Time-Inhomogeneous Diffusion Processes

Let time-dependent drift coefficient $b(t,x)$ be given,[11] and consider a time-dependent diffusion equation with the reflecting boundary condition

$$(3.22) \qquad Bq(t,x) = 0, \quad in \ D,$$

$$(3.23) \qquad \frac{\partial q}{\partial n}(t,x) = 0, \quad on \ \partial D_t,$$

where $D \subset (a,b) \times \mathbf{R}^d$, $D_t = \{x : (t,x) \in D\}$ may depend on t, and B is a time-dependent parabolic differential operator

$$(3.24) \qquad B = B(t) = \frac{\partial}{\partial t} + \frac{1}{2}\Delta + \{b(t,x) + a(t,x)\}\cdot\nabla,$$

with the Laplace-Beltrami operator Δ. In (3.24) the drift coefficient is decomposed into two components $b(t,x)$ and $a(t,x)$, which will find different meanings in applications; namely, $b(t,x)$ is unperturbed basic drift,[12] while $a(t,x)$ represents perturbation of drift which is under the influence of duality or in other words time reversal.

A time inhomogeneous reflecting diffusion process $\{X_t, Q_\mu, t \in [a,b]\}$ is defined in terms of the fundamental solution $q(s,x;t,y)$ of the diffusion equation (3.22) with the reflecting boundary condition (3.23) given above and with an initial distribution μ. We assume the existence of the fundamental solution for the moment and discuss duality in this section. We require the reflecting boundary condition in order to make diffusion processes conservative. However, in typical cases in applications, the diffusion processes which will be discussed in the following chapters do not hit the boundary because of strong (singular) repulsive drift at the boundary. In these cases, the reflecting boundary condition will turn out to be irrelevant. However, if the diffusion processes can reach the boundary, we need the reflecting boundary condition (see examples in Section 7.9). The existence of such singular diffusion processes will be discussed in Chapters 5 and 6.

The reflecting diffusion process $\{X_t, Q_\mu, t \in [a,b]\}$ is characterized in

---

[11] From now on and in Chapter 4 the diffusion coefficient will be assumed to be time-independent. This is a technical assumption. It may be time-dependent. This case can be handled parallel assuming a gauge condition $\operatorname{div} b - (1/\sqrt{\sigma_2})(\partial\sqrt{\sigma_2}/\partial t) = 0$

[12] In applications b is the vector potential of the electromagnetic field and denoted usually with A, cf. e.g., Lorentz (1915)

terms of finite dimensional distributions as explained already in Chapter 2:

$$Q_\mu[f(X_a, X_{t_1}, \ldots, X_{t_{n-1}}, X_b)]$$

$$= \int \mu(dx_0)q(a, x_0; t_1, x_1)dx_1 q(t_1, x_1; t_2, x_2)dx_2 \cdots$$

$$\cdots q(t_{n-1}, x_{n-1}; b, x_n)dx_n f(x_0, x_1, \ldots, x_n),[13]$$

where $a < t_1 < \cdots < t_{n-1} < b$, and $f(x_0, x_1, \ldots, x_n)$ is any bounded measurable function on $(\mathbf{R}^d)^{n+1}$.

As already remarked in Chapter 2, instead of time inhomogeneous diffusion processes X_t, $t \in [a, b]$, $-\infty < a < b < +\infty$, we consider space-time diffusion processes (t, X_t), and apply the theory of time homogeneous processes to them. Since space-time processes are time homogeneous, and any constant time is *the last exit time* for them, we can reverse the space-time processes from the terminal time b,[14] applying Theorem 3.3.

For time-inhomogeneous processes (resp. space-time processes) it is convenient to consider time reversal of diffusion processes with time *running backwards* (not like in (3.9)) decreasing from the terminal time b to the initial time a, namely, we use a clock running backwards.[15] This point must be emphasized to avoid confusion.

Then, a given process and its time reversal are in duality (in the sense of (3.27) below) with respect to the distribution density

$$(3.25) \qquad \mu_t(y) = \int \mu(dx)q(a, x; t, y),$$

where μ is an arbitrary initial distribution at $t = a$ and $q(s, x; t, y)$ denotes a transition probability density. For a space-time process the function $\mu_t(y)$ defined by (3.25) is nothing but a density function of the measure $m(B)$ given in (3.17).

In fact, for the space-time diffusion processes the measure $m(dtdy)$ defined in (3.17) is equal to

---

[13] For simplicity we write dx instead of $\sqrt{\sigma_2(x)}\,dx$

[14] This is for simplicity. We can reverse the process from the last exit time, if it is natural to do so

[15] See (3.52) and what follows

$$(3.26) \qquad m(dtdy) = Q_\mu[\int_a^b dr1_{(dtdy)}(r, X_r)]$$

$$= \int \mu(dx) \int_a^b q(a, x; r, y) \delta_t(dr) \cdot 1_{(dtdy)}$$

$$= \int \mu(dx) q(a, x; t, y) \cdot 1_{(dtdy)}$$

$$= \mu_t(y) dtdy.$$

Then, it is easy to see that we have, corresponding to the (time-homogeneous) duality relation (3.16), **time dependent duality**

$$(3.27) \qquad < g, Q_{s,t}f >_{\mu_s} = < g\widehat{Q}_{s,t}, f >_{\mu_t}, \quad s < t,$$

as time-inhomogeneous processes, where

$$Q_{s,t}f(s, x) = \int q(s, x; t, y) dy f(t, y),$$

$$(3.28)$$

$$g\widehat{Q}_{s,t}(t, y) = \int dx \, g(s, x) \widehat{q}(s, x; t, y),$$

$$\widehat{q}(s, x; t, y) = \mu_s(x) \, q(s, x; t, y) \frac{1}{\mu_t(y)}.$$

Let us define semi-groups of the space-time processes by

$$P_r f(s, x) = Q_{s,s+r}f(s, x), \quad if \quad r \geq 0, \quad (= 0, \quad otherwise),$$

$$(3.29)$$

$$\widehat{P}_r g(t, y) = g\widehat{Q}_{t-r,t}(t, y), \quad if \quad r \geq 0, \quad (= 0, \quad otherwise),$$

where it is enough to consider small $r > 0$ because of the Markov property. Then we have

$$< g, P_r f >_m = \int_a^{b-r} ds < g, Q_{s,s+r}f >_{\mu_s}$$

$$= \int_{a+r}^{b} dt <g, Q_{t-r,t} f>_{\mu_{t-r}}$$

$$= \int_{a+r}^{b} dt <g\widehat{Q}_{t-r,t}, f>_{\mu_t}$$

$$= <\widehat{P}_r g, f>_m ,$$

where the time-dependent duality relation (3.27) has been employed, and hence we have **duality**

(3.30) $$< g, P_r f >_m = < \widehat{P}_r g, f >_m ,$$

which is nothing but the *duality relation (3.16) with arbitrary initial distribution μ for space-time processes (t, X_t).*

Therefore, we can apply a space-time version of the duality formula (3.7):

(3.31) $$\int_{a}^{b} dt \int 1_{D_t} \{(Bf)g - f(\widehat{B}g)\} \mu \sqrt{\sigma_2}\, dx$$

$$= -\frac{1}{2} \int_{a}^{b} dt \int 1_{\partial D_t} \{\frac{\partial f}{\partial n} g - f\frac{\partial g}{\partial n}\} \mu \sqrt{\sigma_2}\, d\tilde{x}$$

$$+ \int_{a}^{b} dt \int 1_{D_t} fg\, (B^{\circ}\mu) \sqrt{\sigma_2}\, dx$$

$$+ \int_{a}^{b} dt \int 1_{\partial D_t} fg \{\frac{1}{2}\frac{\partial \mu}{\partial n} - (b_n + a_n)\mu\} \sqrt{\sigma_2}\, d\tilde{x},$$

$$- \int_{a}^{b} dt \int 1_{D_t} f\{a + \hat{a} - \sigma^T\sigma\nabla \log \mu\}\cdot\nabla g\, \mu \sqrt{\sigma_2}\, dx,$$

where functions $f(t,x)$ and $g(t,x)$ are of compact support in $(a, b) \times \mathbf{R}^d$. The differential operators in (3.31) are defined by

$$Bf = \frac{\partial f}{\partial t} + \frac{1}{2}\Delta f + \{b(t,x) + a(t,x)\}\cdot\nabla f,$$

(3.32)

$$B^\circ\mu = -\frac{\partial\mu}{\partial t} + \frac{1}{2}\Delta\mu - \frac{1}{\sqrt{\sigma_2}}\nabla\{\sqrt{\sigma_2}\,(b(t,x) + a(t,x))\mu\}.$$

Moreover, we define the adjoint operators

$$\widehat{B}g = -\frac{\partial g}{\partial t} + \frac{1}{2}\Delta g + \{-b(t,x) + \widehat{a}(t,x)\}\cdot\nabla g,$$

(3.33)

$$(\widehat{B})^\circ\mu = \frac{\partial\mu}{\partial t} + \frac{1}{2}\Delta\mu - \frac{1}{\sqrt{\sigma_2}}\nabla\{\sqrt{\sigma_2}\,(-b(t,x) + \widehat{a}(t,x))\mu\}.$$

With the help of formula (3.31), as in the proof of Theorem 3.1, we get
the **duality relation of drift coefficients** *for a pair of space-time
reflecting diffusion processes*

(3.34) $$a(t,x) + \widehat{a}(t,x) = \sigma^T\sigma\,\nabla(\log\mu(t,x)), \quad in \ \ D,$$

and

(3.35) $$\frac{\partial\mu}{\partial t} + \frac{1}{\sqrt{\sigma_2}}\nabla(\sqrt{\sigma_2}\,(b + \frac{a - \widehat{a}}{2})\mu) = 0, \quad in \ \ D,$$

$$(b_n + \frac{a_n - \widehat{a}_n}{2})\mu = 0, \quad on \ \ \partial D.$$

The first equation of (3.35) follows from a pair of Fokker-Planck's
equations

$$-\frac{\partial\mu}{\partial t} + \frac{1}{2}\Delta\mu - \frac{1}{\sqrt{\sigma_2}}\nabla\{\sqrt{\sigma_2}\,(b(t,x) + a(t,x))\mu\} = 0,$$

$$\frac{\partial\mu}{\partial t} + \frac{1}{2}\Delta\mu - \frac{1}{\sqrt{\sigma_2}}\nabla\{\sqrt{\sigma_2}\,(-b(t,x) + \widehat{a}(t,x))\mu\} = 0.$$

Since the μ_t satisfies these equations in D at the same time, subtracting one
from another, we get (3.35). We have the second equation of (3.35),
subtracting the boundary conditions

$$\frac{1}{2}\frac{\partial \mu}{\partial n} - (b_n + a_n)\mu = 0, \quad on \ \partial D,$$

$$\frac{1}{2}\frac{\partial \mu}{\partial n} - (-b_n + \widehat{a}_n)\mu = 0, \quad on \ \partial D.$$

In applications in later chapters we assume that the difference of the additional drift coefficients $a(t, x)$ and $\widehat{a}(t, x)$ is given by a potential function $S(t, x)$, namely

(3.36) $a(t, x) - \widehat{a}(t, x) = 2\sigma^T\sigma \nabla S(t, x).$

We will call the system of equations (3.34) with (3.35) the **duality relation for space-time diffusion processes**.

Summarizing, we have shown

Theorem 3.4. *Let* $\{X_t, Q_\mu\}$ *be the reflecting diffusion process which is determined by the fundamental solution of the diffusion equation* (3.22) *with the reflecting boundary condition* (3.23). *Then:*

The additional drift coefficients **a** *and* $\widehat{a}$ *of the diffusion process and its time reversal satisfy the duality relation* (3.34) *and* (3.35).

Moreover under the assumption (3.36)

(3.37)
$$a(t, x) = \sigma^T\sigma \nabla\log \ \phi_t(x), \quad with \quad \phi_t(x) = e^{R(t,x)+S(t,x)},$$

$$\widehat{a}(t, x) = \sigma^T\sigma \nabla\log \ \widehat{\phi}_t(x), \quad with \quad \widehat{\phi}_t(x) = e^{R(t,x)-S(t,x)},[16]$$

where $2R(t, x) = \log \mu_t(x)$ *and* $S(t, x)$ *are assumed to be differentiable outside the zero set* $\{(t, x): \mu_t(x) = 0\}$ *of the distribution density*

(3.38) $\mu_t(x) = \widehat{\phi}_t(x)\phi_t(x)$

of the diffusion process.

We should pay attention to the similarity between formula (3.38) and the factorization $\mu_t(x) = \overline{\psi}_t(x)\psi_t(x)$ in terms of a wave function and its complex conjugate in quantum theory.

---

[16] Compare with (3.6)

In terms of the pair of functions $R(t,x)$ and $S(t,x)$, the duality relations (3.34) and (3.35) can be represented as

(3.34')

$$\frac{a(t,x) + \hat{a}(t,x)}{2} = \sigma^T \sigma \nabla R(t,x),$$

$$\frac{a(t,x) - \hat{a}(t,x)}{2} = \sigma^T \sigma \nabla S(t,x),$$

and

(3.35')
$$\frac{\partial R}{\partial t} + \frac{1}{2} \Delta S + (\sigma \nabla S) \cdot (\sigma \nabla R) + b \cdot \nabla R = 0,$$

$$(b_n + \frac{\partial S}{\partial n}) \mu = 0.$$

The formulae in (3.37) with (3.35) of the drift coefficients a and $\hat{a}$ in terms of the pair of functions $\phi_t(x)$ and $\hat{\phi}_t(x)$, and the formulae (3.34') and (3.35') with the pair of functions $R(t,x)$ and $S(t,x)$ both will find usefulness in later chapters.

3.4. Schrödinger's and Kolmogoroff's Representations

In the context of time reversal of diffusion processes we will represent a diffusion process $\{X_t, Q_\mu\}$ in three ways: in Schrödinger's representation, Kolmogoroff's representation, and in time reversed Kolmogoroff's representation. The identification of the three representations will reveal an important mathematical structure which has been hidden in quantum mechanics, which will be discussed in Chapter 4.

First of all let us formulate a theorem which links diffusion processes $\{X_t, Q_\mu\}$ to Schrödinger's representation.

We denote by L the parabolic differential operator

(3.39)
$$L = \frac{\partial}{\partial t} + \frac{1}{2} \Delta + b(t,x) \cdot \nabla.$$

Let $\phi(t, x)$ be a non-negative continuous function and define

(3.40)
$$c(t, x) = - \frac{L\phi(t, x)}{\phi(t, x)},$$

on a subset

(3.41)
$$D = \{(t, x): \phi(t, x) \neq 0\},$$

where we assume $\phi \in C^{1,2}(D)$. The function $c(t, x)$ defined in (3.40) will be called **creation and killing induced by** $\phi(t, x)$.

Theorem 3.5. *Let $\phi(t, x)$ be a non-negative continuous function such that $\phi \in C^{1,2}(D)$, where D is defined in (3.41), and let c(t, x) be the creation and killing induced by $\phi(t, x)$. Let p(s, x; t, y) be the (weak) fundamental solution* [17] *of the diffusion equation*

(3.42)
$$Lp + c(t, x)p = 0,$$

with the same boundary values as the given $\phi(t, x)$.

Then:

(i) *The function $\phi(t, x)$ satisfies (3.42) and*

(3.43)
$$\phi(s, x) = \int p(s, x; t, y)\phi(t, y)dy, \quad a \leq s < t \leq b,$$

that is, the function $\phi(t, x)$ is a space-time p-harmonic function (notice p(s, x; t, y) is with creation and killing c(t, x) and not a probability density).

(ii) *The function transformed by means of $\phi(t, x)$*

(3.44)
$$q(s, x; t, y) = \frac{1}{\phi(s, x)} p(s, x; t, y)\phi(t, y)$$

is a transition probability density on the subset D defined in (3.41).

Proof. The function ϕ satisfies equation (3.42) trivially because of (3.40) and hence also equation (3.43). Concerning the second assertion, the

---

[17] We assume the existence of the (weak) fundamental solution in this section. But the existence will be shown in Chapters 5 and 6

Chapman-Kolmogoroff equation is clear. Because of equation (3.43) the normalization

$$\int_{D_t} q(s,x;t,y)dy = 1, \quad for \ \forall(s,x) \in D, s < t \le b,$$

holds, where $D_t = \{x: (t,x) \in D\}$, which completes the proof.

In Theorem 3.5 it should be remarked that $p(s,x;t,y)$ satisfies the Chapman-Kolmogoroff equation but is not normalized, and hence it is *not* a transition probability density. Nevertheless, we obtain a *renormalized transition probability* density $q(s,x;t,y)$ through the transformation in terms of a p-harmonic function ϕ. This relation between $p(s,x;t,y)$ and $q(s,x;t,y)$ will be crucial in the following.

Besides the function $\phi(t,x)$ with which we have defined the transition probability density $q(s,x;t,y)$ at (3.44), we assume that we can choose another bounded non-negative measurable function $\widehat{\phi}(a,x)$ on $\mathbf{R}^d$ such that

(3.45) $$\int_{D_a} dx\, \widehat{\phi}(a,x)\phi(a,x) = 1,$$

where $D_a = \{x: (a,x) \in D\}$.[18] Moreover, we extend it for $\forall t \in [a,b]$ by

(3.46) $$\widehat{\phi}(t,y) = \int dx\, \widehat{\phi}(a,x)p(a,x;t,y), \quad for \ \forall (t,y) \in D,$$

where $p(s,x;t,y)$ is the (weak) fundamental solution of (3.42). Therefore, the function $\widehat{\phi}(t,x)$ is a space-time p-coharmonic function. In the following $\widehat{\phi}(t,x)$ and $\phi(t,x)$ will be denoted also as $\widehat{\phi}_t(x)$ and $\phi_t(x)$, respectively, without mentioning it.

Let $\{X_t, Q\}$[19] be a time-inhomogeneous (or space-time) diffusion process on the state space D with the transition probability density $q(s,x;t,y)$ defined by (3.44) (not $p(s,x;t,y)$!) and a prescribed initial distribution density $\mu(x) = \widehat{\phi}_a(x)\phi_a(x)$.

---

[18] If $\widehat{\phi} = \phi$, the requirement reduces to " $\phi \in L^2$ "

[19] Further discussion on the construction of the diffusion process with $q(s,x;t,y)$ will be given in Chapters 5 and 6

Then, the finite dimensional distributions of the diffusion process $\{X_t, Q\}$ is given by

(3.47) $Q[f(X_a, X_{t_1}, \ldots, X_{t_{n-1}}, X_b)]$

$$= \int dx_0 \hat{\phi}_a(x_0)\phi_a(x_0)q(a, x_0; t_1, x_1)dx_1 q(t_1, x_1; t_2, x_2)dx_2 \cdots$$

$$\cdots q(t_{n-1}, x_{n-1}; b, x_n)dx_n f(x_0, x_1, \ldots, x_n)$$

$$= \int dx_0 \hat{\phi}_a(x_0)\phi_a(x_0)\frac{1}{\phi_a(x_0)}p(a, x_0; t_1, x_1)\phi_{t_1}(x_1)dx_1 \frac{1}{\phi_{t_1}(x_1)}p(t_1, x_1; t_2, x_2)\times$$

$$\times \phi_{t_2}(x_2)dx_2 \cdots \frac{1}{\phi_{t_{n-1}}(x_{n-1})}p(t_{n-1}, x_{n-1}; b, x_n)\phi_b(x_n)dx_n f(x_0, x_1, \cdots, x_n),$$

where we denote $\phi_t(x) = \phi(t, x)$ and $a < t_1 < \cdots < t_{n-1} < t_n = b$.

We will call formula (3.47) **Kolmogoroff's representation** (or q-representation) of the diffusion process Q, and denote it as

(3.48) $Q = [\, \hat{\phi}_a \phi_a q \gg.$

Kolmogoroff's representation is nothing but the standard way of defining a diffusion process through the finite dimensional distributions with a given initial distribution density $\mu(x) = \hat{\phi}_a(x)\phi_a(x)$ and the transition probability density given in (3.44). We introduce this terminology to distinguish it from another representation which will be given below.

Cancelling ϕ_{t_i} in (3.47), we get

(3.49) $Q[f(X_a, X_{t_1}, \ldots, X_{t_{n-1}}, X_b)]$

$$= \int dx_0 \hat{\phi}(a, x_0)p(a, x_0; t_1, x_1)dx_1 p(t_1, x_1; t_2, x_2)dx_2 \cdots$$

$$\cdots p(t_{n-1}, x_{n-1}; b, x_n)\phi(b, x_n)dx_n f(x_0, x_1, \ldots, x_n).$$

This representation for diffusion processes was first considered by Schrödinger (1931). Therefore, let us call the formula (3.49) expressed in terms of a transition density $p(s, x; t, y)$ and a pair of functions $\widehat{\phi}(a, x)$ and $\phi(b, x)$ **Schrödinger's representation** (or **p-representation**) of the diffusion process Q, and denote it as

$$(3.50) \qquad\qquad Q = [\, \widehat{\phi}_a p >> << p \, \phi_b \,],$$

which indicates that $\widehat{\phi}_a$ is assigned at the initial time and ϕ_b at the terminal time,[20] and the time-symmetry in formula (3.49).

The Schrödinger representation (3.50) shows clearly the time-symmetry but does not imply the Markov property of the process. As a matter of fact it is *not* a Markov process *with* the transition density $p(s, x; t, y)$. To see the Markov property of the process, we must go into the Kolmogoroff representation (3.48) with the transition probability density $q(s, x; t, y)$ which is the ϕ-transformation of $p(s, x; t, y)$ defined in (3.44). This point should be emphasized and kept in mind.

The notations and terminology introduced above will be convenient, since we shall speak of the representations (3.47) and (3.49) at the same time and must distinguish them.

Starting from (3.49) and applying the same manipulation now with $\widehat{\phi}(t, x)$, we have

$$Q[f(X_a, X_{t_1}, \ldots, X_{t_{n-1}}, X_b)]$$

$$= \int dx_0 \widehat{\phi}(a, x_0) p(a, x_0; t_1, x_1) dx_1 p(t_1, x_1; t_2, x_2) dx_2 \cdots$$

$$\cdots p(t_{n-1}, x_{n-1}; b, x_n) \phi(b, x_n) dx_n f(x_0, x_1, \ldots, x_n)$$

$$= \int dx_0 \widehat{\phi}_a(x_0) p(a, x_0; t_1, x_1) \frac{1}{\widehat{\phi}_{t_1}(x_1)} dx_1 \widehat{\phi}_{t_1}(x_1) p(t_1, x_1; t_2, x_2) \frac{1}{\widehat{\phi}_{t_2}(x_2)} \cdots$$

$$\cdots dx_{n-1} \widehat{\phi}_{t_{n-1}}(x_{n-1}) p(t_{n-1}, x_{n-1}; b, x_n) \frac{1}{\widehat{\phi}_b(x_n)} \widehat{\phi}_b(x_n) \phi_b(x_n) dx_n f(x_0, x_1, \ldots, x_n).$$

Defining the **adjoint transition probability density** $\widehat{q}(s, x; t, y)$ with the formula of $\widehat{\phi}$-transformation

---

[20] Our p (resp. q)-representation has nothing to do with Dirac's "q-number'

(3.51)
$$\hat{q}(s,x;t,y) = \hat{\phi}(s,x)p(s,x;t,y)\frac{1}{\phi(t,y)},$$

which satisfies the Chapman-Kolmogoroff equation and the (time-reversed) normality condition

$$\int dx\hat{q}(s,x;t,y) = 1,$$

we get the third formula

(3.52) $Q[f(X_a, X_{t_1}, \dots, X_{t_{n-1}}, X_b)]$

$$= \int f(x_0, x_1, \dots, x_n)dx_0\hat{q}(a,x_0;t_1,x_1)dx_1\hat{q}(t_1,x_1;t_2,x_2) \cdots$$

$$\cdots dx_{n-1}\hat{q}(t_{n-1},x_{n-1};b,x_n)\hat{\phi}(b,x_n)\phi(b,x_n)dx_n .$$

Formula (3.52) shows that the process $\{X_t, Q\}$ is a diffusion process with the transition probability density $\hat{q}(s,x;t,y)$ and the "initial" (though actually terminal) distribution density $\mu_b(x) = \hat{\phi}(b,x)\phi(b,x)$, if we trace the process backward with time running backward from the terminal time b to the initial time a (read the formula (3.52) from right to left).

Let us denote by

$$<<\hat{q}\,\hat{\phi}_b\phi_b\,]$$

the probability measure of the time reversed process defined through the formula (3.52) and call it **time reversed $\hat{q}$-representation**.

Then, formulae (3.47) and (3.52) together yield

(3.53)
$$[\hat{\phi}_a\phi_a\,q >>$$

$$= <<\hat{q}\,\hat{\phi}_b\phi_b\,],$$

where the meaning of the formula is self-explanatory.

Thus we have shown a theorem which will be extremely important in connection with quantum mechanics.

Theorem 3.6. *Under the same conditions of* Theorem 3.5 *and* (3.45) *the diffusion process* $\{X_t, Q\}$ *can be represented in three ways*

$$(3.54) \qquad Q = [\,\widehat{\phi}_a p >> << p\,\phi_b\,]$$

$$= [\,\widehat{\phi}_a \phi_a q >>$$

$$= << \widehat{q}\,\widehat{\phi}_b \phi_b\,],$$

where $[\,\widehat{\phi}_a p >> << p\,\phi_b\,]$ *is the Schrödinger representation, the second one* $[\,\widehat{\phi}_a \phi_a q >>$ *is the Kolmogoroff representation, and the last one* $<< \widehat{q}\,\widehat{\phi}_b \phi_b\,]$ *is the time-reversed* $\widehat{q}$*-representation of the measure* Q.[21]

Formula (3.54) in the theorem means that one can give two completely different descriptions to a diffusion process $\{X_t, Q\}$ in terms of diffusion equations of a different kind: The first one is a diffusion equation with a *creation and killing* $c(t, x)$

$$(3.55) \qquad \frac{\partial p}{\partial t} + \frac{1}{2}\Delta p + b(t,x)\cdot\nabla p + c(t,x)p = 0,$$

and the second one is a diffusion equation with an *additional drift coefficient* $a(t, x)$

$$(3.56) \qquad \frac{\partial q}{\partial t} + \frac{1}{2}\Delta q + b(t,x)\cdot\nabla q + a(t,x)\cdot\nabla q = 0.$$

In order to use the fundamental solution $p(s, x; t, y)$ of (3.55) one needs *a pair of functions* $\{\widehat{\phi}_a(x), \phi_b(x)\}$, which is an *entrance-exit law* of $p(s, x; t, y)$ (see the equation (3.49)). The function $\phi(t, x)$ satisfies the diffusion equation (3.55), and the function $\widehat{\phi}(t, x)$ satisfies another diffusion equation

$$(3.57) \qquad -\frac{\partial \widehat{p}}{\partial t} + \frac{1}{2}\Delta \widehat{p} - b(t,x)\cdot\nabla\widehat{p} + c(t,x)\widehat{p} = 0,$$

which is the adjoint of equation (3.55).

---

[21] I have called a diffusion process with the representation (3.49) the "Schrödinger process" in Nagasawa (1989,a), since it was first considered by Schrödinger, but I abandon this terminology to avoid confusion, because it is not a diffusion process of a special kind but a special representation of an arbitrary diffusion process

In order to adopt the description in terms of the fundamental solution $q(s,x;t,y)$ of (3.56) one must start with the initial distribution $\mu_a(x) = \widehat{\phi}_a(x)\phi_a(x)$ (see equation (3.47)). In addition, if one traces the diffusion process backward with reversed time, then one needs the terminal distribution $\mu_b(x) = \widehat{\phi}_b(x)\phi_b(x)$ and the time-reversed transition probability density $\widehat{q}(s,x;t,y)$, which is the fundamental solution of

$$(3.58) \qquad -\frac{\partial \widehat{q}}{\partial t} + \frac{1}{2}\Delta\widehat{q} - b(t,x)\cdot\nabla\widehat{q} + \widehat{a}(t,x)\cdot\nabla\widehat{q} = 0,$$

where $a(t,x)$ and $\widehat{a}(t,x)$ satisfy the duality relation (3.36) or (3.34'), since the diffusion equations (3.56) and (3.58) are in duality with respect to the probability distribution $\mu_t(x)dx = \widehat{\phi}(t,x)\phi(t,x)dx$ by Theorem 3.4 (cf. Theorem 3.7 below). Both representations will be needed in Chapter 4 to establish the equivalence of Schrödinger and diffusion equations.

It is worthwhile to give some detail about the crucial difference between the representations in formula (3.54). The q-representation $[\widehat{\phi}_a\phi_a q >>$ describes the transition of the diffusion process $\{X_t, Q\}$ in the "real world", since $\mu_a(x) = \widehat{\phi}_a(x)\phi_a(x)$ stands for the "real", namely *observable* initial distribution density of the diffusion process $\{X_t, Q\}$ and $q(s,x;t,y)$ governs the *real transition* of the process. This argument also applies to the reversed $\widehat{q}$-representation $<< \widehat{q}\,\widehat{\phi}_b\phi_b\,]$, even though time is reversed.

In contrast, the description in the p-representation $[\widehat{\phi}_a p >><< p\,\phi_b]$ belongs to not the real but an "imaginary" or "fictitious" world, because the function $\widehat{\phi}_a(x)$ is *not* observable, namely not an initial distribution but an *entrance law* for the $p(s,x;t,y)$ which belongs also to a "fictitious" world. In fact, the transition (*not* probability!) density $p(s,x;t,y)$ *does not* describe the real transition of the process $\{X_t, Q\}$. Instead, it governs the evolution of the entrance-exit law, and hence it is *not observable* either. This is why I say the description in the p-representation belongs to a "fictitious" world, or is of a "fictitious description".

One of the important characteristics of the p-representation is this: we can always *recover* the description in the real world with it, i.e., the distribution density $\mu_t(x)$ of the *real* diffusion process can be obtained through the multiplication of the pair of functions $\widehat{\phi}(t,x)$ and $\phi(t,x)$ which satisfy (3.57) and (3.55), respectively, even though each single one is "fictitious".

Quantum mechanics in terms of a pair of wave functions $\overline{\psi}$ and ψ can be regarded as a complex-valued counterpart of the *fictitious* description: namely, the pair of wave functions belongs to an "*imaginary or fictitious*" world exactly as the pair of functions $\hat{\phi}_t$ and ϕ_t does in diffusion theory. Nevertheless, the product $\overline{\psi}\psi$ (= probability density !) recovers the *real world* exactly as the product $\hat{\phi}\phi$ does in diffusion theory. It will be shown in Chapter 4 that the pair of wave functions $\{\overline{\psi}, \psi\}$ is exactly the complex-valued counterpart of the pair $\{\hat{\phi}, \phi\}$ of real-valued functions in diffusion theory.

The last but not the least important remark is on systems of diffusion particles. To be concrete we consider a system of two interacting diffusion particles in $\mathbf{R}^3$, which we denote by x_t^1 and x_t^2. The system defines a diffusion process

$$X_t = (x_t^1, x_t^2), \;\; in \;\; \mathbf{R}^6,$$

the distribution of which

$$\mu_t = Q \circ X_t^{-1}$$

is, therefore, a probability distribution in $\mathbf{R}^6$. I have said that the distribution is *observable* and hence belongs to the real world, but it was not meant to be observable in the three-dimensional space. To obtain the distribution of the system in the three-dimensional space, we need the empirical distribution of the system (x_t^1, x_t^2):

$$U_t^x = \frac{1}{2} \{ \delta_{x_t^1} + \delta_{x_t^2} \},$$

or what is the same

$$U_t^x(B) = \frac{1}{2} \{ 1_B(x_t^1) + 1_B(x_t^2) \},$$

where 1_B denotes the indicator function of a measurable subset B. Then, the expectation of the empirical distribution U_t^x :

$$\mu_t^x(B) = Q[U_t^x(B)]$$

is the probability distribution of the system in the three-dimensional space. Therefore, this is a (not only mathematical but) experimentally observable distribution in the three-dimensional space.

If we are interested in the distribution of a single particle in the system, the marginal distribution

$$\mu_t^i = Q \circ (x_t^i)^{-1}, \quad i = 1, 2,$$

should be considered.

An important implication of the fact explained above will be discussed in Chapter 4 in connection with quantum mechanics.

It is easy to see that the transition probability densities $q(s, x; t, y)$ and $\widehat{q}(s, x; t, y)$ are in duality (time dependent) with respect to the measure $\mu_t(x)dx = \widehat{\phi}(t, x)\phi(t, x)dx$. In fact, let us set

$$Q_{t-s}f(s, x) = \int q(s, x; t, y)f(t, y)dy, \quad s < t,$$

(3.59)

$$g\widehat{Q}_{t-s}(t, y) = \int dx g(s, x)\widehat{q}(s, x; t, y), \quad s < t.$$

Then

$$\int dx\mu_s(x)g(s, x)Q_{t-s}f(s, x)$$

$$= \int dx g(s, x)\phi_s(x)\widehat{\phi}_s(x)\frac{1}{\phi_s(x)}p(s, x; t, y)\phi_t(y)f(t, y)dy$$

$$= \int dx g(s, x)\widehat{\phi}_s(x)p(s, x; t, y)\frac{1}{\widehat{\phi}_t(y)}f(t, x)\widehat{\phi}_t(y)\phi_t(y)dy$$

$$= \int dx g(s, x)\widehat{q}(s, x; t, y)f(t, y)\widehat{\phi}_t(y)\phi_t(y)dy$$

$$= \int dx g\widehat{Q}_{t-s}(t, y)f(t, y)\mu_t(y)dy.$$

Hence Q_{t-s} and $\widehat{Q}_{t-s}$ are in duality with respect to the measure $\mu_t(x)dx$.

Summarizing, we have

Theorem 3.7. *Assume a gauge condition*

$$(3.60) \qquad\qquad \text{div } \boldsymbol{b} = \frac{1}{\sqrt{\sigma_2}}\nabla(\sqrt{\sigma_2}\,\boldsymbol{b}) = 0.^{22}$$

Let Q_{t-s} and $\widehat{Q}_{t-s}$ be defined in (3.59) in terms of $q(s,x;t,y)$ and $\hat{q}(s,x;t,y)$ given in (3.44) and (3.51), respectively.

Then:

(i) *Time dependent duality*

$$(3.61) \qquad\qquad < g, Q_{t-s}f>_{\mu_s} = < g\widehat{Q}_{t-s},f>_{\mu_t}, \quad s < t,$$

holds, which coincides with the time dependent duality relation given in (3.27), and hence (3.30) holds as space-time processes.

Consequently:

(ii) *The diffusion process $[\hat{\phi}_a \phi_a q>>$ and its time reversal $<<\hat{q}\,\hat{\phi}_b \phi_b]$ have additional drift terms with*

$$\boldsymbol{a}(t,x) = \sigma^T\sigma\nabla\log\phi(t,x),$$

$$(3.62)$$

$$\hat{\boldsymbol{a}}(t,x) = \sigma^T\sigma\nabla\log\hat{\phi}(t,x),$$

respectively, and $\mu = \hat{\phi}\,\phi$ satisfies (3.35).

(iii) *The transition probability densities $q(s,x;t,y)$ and $\hat{q}(s,x;t,y)$ are the fundamental solutions of*

$$Lq + \sigma^T\sigma\nabla\log\phi(t,x)\cdot\nabla q = 0,$$

$$(3.63)$$

$$\widehat{L}\hat{q} + \sigma^T\sigma\nabla\log\hat{\phi}(t,x)\cdot\nabla\hat{q} = 0,$$

respectively, where L is defined in (3.39) and $\widehat{L}$ is the formal adjoint of the operator L,

$$(3.64) \qquad\qquad \widehat{L} = -\frac{\partial}{\partial t} + \frac{1}{2}\Delta - \boldsymbol{b}(t,x)\cdot\nabla.$$

---

[22] This is technically necessary but harmless

Proof. The first assertion has been shown already. The second one follows from Theorem 3.4 combined with Theorem 3.5. When we apply Theorem 3.5, we need

$$(3.65) \qquad c(t,x) = -\frac{L\phi(t,x)}{\phi(t,x)} = -\frac{\widehat{L}\widehat{\phi}(t,x)}{\widehat{\phi}(t,x)} .$$

This can be shown easily as follows: Let us denote $\mu = \widehat{\phi}\,\phi$. Then

$$(3.66) \qquad \frac{\widehat{L}\widehat{\phi}}{\widehat{\phi}} = \frac{\phi}{\mu}\widehat{L}(\frac{\mu}{\phi}) = \frac{L\phi}{\phi} + \frac{1}{\phi}B^\circ\mu,$$

where $B^\circ\mu$ is defined in (3.32), and it vanishes because $\mu = \widehat{\phi}\,\phi$ is the distribution density of the process $\{X_t, Q\}$. In (3.66) we have used the gauge condition (3.60). The third assertion is just a rephrase of the second one under the gauge condition which is employed to get (3.64).

In applications which will be discussed in the following chapters, the function $\phi(t,x)$ vanishes on a subset. Then, the formulae in (3.62) show that the diffusion process $\{X_t, Q\}$ has singular drift both in the q-representation $[\widehat{\phi}_a\phi_a q >>$ and its time reversed $\widehat{q}$-representation $<< \widehat{q}\,\widehat{\phi}_b\phi_b\,]$, while the formula in (3.40) shows that it has a singular creation and killing $c(t,x)$ induced by $\phi(t,x)$ in the p-representation $[\widehat{\phi}_a p >><< p\,\phi_b\,]$. The existence of such singular diffusion processes is non-trivial and will be discussed in Chapters 5 and 6.

3.5. Some Remarks

(i) The formula (3.44) is called "harmonic transformation" in the context of potential theory. The harmonic transformation of Brownian motions was discussed by Doob in the 1950's in developing probabilistic treatments of potential theory (cf. Doob (1984)). According to Doob the h-transformed process has an additional drift term

$$(3.67) \qquad a(t,x) = \sigma^T\sigma\nabla(\log\,\phi_t(x)),$$

which follows from a simple formula

$$(3.68) \qquad L\,q + \sigma^T\sigma\nabla(\log\,\phi)\cdot\nabla q = \phi^{-1}(Lp + cp) - \phi^{-2}p(L\phi + c\phi),$$

for $q = p/\phi$, where $c(t,x)$ is arbitrary.[23] If $Lp + cp = 0$ and $L\phi + c\phi = 0$, then the right-hand side vanishes. Theorem 3.5 and 3.7 together generalize Doob's h-transformation to a wider class of transition densities with creation and killing.

(ii) *The Zero Set and Segregation.* Let $\{X_t, Q\}$ be the diffusion process which we have considered in this chapter. It will be shown in Chapters 5 and 6 that this diffusion process cannot cross over the zero set of its distribution density, namely, ergodic decomposition occurs. Therefore, in order to recover the distribution, we must have one diffusion particle in each decomposed region. This will turn out to be a serious problem, when excited states of, say, a hydrogen atom, will be considered. To overcome the difficulty, as I have mentioned already, *we must consider many, or infinitely many particles, when* $\mu_t(x)$ *vanishes on a subset* (we say *segregation occurs*), and interpret the distribution as the statistical spatial distribution of infinitely many particles, and hence we must consider statistical mechanics for Schrödinger equations (cf. Chapters 7, 8 and 9).

(iii) *Schrödinger and Kolmogoroff.* Schrödinger's representation gives a time-symmetric picture

$$Q = [\, \hat{\phi}_a p >><< p\, \phi_b \,]$$

of a diffusion process[24] though it describes the evolution of an entrance-exit law in a "fictitious world"; nevertheless it is useful in constructing our singular diffusion processes in terms of the variational principle, which will be discussed in Chapter 5. Schrödinger (1931) emphasized the importance of the symmetry in time reversal in connection with the time reversibility of Schrödinger equations.

However, to establish diffusion theory for quantum mechanics, not only the p-representation but also Kolmogoroff's *duality relation* (3.34) will play an indispensable role, as will be seen in the next chapter. In order to establish the duality relation (3.34) we need Kolmogoroff's characterization of time reversal

$$[\, \hat{\phi}_a \phi_a q >> = << \hat{q}\, \hat{\phi}_b \phi_b \,]$$

in the q-representation, as we have seen in Theorems 3.6 and 3.7. The

---

[23] Cf. Lemma 3.2 of Nagasawa (1989, a)

[24] The Schrödinger's representation was generalized by Kuznetzov (1973) to processes with random birth and death motivated by Hunt (1960). For further discussions see Getoor-Glover (1984), Getoor-Sharpe (1984)

construction of our diffusion processes in the form of the q-representation will be discussed in Chapter 6.

As we have mentioned, a decisive role of both representations will be seen in Chapter 4 in establishing the equivalence of Schrödinger and diffusion equations: Schrödinger's representation is enough to claim the identification $\overline{\psi}\psi = \hat{\phi}\,\phi$ and this formula shows the most important interrelation between Schrödinger equations and diffusion equations. However, in order to establish the *equivalence* between them, we need the q-representation and Kolmogoroff's duality relation of time reversal.

(iv) *Diffusion Processes with Singular Drift.* We have seen that if we consider duality of diffusion processes with respect to a probability density which vanishes on a subset we must necessarily handle the case of singular drift. There is a well-known method of drift transformation due to Maruyama (1954) and Girsanov (1960) to treat singular drift as explained in Chapter 2. To apply it one must usually verify the Novikov or Kazamaki condition (cf., e.g. Revuz-Yor (1991)). However, it is clear that the singularity of the drift coefficient $\nabla \log \phi_t(x)$ is too strong to be controlled by the Novikov or Kazamaki condition, since the absolute continuity fails. This point was paid attention to after the articles of Nelson (1966, 67) (cf. Appendix), and the existence problem of such singular diffusion processes have been solved by many probabilists with various methods. Contributions are, under different assumptions, Albeverio- Høegh-Krohn (1974), Nagasawa (1980), Carlen (1984), Zheng-Meyer (1984/85), Carmona (1979, 85), Blanchard-Golin (1987), Föllmer (1988), Norris (1988), Nagasawa (1989, a; 90), Aebi (1989, 93) among others.

(v) *Time Reversal.* For further discussions and applications on time reversal of Markov processes, see (besides Schrödinger (1931)[25] and Kolmogoroff (1936, 37)) Hunt (1960), Nagasawa (1961, 64), Ikeda-Nagasawa-Sato (1964), Kunita-Watanabe (1966), Chung-Walsh (1969), Meyer (1971), Azema (1973), Kuznezov (1973), Nagasawa-Maruyama (1979), Getoor-Glover (1984), Getoor-Sharpe (1984), Dynkin (1985), Föllmer (1985, 86), and Föllmer-Wakolbinger (1986), especially for applications to infinite dimensional diffusion processes. See also Revuz-Yor (1991), in which interesting examples are discussed.

---

[25] Motivated by Schrödinger (1931), Bernstein (1932) introduced an interesting class of conditional processes. His formulation plays, however, no role in this monograph. For an approach based on Bernstein (1932) cf. Jamison (1974,a; 1975). Influenced by Schrödinger (1931), Fényes (1952) discussed diffusion theory of quantum mechanics (cf. Appendix, and also Jammer (1974) for further literature)

Chapter IV

Equivalence of Diffusion and Schrödinger Equations

The equivalence of Schrödinger's representation and Kolmogoroff's representation of a diffusion process established in Theorem 3.6 will play a crucial role in this chapter. One of the consequences is the duality formula, especially (3.35'), which will be renumbered as (4.7). Accepting the duality formula, one can regard Theorem 4.1 as an assertion on transformations between partial differential equations of different types. Referring to the Appendix is recommended in connection with the theme of this chapter.

4.1. Change of Variables Formulae

Corresponding to the Schrödinger equation and its complex conjugate;

$$i\frac{\partial \psi}{\partial t} + \frac{1}{2}\Delta \psi + i\boldsymbol{b}(t,x)\cdot\nabla\psi - V(t,x)\psi = 0,$$

(4.1)

$$-i\frac{\partial \overline{\psi}}{\partial t} + \frac{1}{2}\Delta \overline{\psi} - i\boldsymbol{b}(t,x)\cdot\nabla\overline{\psi} - V(t,x)\overline{\psi} = 0,$$

we take a pair of diffusion equations

$$\frac{\partial \phi}{\partial t} + \frac{1}{2}\Delta \phi + \boldsymbol{b}(t,x)\cdot\nabla\phi + c(t,x)\phi = 0,$$

(4.2)

$$-\frac{\partial \widehat{\phi}}{\partial t} + \frac{1}{2}\Delta \widehat{\phi} - \boldsymbol{b}(t,x)\cdot\nabla\widehat{\phi} + c(t,x)\widehat{\phi} = 0,$$

because of the formal "resemblance" between (4.1) and (4.2), which will turn out to be crucial.[1]

Namely, we will show that there is a simple transformation between the pairs $\{\psi, \overline{\psi}\}$ and $\{\phi, \hat{\phi}\}$ of solutions of (4.1) and (4.2), respectively, and hence the pair of Schrödinger equations (4.1) and the pair of diffusion equations (4.2) will turn out to be equivalent.

We assume a gauge condition

$$\mathrm{div}\, b = \frac{1}{\sqrt{\sigma_2}} \nabla(\sqrt{\sigma_2}\, b) = 0,$$

which is harmless. Then the second equation of (4.2) is the formal adjoint of the first one.[2]

As shown in the preceding chapter, a diffusion processes $\{X_t, Q\}$ has the p-representation,

$$Q = [\, \hat{\phi}_a p >> << p\phi_b\,],$$

where $p = p(s, x; t, y)$ is the (weak) fundamental solution[3] of the diffusion equations in (4.2) with creation and killing $c(t, x)$, and the pair $\{\hat{\phi}_a(x), \phi_b(x)\}$ is an entrance-exit law satisfying the normalization condition

$$\int \hat{\phi}_a(x) p(a, x; b, y) \phi_b(y)\, dx dy = 1.$$

We take the "fictitious" p-representation here instead of the "real" q-representation. The reason is the formal resemblance between the equations in (4.1) and (4.2). This idea is due to Schrödinger (1931), who was motivated by the fact that one obtains the distribution density μ as *a product*

$$\mu = \hat{\phi}\phi$$

of solutions ϕ and $\hat{\phi}$ of the diffusion equations in duality given in (4.2).

---

[1] This choice of the pair of diffusion equations will give the simplest dependence between V and c (cf. Theorem 4.1). See Eddington's remark on (4.1) quoted in Section 8.4

[2] We can, therefore, regard the equations in (4.2) as time reversal of each other

[3] We do not discuss the existence in this chapter. Cf. Chapters 5 and 6

One can identify the distribution $\mu = \hat{\phi}\phi$ with the product of solutions ψ and $\overline{\psi}$ of the Schrödinger equation and its complex conjugate given in (4.1), namely

(4.3) $$\hat{\phi}\phi = \overline{\psi}\psi.$$

In order to show the equivalence between the Schrödinger equations (4.1) and the pair of diffusion equations (4.2), we need, not only the identification of the product representation such as given in (4.3), but also a direct correspondence between the pair $\{\overline{\psi}, \psi\}$ of the complex-valued wave functions satisfying (4.1) and the pair $\{\hat{\phi}, \phi\}$ of the real-valued solutions of equation (4.2).

One of the keys for this is an application of Theorem 3.6, namely we employ Kolmogoroff's representation with time reversal

$$Q = [\,\hat{\phi}_a\phi_a\, q>> = <<\hat{q}\,\hat{\phi}_b\phi_b\,],$$

where q and $\hat{q}$ are (weak) fundamental solutions of

(4.4)
$$\frac{\partial q}{\partial t} + \frac{1}{2}\Delta q + b(t,x)\cdot\nabla q + a(t,x)\cdot\nabla q = 0,$$

$$-\frac{\partial \hat{q}}{\partial t} + \frac{1}{2}\Delta \hat{q} - b(t,x)\cdot\nabla \hat{q} + \hat{a}(t,x)\cdot\nabla \hat{q} = 0,$$

respectively. Then the distribution density $\mu = \hat{\phi}\phi$ satisfies a pair of Fokker-Plank's equations

(4.5)
$$-\frac{\partial \mu}{\partial t} + \frac{1}{2}\Delta\mu - \frac{1}{\sqrt{\sigma_2}}\nabla\{\sqrt{\sigma_2}\,(b(t,x) + a(t,x))\mu\} = 0,$$

$$\frac{\partial \mu}{\partial t} + \frac{1}{2}\Delta\mu - \frac{1}{\sqrt{\sigma_2}}\nabla\{\sqrt{\sigma_2}\,(-b(t,x) + \hat{a}(t,x))\mu\} = 0,$$

at the same time, and hence it satisfies a "continuity equation"

(4.6) $$\frac{\partial \mu}{\partial t} + \frac{1}{\sqrt{\sigma_2}}\nabla\{\sqrt{\sigma_2}\,(b + \frac{a - \hat{a}}{2})\mu\} = 0, \quad in \ \ D,$$

(cf. (3.35) in Chapter 3). Because of Theorem 3.4 the drift coefficients

$a(t,x)$ and $\hat{a}(t,x)$ satisfy the duality formula (3.4) with (3.5), which can be represented under (3.6) in terms of a pair of functions $R(t,x)$ and $S(t,x)$ as

$$\frac{a(t,x) + \hat{a}(t,x)}{2} = \sigma^T \sigma \nabla R(t,x),$$

$$\frac{a(t,x) - \hat{a}(t,x)}{2} = \sigma^T \sigma \nabla S(t,x).$$

Therefore, since $\mu = e^{2R}$, the continuity equation (4.6) becomes

$$(4.7) \qquad \frac{\partial R}{\partial t} + \frac{1}{2}\Delta S + (\sigma\nabla S)\cdot(\sigma\nabla R) + b\cdot\nabla R = 0,$$

which relates the pair of functions $R(t,x)$ and $S(t,x)$.

For the equivalence of Schrödinger and diffusion equations the formula (4.7) will play a crucial role, since it follows from the Schrödinger equation as well, as will be shown in the following Lemma 4.1.

Besides formula (4.7), another key to establishing the equivalence is an application of the following simple lemmas on change of variable formulae.

Lemma 4.1. (Nelson (1966))   *Let $\psi(t,x)$ be a complex-valued function and set $D = \{(t,x): 0 < |\psi(t,x)| < \infty\} \subset [a,b]\times\mathbf{R}^d$.*

Assume $\psi(t,x)$ is differentiable once in t and twice in x and given in the exponential form

$$\psi = e^{R + iS},$$

in terms of real-valued functions $R(t,x)$ and $S(t,x)$.   Then, $\psi(t,x)$ satisfies the Schrödinger equation in (4.1) in D, if and only if R and S satisfy in D a system of non-linear equations

$$(4.8) \qquad V = -\frac{\partial S}{\partial t} + \frac{1}{2}\Delta R + \frac{1}{2}(\sigma\nabla R)^2 - \frac{1}{2}(\sigma\nabla S)^2 - b\cdot\nabla S,$$

$$(4.9) \qquad 0 = \frac{\partial R}{\partial t} + \frac{1}{2}\Delta S + (\sigma\nabla S)\cdot(\sigma\nabla R) + b\cdot\nabla R.$$

Proof. Substitute $\psi = e^{R + iS}$ in (4.1) and divide by ψ. Then, equations (4.8) and (4.9) can be obtained from the real and imaginary parts, respectively. This completes the proof.

Corresponding to the preceding lemma for complex-valued wave functions, we have a similar change of variable formulae for a pair of real-valued functions $\{\hat{\phi}, \phi\}$:

Lemma 4.2. (Nagasawa (1989, a)) *Let* $\{\hat{\phi}, \phi\}$ *be a pair of non-negative functions and* $D = \{(s,x): 0 < \hat{\phi}(s,x)\phi(s,x) < \infty\} \subset [a,b] \times \mathbf{R}^d$. *Define*

$$(4.10) \qquad R = \tfrac{1}{2}\log \hat{\phi}\,\phi \quad and \quad S = \tfrac{1}{2}\log \phi/\hat{\phi}\,,$$

and assume they are differentiable once in t and twice in x in D. Then:

(i) $\phi = e^{R + S}$ *satisfies the first diffusion equation of (4.2) in D if and only if R and S satisfy in D*

$$(4.11) \qquad -c = \{-\frac{\partial S}{\partial t} + \tfrac{1}{2}\Delta R + \tfrac{1}{2}(\sigma\nabla R)^2 - \tfrac{1}{2}(\sigma\nabla S)^2 - \boldsymbol{b}\cdot\nabla S\}$$

$$+\{\frac{\partial R}{\partial t} + \tfrac{1}{2}\Delta S + (\sigma\nabla S)\cdot(\sigma\nabla R) + \boldsymbol{b}\cdot\nabla R\}$$

$$+\{2\frac{\partial S}{\partial t} + (\sigma\nabla S)^2 + 2\,\boldsymbol{b}\cdot\nabla S\}.$$

(ii) $\hat{\phi} = e^{R - S}$ *satisfies the second diffusion equation of (4.2) in D if and only if R and S satisfy in D*

$$(4.12) \qquad -c = \{-\frac{\partial S}{\partial t} + \tfrac{1}{2}\Delta R + \tfrac{1}{2}(\sigma\nabla R)^2 - \tfrac{1}{2}(\sigma\nabla S)^2 - \boldsymbol{b}\cdot\nabla S\}$$

$$-\{\frac{\partial R}{\partial t} + \tfrac{1}{2}\Delta S + (\sigma\nabla S)\cdot(\sigma\nabla R) + \boldsymbol{b}\cdot\nabla R\}$$

$$+\{2\frac{\partial S}{\partial t} + (\sigma\nabla S)^2 + 2\,\boldsymbol{b}\cdot\nabla S\}.$$

Proof. Substituting $\phi = e^{R + S}$ and $\hat{\phi} = e^{R - S}$ into the equations in (4.2), and dividing by ϕ and $\hat{\phi}$, we get equations (4.11) and (4.12).

Remark 4.1. Functions $\hat{\phi}(t,x)$ and $\phi(t,x)$ are not necessarily non-negative. They may take negative values as well, which will be seen in the following chapters in connection with ergodic decomposition of the state space. In fact we can multiply $(-1)^{K(t,x)}$, where $K(t,x)$ takes values 0 or 1 and may change its values only at the zero set of the product $\hat{\phi}(t,x)\phi(t,x)$ (see Lemma 5.2 and Remark 5.2), but both $\hat{\phi}(t,x)$ and $\phi(t,x)$ take the same positive or negative sign at the same time, and hence their product $\hat{\phi}(t,x)\phi(t,x)$ remains always non-negative.

4.2. Equivalence Theorem

We will prove an equivalence relation between Schrödinger and diffusion equations, applying the formula (4.7), Lemma 4.1, and Lemma 4.2. We assume the gauge condition div $b = 0$ on the drift coefficient b.

Theorem 4.1. (Nagasawa (1989, a)[4])
(i) *Let a pair $\{\hat{\phi}, \phi\}$ of real-valued functions satisfy diffusion equations in (4.2) in D, and define R and S by (4.10). Define a complex-valued function $\psi(t,x)$ by*

$$(4.13) \qquad \psi(t,x) = e^{R(t,x) + iS(t,x)}.$$

Then, it is a (weak) solution of the Schrödinger equation (4.1) in D with a potential function

$$(4.14) \qquad V(t,x,\psi) = -c(t,x) - 2\frac{\partial S}{\partial t}(t,x) - (\sigma\nabla S)^2(t,x) - 2\,b\cdot\nabla S(t,x).^5$$

(ii) *Conversely, let a complex-valued function $\psi(t,x)$ be a (weak) solution of the Schrödinger equation (4.1) in D of the form (4.13) and define ϕ and $\hat{\phi}$ by*

$$\phi(t,x) = e^{R(t,x) + S(t,x)},$$
$$(4.15)$$
$$\hat{\phi}(t,x) = e^{R(t,x) - S(t,x)}.$$

---

[5] The relation between V and c depends on our choice of the pair of diffusion equations in (4.2). Another choice gives more complicated dependence, and hence we avoid it

Then, they are (weak) solutions of the diffusion equations (4.2) *in D with creation and killing*

$$(4.16) \qquad c(t, x, \phi) = - V(t, x) - 2 \frac{\partial S}{\partial t}(t, x) - (\sigma \nabla S)^2(t, x) - 2 \, \boldsymbol{b} \cdot \nabla S(t, x).$$

Proof. We will give a proof in the case of $\phi \in C^{1,2}(D)$. First of all, we remark that, if ϕ and $\hat{\phi}$ satisfy equations (4.2), then formula (4.7) holds, which is exactly the equation appeared in (4.9).

(i) Let $\phi(t, x) = e^{R(t,x) + S(t,x)}$ satisfy the first equation of (4.2). Then equality (4.11) holds by Lemma 4.2. Since the equation (4.7) holds as remarked above, the second line of the right-hand side of (4.11) vanishes. Hence, equation (4.8) holds with $V(t, x, \psi)$ defined in (4.14). Consequently, $\psi(t, x) = e^{R(t, x) + i S(t, x)}$ satisfies the Schrödinger equation (4.1) with this potential $V(t, x, \psi)$ by Lemma 4.1.

For the converse statement:
(ii) Let $\psi(t, x) = e^{R(t, x) + i S(t, x)}$ satisfy the Schrödinger equation in (4.1); then the first line of the right-hand side of (4.11) in Lemma 4.2 is equal to V, because of equation (4.8). Moreover, the second line vanishes because of (4.9). Hence, Lemma 4.2 implies that $\phi(t, x) = e^{R(t, x) + S(t, x)}$ satisfies the first equation of (4.2) with the creation and killing $c(t, x, \phi)$ defined in (4.16).

The same argument applies to $\hat{\phi}(t, x)$. This completes the proof.

Remark 4.2. In the assertion (i) of Theorem 4.1, an application of formula (4.7) is indispensable, but it should be emphasized the fact that if ϕ and $\hat{\phi}$ satisfy (4.2) then formula (4.7) holds automatically as a consequence of the equivalence of the p-representation and the q-representation. Moreover, it should be remarked, though it is not stated in the assertion (ii) of Theorem 4.1, that there exists a diffusion process $\{X_t, Q\}$ in the p-representation corresponding to the functions $\phi(t, x)$ and $\hat{\phi}(t, x)$ given in (4.15), and also in the q-representation, cf. Chapters 5 and 6.

Theorem 4.2. *In terms of the q-representation, the transition probability density $q(s, x; t, y)$ of the diffusion process $\{X_t, Q\}$ determined by the Schrödinger equation* (4.1) *is the weak fundamental solution of*

$$(4.17) \qquad \frac{\partial q}{\partial t} + \tfrac{1}{2} \Delta q + \{ \boldsymbol{b} + \sigma^T \sigma \nabla(\log \phi) \} \cdot \nabla q = 0, \ in \ D.$$

Moreover, the distribution density $\mu(t, x) = \hat{\phi}(t, x)\phi(t, x)$ of the diffusion process is a weak solution of the formal adjoint equation of (4.17)

$$(4.18) \quad -\frac{\partial \mu}{\partial t} + \frac{1}{2}\Delta \mu - \frac{1}{\sqrt{\sigma_2}}\nabla\{\sqrt{\sigma_2}(b + \sigma^T\sigma\nabla(\log \phi))\mu\} = 0, \quad in \quad D,$$

which is a non-linear equation of $\mu = \hat{\phi}\,\phi$.

Consequently, the equations in (4.1), (4.2), and (4.4) (*i.e.* (4.17)) are equivalent with each other.

Proof. Cf. Theorem 3.7.

Theorem 4.1 provides a way of transforming a pair of Schrödinger equations to a pair of diffusion equations, and vice versa. This transformation is nothing other than the simple relation of two-dimensional vectors in the polar coordinate on the one hand, namely e^{R+S} and e^{R-S}, and in the complex plane, namely e^{R+iS} and e^{R-iS}, on the other hand.

Remark 4.3. It should be noted that our equivalence relation established in Theorem 4.1 has nothing to do with the so-called "imaginary time argument". In the so called "imaginary time argument" one substitutes "$t \to it$" in formulae (*e.g.* of action functionals), and after manipulations applies analytic continuation. This is often employed to get a corresponding diffusion equation from a Schrödinger equation. But it is too naive and obscure to be a justifiable transformation, and in fact it does not work for the Schrödinger equation (4.1) with a vector potential $b(t, x)$. Therefore, it would be better to avoid it.

4.3. Discussion of the Non-Linear Dependence

In stationary cases with $\sigma^T\sigma = \delta_{ij}$, $b \equiv 0$, and $S(t, x) = -\lambda \cdot t$, where λ is a properly chosen constant, the non-linear dependence (4.14) or (4.16) in Theorem 4.1 disappears, because

$$V(x) + c(x) = 2\lambda.$$

In this case, substituting

$$\psi(t, x) = e^{-i\lambda t}\varphi(x) \quad and \quad \phi(t, x) = e^{-\lambda t}\varphi(x)$$

into the Schrödinger equation (4.1) and the diffusion equation (4.2), respectively, we have in both cases the same eigenvalue problem

$$-\frac{1}{2}\Delta\varphi + V(x)\varphi = \lambda\varphi,$$

and $\mu = \overline{\psi}\psi = \hat{\phi}\,\phi = \varphi^2$, through which Schrödinger achieved great success in his first paper (cf. Schrödinger (1926, I)) on the subject of quantum mechanics.

Remark 4.4. Excited states are degenerated in higher dimensions in general. Therefore, through superposition we get mixed states. The diffusion processes determined by the mixed states are not symmetric anymore with respect to time reversal (cf. examples in Nagasawa (1980)).

Schrödinger proposed a time-dependent equation such as given in (4.1) in his fourth paper (cf. Schrödinger (1926, IV)) on wave mechanics. He deduced his equation through the factorization of a "physically plausible" wave equation of fourth order

$$\left\{\frac{\partial^2}{\partial t^2} + (\tfrac{1}{2}\Delta - V)^2\right\}\psi = 0,$$

from which he gets a pair of equations

$$\{-i\frac{\partial}{\partial t} + \frac{1}{2}\Delta - V(t,x)\}\psi = 0,$$

and

$$\{+i\frac{\partial}{\partial t} + \frac{1}{2}\Delta - V(t,x)\}\overline{\psi} = 0.$$

Then, he took one of them as his fundamental equation in quantum mechanics. Consequently, his deduction of the wave (Schrödinger) equation was heuristic, though he showed soon afterwards that his equation could also be obtained through an application of Heisenberg-Born-Jordan's commutation relation to a prescribed Hamiltonian (cf. Schrödinger (1926)), through which he proved a connection between two theories.

Our deduction of the Schrödinger equation from the diffusion equations in duality is not heuristic, nor does it depend on any additional postulate (this point should be paid attention). Therefore, we can claim that the Schrödinger equation is *intrinsically* in the theory of diffusion processes as an equation in the context of time reversal of diffusion processes (namely diffusion equations in duality), and the transformation

$$\psi = e^{R + iS} \qquad\qquad \phi = e^{R + S}$$
$$\Leftrightarrow \; (R, S) \; \Leftrightarrow$$
$$\overline{\psi} = e^{R - iS} \qquad\qquad \hat{\phi} = e^{R - S}$$

with

$$\overline{\psi}\psi = \hat{\phi}\phi$$

establishes a simple correspondence between quantities in quantum mechanics and diffusion theory (cf. Theorem 4.1).

However, in this correspondence, we must be concerned with the non-linear relation between two functions $c(t, x)$ and $V(t, x)$.[6]

Let us write the equations (4.14) and (4.16) as

$$(4.19) \qquad\qquad c + V + 2\frac{\partial S}{\partial t} + (\sigma\nabla S)^2 + 2b\cdot\nabla S = 0.$$

This formula shows that in treating the non-linear relation, we have freedom of how much portion of the non-linear terms of (4.19) are to be included in the function c or V. This freedom is mathematically a technical problem, but it induces completely different physical pictures, as will be seen. Equations (4.14) and (4.16) are two extreme cases: In (4.14) we regard c is given, *i.e.*, we consider that the diffusion equation with c is linear, so that the Schrödinger equation with V becomes non-linear, while in (4.16) we consider V is given, *i.e.*, the Schrödinger equation with V is linear but the diffusion equation with c is non-linear. It should be kept in mind that the function c and V cannot be assigned freely (cf. Appendix).

If one were to be faithful to the original equation introduced by Schrödinger, one should assume that V is given, although there is no evidence for the Schrödinger equation to be linear; also we must keep the possibility that the Schrödinger equation might be non-linear in general. Then the diffusion process corresponding to the Schrödinger equation turns out to have non-linear drift caused by the nonlinear creation and killing c given in (4.16). To understand this point we must find a plausible probabilistic reasoning (or mathematical structure) of this non-linearity of diffusion processes.

As far as I know, there are two cases in which non-linearity arises naturally in the theory of diffusion processes.

---

[6] The non-linear dependence of Schrödinger and diffusion equations is caused by the non-linearity of the change of variables. For the statistical mechanical nature of the non-linear relation, cf. Chapters 7 and 8

The first one is (generalized) branching diffusion processes which can be described by diffusion equations with nonlinear creation and killing (and in general also non-linear drift) terms

$$\sum_{n,m} c_{n,m} u(t,x)^n \left(\frac{\partial u(t,x)}{\partial x}\right)^m,$$

(cf. Chapter 12). This non-linearity is a consequence of branching, namely, creation and annihilation of particles.

The second one appears in connection with statistical mechanics. Let us consider the propagation of chaos for systems of interacting diffusion processes (cf. McKean (1966, 67)), which induces McKean-Vlasov's equation

$$\frac{\partial u}{\partial t} = \frac{1}{2}\frac{\partial^2 u}{\partial x^2} - \frac{\partial}{\partial x}(b[x,u]u),$$

in the simplest case of one dimension, where the drift coefficient depends on u and is given by

$$b[x,u] = \int b(x,y)u(t,y)dy,$$

with a pair interaction $b(x,y)$ between particles in the system we consider.

Both non-linearities do not resemble the one given in (4.19) or (4.16). Therefore, it is natural that one might regard the non-linearity appearing in (4.19) as the third kind. If it is the case, one must discover a reasonable mathematical structure behind this.

However, I would like to consider, as an attempt, "the third kind of non-linearity (4.19)", something like the second one. That is, I regard the non-linear diffusion equation induced by a Schrödinger equation as the statistical equation of a system of interacting diffusion particles. This will be discussed in Chapters 7 and 8, applications of which will be given in Chapter 9.

It should be remarked that even if we would assume c is given and hence the Schrödinger equation has the non-linear potential $V(t,x,\psi)$ given in (4.14), the situation we have encountered (with the non-linear c) would not change at all, and the discussion above applies also to this case.

4.4. A Solution to Schrödinger's Conjecture

Schrödinger closes his article (1931) on time reversal of Brownian motions by claiming :

> "Die Überlegungen der ersten drei Paragraphen sind mit wenig Änderung auch auf viel kompliziertere Fälle anwendbar - mehrere räumliche Koordinaten, variabler Diffusionskoeffizient, äußere Kräfte, die irgendwelche Ortsfunctionen sind. Für die Wahrscheinlichkeitsdichte erhält man immer das Product der Lösungen zweier adjungierter Gleichungen, welche sich im allgemeinen nicht nur durch das Vorzeichen der Zeit, sondern auch in anderen Gliedern unterscheiden. ⋯⋯ Ich möchte aber auf diese Dinge nicht näher eingehen, bevor sich herausgestellt hat, ob man sie etwa wirklich zu einem besseren Verständnis der Quantenmechanik ausnützen kann."

It seems that Schrödinger was convinced, but he could not conclude whether the diffusion theory would provide really a "better understanding of quantum mechanics" in 1931. He postponed further discussion without giving a clear perspective in his article. Therefore, let us regard this as a conjecture.

However, we now have a clear answer to his conjecture and can claim: *The diffusion theory in terms of diffusion processes certainly provides a better understanding of quantum mechanics*. In fact, as a corollary of Theorem 4.1, we can formulate a proposition,[7] which gives a solution to Schrödinger's conjecture:

Proposition 4.1. *Non-relativistic quantum mechanics is a diffusion theory which contains the Schrödinger equation intrinsically as an equation in the context of time reversal (or duality).*[8]

It is not so clear what Schrödinger meant by "providing a better understanding of quantum mechanics". Nevertheless, Proposition 4.1 is certainly a definitive answer to it. Thus Schrödinger's program in 1931 is completed.

---

[7] The word "Proposition" in this monograph is used differently. In mathematical literature, it usually means a "small theorem"

[8] Needless to say, but this proposition does not deny the importance of wave functions and Schrödinger equations. See the following sections

4.5. A Unified Theory

It is worthwhile to make some additional remarks and rephrase Proposition 4.1. As remarked in Chapter 1, both Born-Heisenberg-Jordan's algebraic theory and Schrödinger's wave theory have no built-in mathematical structure to conclude Born's "statistical interpretation". This has been a main source of the conceptual confusion in quantum mechanics. The confusion is now solved completely through the equivalence of the Schrödinger and diffusion equations. The equivalence has revealed a mathematical structure with which we can formulate a unified theory, in which Born-Heisenberg-Jordan's algebraic theory, Schrödinger's wave theory, and Born's statistical interpretation find their own right places; namely, elaborating on Proposition 4.1, we can claim

Proposition 4.2. *Born-Heisenberg-Jordan's algebraic theory, Schrödinger's wave theory, and diffusion theory are unified in a theory on a firm mathematical ground based on the equivalence of the Schrödinger and diffusion equations*:

$$L^2\text{-theory} \qquad Transformation$$
$$\Downarrow \qquad\qquad \Downarrow$$

Algebraic Theory $\Leftrightarrow$ Wave Theory $\Leftrightarrow$ Diffusion Theory

Commutation	Schrödinger	Diffusion equations
relation	equation	in duality;
		Diffusion processes
		with time reversal

As L^2-theories the algebraic theory with commutation relation (1.1) and Schrödinger's wave theory with Schrödinger equation (4.1) are equivalent. This equivalence is well-known since Schrödinger (1926). The wave theory with the Schrödinger equation and the diffusion theory with time reversal (diffusion equations in duality) are equivalent through the transformation

$$\psi = e^{R+iS} \qquad\qquad \phi = e^{R+S}$$
$$\Leftrightarrow (R, S) \Leftrightarrow$$
$$\overline{\psi} = e^{R-iS} \qquad\qquad \widehat{\phi} = e^{R-S}$$

This part of the equivalence in the above diagram is new.

I would like to emphasize the fact that *nothing new* (namely, *no new postulate*) *is added* in our unified theory,[9] but just a simple mathematical structure which had been hidden is revealed.

It should be remarked that "Born's interpretation" is no longer an "interpretation", but a "formula"

$$\overline{\psi}_t(x)\,\psi_t(x)dx = \hat{\phi}_t(x)\phi_t(x)dx = Q \circ X_t^{-1},$$

in our unified theory.

The unified theory claims that a "quantum particle" moves with specific noise and drift; namely, the movement of the particle is described mathematically in terms of the theory of diffusion processes.[10] The movement shows wave-like interference (cf. Section 4.8). The wave-like character of the movement can be well-described in terms of the Schrödinger (wave) equation. To specify (or find) the diffusion and drift coefficients for mechanical systems, we can adopt the commutation relation in the algebraic theory as a technical tool. This point will be discussed in the next section.

If we treat systems of, say, two particles in $\mathbf{R}^3$, then $\overline{\psi}_t(x)\psi_t(x)$ is a distribution density in $\mathbf{R}^6$. Based on this fact, it has been claimed that the distribution density $\overline{\psi}_t(x)\psi_t(x)$ must be understood as something not concerning physical substance but just a mathematical device in Born's statistical interpretation. But this claim should be reconsidered carefully.

Consider a system of interacting k-particles x_t^i, $i = 1, 2, \ldots, k$, in $\mathbf{R}^3$. Then our diffusion process

$$X_t = (x_t^1, \ldots, x_t^k)$$

moves in $\mathbf{R}^{3k}$, and its probability distribution

$$P^X = Q \circ X_t^{-1}$$

is defined in $\mathbf{R}^{3k}$ accordingly. When we discuss the distribution of the

---

[9] This is a crucial point which distinguishes our theory from the other ones explained in the Appendix

[10] Our diffusion process has nothing to do with the so-called "hidden variables", since it follows from (actually equivalent to) the Schrödinger equation

system in the three-dimensional space, we should naturally take the empirical distribution

$$U_t^x = \frac{1}{k} \sum_{i=1}^{k} \delta_{x_t^i}$$

of the k-particles $(x_t^1, \dots, x_t^k)$ in the system. Then the expectation of the empirical distribution

$$\eta_t^x = Q[U_t^x]$$

is the probability distribution of the system *in the three-dimensional space and hence experimentally observable*. Consequently our diffusion process $X_t = (x_t^1, \dots, x_t^k)$ is not just a mathematical device but physically substantial.

If a system contains particles of different types, we must generalize the above argument. Let us consider a system

$$X_t = (x_t^1, \dots, x_t^m, y_t^1, \dots, y_t^n)$$

of particles of two different types in $\mathbf{R}^3$. Then in this case we define a pair of empirical distributions

$$U_t^x = \frac{1}{m} \sum_{i=1}^{m} \delta_{x_t^i}, \qquad U_t^y = \frac{1}{n} \sum_{i=1}^{n} \delta_{y_t^i},$$

the expectations of which

$$\eta_t^x = Q[U_t^x], \qquad \eta_t^y = Q[U_t^y],$$

are the probability distributions in the three-dimensional space of the particles $(x_t^1, \dots, x_t^m)$ and $(y_t^1, \dots, y_t^n)$, respectively. In the simplest case of two particles (x_t, y_t) of different types

$$\eta_t^x = Q[U_t^x] = Q[\delta_{x_t}], \qquad \eta_t^y = Q[U_t^y] = Q[\delta_{y_t}],$$

reduce to the marginal distributions of $X_t = (x_t, y_t)$. It is routine to extend the above argument to systems of many types.

It should be remarked that what has been explained above is well-known in diffusion theory but new in quantum mechanics.[11]

---

[11] In conventional Born's statistical interpretation random variables x_t^i were not introduced and hence it was not possible to consider the empirical distributions

The so-called "semi-classical" limit of diffusion processes provides a way of regarding the drift coefficient approximately as the classical velocity field when the noise (the diffusion coefficient) is negligible relative to the drift coefficient. However, it should be emphasized that the drift field cannot be regarded as the classical velocity field in general (cf. Appendix A.6 on the so-called Bohmian mechanics).

Moreover, the theory of diffusion processes provides an important device, namely the notion of *sample paths*, which is lacking in the traditional quantum mechanics. This notion helps us, when we discuss the "problem of observation (measurement)", and it resolves the so-called "paradoxes" in quantum mechanics, and so on. In fact, the theory of stochastic processes claims that a set of sample paths is realized with a probability according to a given probability law Q. This means that we observe actually a sample path(s) in experiments.[12] Moreover, since the probability law Q has the Schrödinger representation which contains $\hat{\phi}_a$ and ϕ_b, we must think of Eddington-Schrödinger's (namely Markov field type) prediction to understand the experiment.[13] Therefore, the non-locality of quantum mechanics should not be a surprise. According to Proposition 4.1 or 4.2, it is clear that quantum mechanics is non-local, because it is equivalent to the theory of diffusion processes. One should perhaps argue this the other way round, namely, the non-locality property of quantum mechanics has revealed its character as a diffusion theory.

4.6. On Quantization

In conventional quantum mechanics the so-called "quantization" (commutation relation) is postulated as an axiom which is applied to plausible prescribed classical Hamiltonians through which the Schrödinger equation is deduced. It was a great discovery, but people have never stopped wondering what "quantization" means, so that it has been a constant source of "mysticism of quantum mechanics". Now we can begin with a diffusion equation (or process), which induces (actually is equivalent to) the Schrödinger equation. Therefore, we need no "quantisation" *as a postulate*, and we can regard the diffusion equation (or what is the same the Schrödinger equation) as the fundamental *equation of motion* in quantum mechanics.

---

[12] In the case of famous Schrödinger's cat (cf. Schrödinger (1935)) there are two sample paths, on one of which the cat is killed, and along another path the cat survives. The probability of each event is 1/2. The wave function $\psi = (\psi_{dead} + \psi_{alive})/\sqrt{2}$ itself belongs to a fictitious description as explained already

[13] Cf. Eddington's comment on $\psi\bar{\psi}$ quoted in Section 8.4

In the unified theory we must specify diffusion and drift coefficients $\sigma^T \sigma(t,x)$ and $b(t,x)$, and a killing and creation $c(t,x)$, or a potential function $V(t,x)$. This is exactly like specifying the "mass and force" in Newton's equation of classical mechanics, assuming the acceleration is well-defined (or in other words, a classical Hamiltonian or Lagrangian). By this I mean that *the diffusion equations* (4.2) *in the p-representation* (*resp.* (4.4) *in the q-representation* or *the stochastic differential equation in the form of* (6.13)) *should be regarded as "the equation of motion" in the unified theory*. The diffusion equation is equivalent to the Schrödinger equation and defines the Schrödinger process. As a matter of fact, the Schrödinger process follows directly from the variational principle with the Lagrangian (5.55), as will be shown in Chapter 5.

When we consider a mechanical system in quantum theory and build a model, we usually employ a similar system of classical mechanics and transform it into quantum mechanics (namely diffusion processes). To do this the "quantization (commutation relation)" is an amazing technical tool (although the true problem is to fix how much noise should be involved).

As Schrödinger (1926) demonstrated, the theory of operator algebra with the commutation relation (quantization) induces the Schrödinger equation. In fact, substituting

$$q^k = x^k \quad and \quad p_k = \frac{1}{i}\frac{h}{2\pi}\frac{\partial}{\partial x^k},$$

which satisfy the commutation relation (1.1),[14] into a classical Hamiltonian

(4.20) $$H(p,q) = \frac{1}{2}\sum_{k=1}^{d}(p_k - b_k)^2 + v(q),$$

one can get

$$-\frac{1}{i}\frac{h}{2\pi}\frac{\partial\psi}{\partial t} + \frac{1}{2}\sum_{k=1}^{d}(\frac{1}{i}\frac{h}{2\pi}\frac{\partial}{\partial x^k} - b_k)^2\psi + v\,\psi = 0,$$

which coincides with the Schrödinger equation (4.1) under the gauge condition div $b = 0$ and with $V = v + b^2$. Therefore, as an abstraction, the theory of operator algebra with the commutation relation (quantization) has been considered as quantum mechanics. As a matter of fact the operator theory is a generalization of the theory of diffusion processes which contains intrinsically Schrödinger equation (4.1) as an equivalent equation of the

[14] Cf. Schrödinger (1926). Cf. also Mackey (1968, 78) for the commutation relation

diffusion equations (4.2).   Historically, the operator theory of quantum mechanics was formulated just by chance one year before the Schrödinger equation came out, and moreover, Born's statistical interpretation had to be attached artificially. This historical accident has caused a lot of confusion[15] in quantum mechanics, as is well known.

Such an interrelation as the one between the Schrödinger equation and the theory of operator algebra can be seen rather often. Let us take the theory of Markov processes as one such case. It is well-known that the theory can be formulated in terms of more abstract one, say, the theory of semi-groups. In this case there is no doubt that semi-group theory is a technical tool for Markov processes, more concretely, for diffusion processes. Moreover, it is also well-known that there are many diffusion processes which cannot be well-treated by means of the theory of semi-groups.

To clarify things, imagine a (fictitious) story which is parallel to the one in quantum mechanics: A substitution of

$$(4.21) \qquad\qquad q^k = x^k \quad and \quad p_k = -\sigma\frac{\partial}{\partial x^k},$$

into a prescribed Hamiltonian (4.20) yields the (formal) generator of a semi-group. Thus, an operator algebra on $C(\mathbf{R}^d)$ (or $L^2(\mathbf{R}^d)$) with an additional *postulate*, namely a "commutation relation",

$$(4.22) \qquad\qquad pq - qp = -\sigma I,$$

provides a diffusion equation

$$\frac{\partial\phi}{\partial t} + \frac{1}{2}\sum_{k=1}^{d}(-\sigma\frac{\partial}{\partial x^k} - b_k)^2\phi + v\phi = 0,$$

which coincides with diffusion equation (4.2) under the gauge condition div $b = 0$ and $c = v + b^2$.

Imagine that someone would claim, based on the above story, that the operator algebra with the commutation relation (4.22) is a completely new theory which describes the motion of something like a non-Newtonian particle, but what we are treating should not be the same as a diffusion process, because the physical meaning of the commutation relation (4.22) is

---

[15] Since it has become clear that quantization is just a technical tool, there remains no corner any more for mysticism based on "quantization"

not clear and should be postulated as an important fundamental axiom of the theory. It would be hardly agreeable for many people, especially for probabilists, if the person claims in addition that one should avoid considering a diffusion process corresponding to the non-negative semi-group in his new theory, and that one must always regard the abstract operator theory as a fundamental starting point.

The abstract operator theory of quantum mechanics with the commutation relation appears to be self-standing, apart from diffusion theory. However, by exactly the same reasoning as explained above for semi-group theory, it must come back time after a time to diffusion theory, since diffusion theory is the root of the general theories of quantum mechanics.

It is clear that if we begin with diffusion equations (or processes), we need no "quantization", since we have the equivalence between diffusion and Schrödinger equations, as we have shown in Theorem 4.1, and the diffusion equation itself is the equation of motion, as remarked already. However, various kinds of the so-called equation of motion were formulated in diffusion theory, from which the Schrödinger equation was deduced, and they were considered as "quantization" (cf. Appendix for this point).

4.7. As a Diffusion Theory

Theorem 4.1 claims the equivalence of Schrödinger and diffusion equations, based on which a unified theory is formulated in Proposition 4.2. In connection with this, it is important not only theoretically but also practically to emphasize the fact that the Schrödinger equation is intrinsically in diffusion theory as formulated in the following

Proposition 4.3. *The applicability of the Schrödinger equation is not confined in conventional quantum mechanics, but can be extended to other fields as an equation in the theory of diffusion processes (or the unified theory).*

In other words, the Schrödinger equation is liberated from the conventional framework of quantum mechanics. Non-relativistic quantum mechanics is just a special case, in which the diffusion coefficient contains *Planck's constant*. As is known, "Schrödinger equations" have been applied to problems arising from various fields which are not considered proper (or appropriate) for "quantum mechanics" to be applied; for example, problems of cosmology, the solar system, distribution of animal populations, distribution of molecules in a cell, distribution of glueons in mesons, and so on. One of the characteristics of these problems is the random movement (more precisely Brownian movement) of involved particles. It is now clear: what was applied is not "quantum mechanics" but "diffusion theory". In connection with this, detailed mathematical aspects of systems of interacting diffusion particles will be discussed in Chapters 7 and 8, and some applications will be given in Chapter 9.

Naturally there is no built-in mechanism in diffusion theory to determine diffusion and drift coefficients a priori. Therefore, when we apply diffusion theory to quantum mechanics we have to choose the diffusion coefficient $\sigma^T \sigma$ and drift coefficient b properly so that we get the right coefficients of the diffusion and hence Schrödinger equations. A well-known heuristic way of guessing the coefficients in quantum mechanics is to apply the commutation relation (4.21) or (1.1) to plausible Hamiltonians. In this sense the commutation relation is an important technical tool as already noted. In general, we must specify diffusion coefficients appropriately based on given experimental data (cf. Chapter 9).

4.8. Principle of Superposition

It has been claimed that the superposition principle in quantum theory is one of its conspicuous characteristics; namely the principle yields the typical "wave effect", to which classical theories including probability theory are unable to give any reasonable explanation.

The principle of superposition means that if ψ_k are arbitrary wave functions, then

$$\psi = \sum \alpha_k \psi_k$$

defines a new wave function, which is called the superposition of ψ_k.

This is nothing but the linearity of the space of wave functions,[16] and gives no trouble to anybody. However, if the superposition of states is combined with the formula $\mu = |\psi|^2$, then it turns out to be an almost unsolvable problem for probabilists.

To make things clear, let us consider the simplest case of two wave functions $\psi^{(1)}$, $\psi^{(2)}$, and their superposition

---

[16] To claim the linearity it is assumed implicitly that Schrödinger equations are linear

$$\psi = \psi^{(1)} + \psi^{(2)},$$

where we ignore the normalizing constant for simplicity. Then, we have

(4.23) $$|\psi|^2 = |\psi^{(1)}|^2 + |\psi^{(2)}|^2 + 2\mathcal{R}e(\psi^{(1)}\overline{\psi}^{(2)}),$$

where the real part $\mathcal{R}e(\psi^{(1)}\overline{\psi}^{(2)})$ of cross-terms produces "interference" of the two wave functions $\psi^{(1)}$ and $\psi^{(2)}$. This term represents the typical *wave effect* in quantum mechanics.

It has been a long-standing open problem, as whether there is such a principle of superposition in diffusion theory, with which one gets "interference" of diffusion processes like (4.23) as in quantum mechanics. In fact, the so-called principle of superposition as described above must and actually does exist in the theory of diffusion processes.[17]

First of all it must be emphasized that in quantum theory we do not add distribution densities $|\psi^{(i)}|^2$. Instead we add wave functions $\psi^{(i)}$. Therefore, in diffusion theory we should add those quantities that correspond to the wave functions, namely we should superpose the functions $\hat{\phi}^{(i)}$ and $\phi^{(i)}$, because they are the real-valued counterpart of the wave functions $\overline{\psi}^{(i)}$ and $\psi^{(i)}$.

Let

(4.24)
$$\overline{\psi}_t^{(1)}\psi_t^{(1)} = \hat{\phi}_t^{(1)}\phi_t^{(1)},$$

$$\overline{\psi}_t^{(2)}\psi_t^{(2)} = \hat{\phi}_t^{(2)}\phi_t^{(2)},$$

be the factorizations (the p-representation) in terms of corresponding diffusion processes Q_1 and Q_2, and define superposition:

(4.25)
$$\hat{\phi}_t = \hat{\phi}_t^{(1)} + \hat{\phi}_t^{(2)},$$

$$\phi_t = \phi_t^{(1)} + \phi_t^{(2)},$$

for $t \in [a, b]$. Then

---

[17] Cf. Nagasawa (1992)

$$(4.26) \qquad \hat{\phi}_t \phi_t = \{\hat{\phi}_t^{(1)} + \hat{\phi}_t^{(2)}\}\{\phi_t^{(1)} + \phi_t^{(2)}\}$$

$$= \hat{\phi}_t^{(1)} \phi_t^{(1)} + \hat{\phi}_t^{(2)} \phi_t^{(2)} + \{\hat{\phi}_t^{(1)} \phi_t^{(2)} + \hat{\phi}_t^{(2)} \phi_t^{(1)}\}.$$

Thus we have "interference"

$$(4.27) \qquad \hat{\phi}_t^{(1)} \phi_t^{(2)} + \hat{\phi}_t^{(2)} \phi_t^{(1)},$$

where we have ignored the normalization. After normalizing $\hat{\phi}_t \phi_t$, we will call the diffusion process Q, which has $\hat{\phi}_t \phi_t$ as its distribution density, *the superposition of the diffusion processes Q_1 and Q_2.*

In general, let $[\hat{\phi}_a^{(k)} p^{(k)} >><< p^{(k)} \phi_b^{(k)}]$ be the p-representations of diffusion processes $Q^{(k)}$, $k = 1, 2, \dots$. With $\alpha_k, \beta_k \geq 0$ we set

$$(4.28) \qquad \hat{\phi} = \sum_k \alpha_k \hat{\phi}^{(k)} \quad and \quad \phi = \sum_k \beta_k \phi^{(k)},$$

where α_k and β_k should be chosen so that

$$(4.29) \qquad \int dx \hat{\phi}_t(x) \phi_t(x) = 1.$$

If there exists a diffusion process Q with the p-representation $[\hat{\phi}_a p >><< p \phi_b]$, then we call the process Q the **superposition of the diffusion processes** $\{Q^{(k)}: k = 1, 2, \dots \}$.

As an example, let us consider an inverse problem; we decompose a diffusion process Q into, say, two diffusion processes. Let us decompose the entrance-exit law $\hat{\phi}_a$ and ϕ_b, ignoring normalization for simplicity, as

$$\hat{\phi}_a = \hat{\phi}_a^{(1)} + \hat{\phi}_a^{(2)}$$
$$(4.30)$$
$$\phi_b = \phi_b^{(1)} + \phi_b^{(2)},$$

and assume that there exists the p-representations

$$(4.31) \qquad Q^{(i)} = [\hat{\phi}_a^{(i)} p^{(i)} >><< p^{(i)} \phi_b^{(i)}], \quad i = 1, 2,$$

where $p^{(i)}$, $i = 1, 2$, denote transition densities in the p-representation. If Q

is the superposition of $Q^{(1)}$ and $Q^{(2)}$, then we must have, for $\forall t \in [a,b]$,

(4.32)

$$\widehat{\phi}_a^{(1)} p^{(1)}(a,t) + \widehat{\phi}_a^{(2)} p^{(2)}(a,t) = \widehat{\phi}_a p(a,t),$$

$$p^{(1)}(t,b)\phi_b^{(1)} + p^{(2)}(t,b)\phi_b^{(2)} = p(t,b)\phi_b,$$

where the notation is self-explanatory. System (4.32) is a requirement on $p^{(i)}$, $i = 1, 2$, and p, for Q to be the superposition of Q_1 and Q_2.

As a special case, if we assume $p = p^{(1)} = p^{(2)}$, then the system of equations (4.32) reduces to the system of equations (4.30) at the initial and terminal times, and hence we can formulate

Theorem 4.3. *If $p = p^{(1)} = p^{(2)}$, then a diffusion process Q can be decomposed into two diffusion processes $Q^{(1)}$ and $Q^{(2)}$ corresponding to the decomposition of $\widehat{\phi}_a$ and ϕ_b at the initial and terminal times, as in (4.30). Moreover, the diffusion processes $Q^{(1)}$ and $Q^{(2)}$ interfere with each other with the correlation $\widehat{\phi}_t^{(1)} \phi_t^{(2)} + \widehat{\phi}_t^{(2)} \phi_t^{(1)}$ (cf. (4.26) and (4.27)).*

This corresponds to the so-called "two slits problem" in quantum mechanics.

Remark 4.5. If we require $p = p^{(1)} = p^{(2)}$, as in Theorem 4.3, then the decomposition (4.30) becomes a delicate problem in connection with admissibility of the initial and final distributions. This point will be discussed in the following chapters.

More generally we have

Theorem 4.4. *(Principle of Superposition) Let $Q^{(k)}$ be diffusion processes with the p-representations $[\widehat{\phi}_a^{(k)} p^{(k)} >> << p^{(k)} \phi_b^{(k)}]$, $k = 1, 2, \dots$. Define*

(4.33)

$$\widehat{\phi}_t = \sum_k \alpha_k \widehat{\phi}_t^k \quad and \quad \phi_t = \sum_j \beta_j \phi_t^j.$$

with α_k and β_j such that

(4.34)

$$\int dx \widehat{\phi}_t(x) \phi_t(x) = 1.$$

Consider a flow $\mu_t = \widehat{\phi}_t \phi_t$ *of probability densities and determine a drift coefficient* $\overline{b}(t, x)$ *through*

(4.35) $$\frac{\partial R}{\partial t} + \frac{1}{2}\Delta S + (\sigma\nabla S)\cdot(\sigma\nabla R) + \overline{b}\cdot\nabla R = 0,$$

where R and S are defined by $R = \frac{1}{2}\log\mu_t$ *and* $S = \frac{1}{2}\log(\phi_t/\widehat{\phi}_t)$. *Let* $\overline{L}$ *be a parabolic differential operator*

$$\overline{L} = \frac{\partial}{\partial t} + \frac{1}{2}\Delta + \overline{b}(t, x)\cdot\nabla$$

with the $\overline{b}(t, x)$, *and assume the integrability condition* (6.4) *on*

$$\overline{c}(t, x) = -\frac{\overline{L}\phi_t(x)}{\phi_t(x)}$$

with $\phi_t(x)$ *given in* (4.33).[18]

Then, there exists[19] *a diffusion process* Q *with the p-representation:*

(4.36) $$Q = [\sum_k \alpha_k \widehat{\phi}_a^{\ k} p \gg \ll p \sum_k \beta_k \phi_b^k],$$

which is the superposition of the diffusion processes $Q^{(k)}$; *i.e., it holds that for a bounded measurable f and* $\forall t \in (a, b)$

(4.37) $$Q[f(X_t)] = \sum_{k, j} \int dy\, \alpha_k \widehat{\phi}_t^{(k)}(y)\, \beta_j\, \phi_t^{(j)}(y) f(y),$$

where

(4.38) $$\widehat{\phi}_t^{(k)}(y) = \int dx \widehat{\phi}_a^{\ k}(x) p(a, x; t, y),$$

$$\phi_t^{(j)}(y) = \int p(t, y; b, z) dz \phi_b^j(z).$$

The cross terms in (4.37) *represent "interference" (namely, the "wave effect") of the diffusion processes* $\{X_t, Q^{(k)}\}$.

---

[18] We can assume the drift coefficient $\overline{b}$ is good enough so that we get a diffusion process corresponding to the parabolic operator $\overline{L}$

[19] Cf. Theorem 5.4, and Corollary 6.2

Because of the non-linear dependence (4.19), if we assume that the Schrödinger equation is linear with a given potential $V(t, x)$, then the corresponding diffusion processes in general are non-linear. Therefore, it is not reasonable to apply the real-valued superposition in terms of ϕ_i, and hence we should apply the complex-valued superposition, namely, defining wave functions ψ_i in terms of R_i and S_i of ϕ_i, and then superpose the wave functions. In the simplest case of (4.23), we get the interference

$$(4.39) \qquad 2\mathcal{R}e(\psi_1 \overline{\psi_2}) = 2e^{R_1 + R_2} \cos(S_1 - S_2).$$

On the other hand, if a rate of creation and killing $c(t, x)$ is given, then the diffusion equation is linear but the corresponding Schrödinger equation in general becomes non-linear. Therefore, it is reasonable to apply the real valued superposition, namely, defining ϕ_i in terms of R_i and S_i, we superpose them. In the simplest case of (4.27) we get the interference

$$(4.40) \qquad \phi_1 \widehat{\phi}_2 + \phi_2 \widehat{\phi}_1 = 2e^{R_1 + R_2} \cosh(S_1 - S_2).$$

If we treat stationary states, we can apply both complex and real superposition, since the non-linear dependence disappears.

Remark 4.6. We have assumed ϕ_i and $\widehat{\phi}_i$, $i = 1, 2$, are non-negative in (4.40). However, they can take positive and negative values in general (cf. Remark 5.2). Therefore, the interference (4.40) turns out to be

$$(4.41) \qquad (-1)^{K(t, x)} 2e^{R_1 + R_2} \cosh(S_1 - S_2),$$

where $K(t, x) = 0$ *or* 1, which may vary only at the zero set of the function $2e^{R_1 + R_2} \cosh(S_1 - S_2)$.

Based on the principle of superposition in quantum theory, it has been claimed often that quantum theory is the third way of describing natural laws besides deterministic and stochastic ways, since probability theory does not provide such a mathematical structure. As a matter of fact the "principle of superposition in quantum mechanics" has been a nightmare for probabilists. Since this problem has been solved in diffusion theory as we have seen, we can now fully rely on the theory of stochastic processes, even when we consider quantum theory. There are a lot of interesting mathematical problems in quantum mechanics, which have been not yet treated in terms of the theory of diffusion processes.[20] They will be challenging problems.

---

[20] P.A. Meyer kindly indicated to me such problems: Quantum field theory, laser theory, quantum Hall effect, superfluidity, superconductivity, fermions, phonons, quantum chemistry, squeezed light, Bohm-Aharonov effect, and so on

4.9. Remarks

(i) Concerning the relation between the Schrödinger equation and diffusion processes, motivated by Schrödinger (1931), Féniyes (1952) discussed the least action principle and formulated "the equation of motion" of probability densities,[21] from which he deduced the Schrödinger equation and moreover showed that the "uncertainty principle" of Heisenberg is of a stochastic nature. Nelson (1966, 67) introduced and discussed "stochastic mechanics" ("stochastic Newtonian equation"). Nagasawa (1980) treated the law of equilibrium distributions and segregation of a population, some applications of which were discussed in Nagasawa (1981) and Nagasawa-Yasue (1982) (cf. also Nagasawa (1985)). Yasue (1981) formulated stochastic calculus of variation and proved that it implies Nelson's "stochastic mechanics".[22] Carmona (1985) showed the existence of a singular diffusion process for the Schrödinger equation, where he applied the fact that the diffusion process can be given with the modified (non-linear) potential (4.16) (in the case of the vector potential $b(t, x) \equiv 0$).   Based on Jamison (1974, a), Zambrini (1985, 86 a, 87) discussed the relation between the diffusion process with the modified potential and Nelson's "stochastic mechanics", and then formulated "Euclidean quantum mechanics".   For details of the above mentioned contributions, refer to the Appendix. The equivalence of the Schrödinger equation and a pair of diffusion equations in duality (diffusion processes with time reversal) was first remarked in Nagasawa (1989, a), and discussed in Nagasawa (1991) in connection with other theories.

(ii) The theory of (special) relativity[23] can hardly be consistent with diffusion theory, as is clear if we consider sample paths of diffusion processes.   This might be one of the sources of the well-known conflict between quantum mechanics and the theory of relativity.

(iii) For a completely different approach in investigating *mathematical structures* of quantum mechanics in terms of probability theory on operator algebras[24] see Umegaki (1954) and references in Takesaki (1979), and for recent development, cf. Meyer (1992), Parthasarathy (1992).

---

[21] For further literatures on the subject cf. Jammer (1974)

[22] For further references on the variational principle, see Section 5.7

[23] Cf. Einstein (1905, c)

[24] It does not so much concern the problem which we are considering, namely, what the Schrödinger equation is

Chapter V

Variation Principle

In the formulation of the conventional variation principle (or the least action principle) one minimizes the so-called action functional defined in terms of a given Lagrangian to characterize extremal processes. In this chapter, however, the problem will be formulated in a different way which is suitable to the p-representation of diffusion processes.

5.1. Problem Setting in the p-Representation

When one treats diffusion processes in the p-representation, one must be aware of what is given in order to construct diffusion processes. If an explicit (such as Brownian) transition density $p(s, x; t, y)$ which satisfies Kolmogoroff's continuity condition, and p-harmonic and p-coharmonic functions $\phi(t, x)$ and $\hat{\phi}(t, x)$ are given beforehand, one can apply Kolmogoroff's method based on Schrödinger's representation (3.49), *i.e.*,

(5.1) $Q[f(X_a, X_{t_1}, \dots, X_{t_{n-1}}, X_b)]$

$$= \int dx_0 \hat{\phi}(a, x_0) p(a, x_0; t_1, x_1) dx_1 p(t_1, x_1; t_2, x_2) dx_2 \cdots$$

$$\cdots p(t_{n-1}, x_{n-1}; b, x_n) \phi(b, x_n) dx_n f(x_0, x_1, \dots, x_n),$$

and get a probability measure $Q = [\hat{\phi}_a p >><< p \phi_b]$.

However, this happens rarely in serious applications, because we must handle, first of all, a transition density which is the fundamental solution of a diffusion equation

$$(5.2) \qquad \frac{\partial p}{\partial t} + \frac{1}{2}\Delta p + \boldsymbol{b}(t,x)\cdot\nabla p + c(t,x)p = 0,$$

with singular creation and killing $c(t,x)$. The existence of the fundamental solution $p(s,x;t,y)$ of (5.2) is not evident. Secondly, p-harmonic and p-coharmonic functions $\phi(t,x)$ and $\hat{\phi}(t,x)$ are not given beforehand, because they are fictitious and are determined not directly through experimental data. What usually occurs is this: Imagine we are releasing an electron gun at $t = a$ and observing at $t = b$ an electron detector or a screen.[1] Then, the experimental data we get is not an entrance-exit law $\{\hat{\phi}_a(x),\ \phi_b(x)\}$ but a pair of probability densities $\mu_a(x)$ and $\mu_b(x)$ at $t = a, b$, respectively. Based on the data $\mu_a(x)$ and $\mu_b(x)$ at $t = a, b$, we predict intermediate states.[2]

Now, let us assume that we already know the explicit form of a "transition law" $p(s,x;t,y)$, and that a pair of distribution densities $\mu_a(x)$ and $\mu_b(x)$ at $t = a, b$ is given as experimental data. Then, if one attempts to construct a diffusion process by formula (5.1), one must find an entrance-exit law $\{\hat{\phi}_a(x),\ \phi_b(x)\}$ out of the given $p(s,x;t,y)$ and $\{\mu_a(x), \mu_b(x)\}$. This problem was first formulated by Schrödinger (1931):

Given a pair of probability densities $\mu_a(x)$ and $\mu_b(x)$ at $t = a, b$, and a "transition function" $p(s,x;t,y)$, find a pair of functions $\{\hat{\phi}_a(x),\ \phi_b(x)\}$ such that

$$(5.3) \qquad \begin{aligned} \mu_a(x) &= \hat{\phi}_a(x)\phi_a(x), \\[4pt] \mu_b(x) &= \hat{\phi}_b(x)\phi_b(x), \end{aligned}$$

where $\phi_a(x)$ and $\hat{\phi}_b(x)$ are defined by

$$(5.4) \qquad \begin{aligned} \phi_a(x) &= \int p(a,x;b,y)dy\,\phi_b(y), \\[4pt] \hat{\phi}_b(x) &= \int \hat{\phi}_a(y)dy\,p(a,y;b,x). \end{aligned}$$

Let us call the system of equations (5.3) with (5.4) *Schrödinger's problem*. Schrödinger (1931) assumed the existence of solutions to the problem (5.3) for the Brownian transition probability density $p(s,x;t,y)$.

---

[1] It is perhaps realistic in this case to take the first hitting time to the screen instead of a constant time b

[2] This is a Markov field type prediction, which will be called "Eddington-Schrödinger prediction"

Then he remarked that there would be no difficulty, at least formally, to extend his arguments to the case of variable coefficients and with potential (creation and killing) functions.

Remark 5.1. Schrödinger's problem (5.3) was investigated by Bernstein (1932), Fortet (1940), Beurling (1960), Jamison (1974, a) assuming the positivity of $p(s, x; t, y)$ besides some analytic conditions. Föllmer (1988) and Nagasawa (1990) have applied Csiszar's projection theorem, which allows us to treat Schrödinger's problem without analytic conditions in the general case of $p(s, x; t, y)$ which vanishes on a subset.

In our formulation of variation principle, which will be discussed in this chapter, we will *not apply* formula (5.1) in constructing diffusion processes because of the reasons explained above, and hence we need not solve Schrödinger's problem (5.3) as a necessary step, although it will be solved later on under a mild integrability condition and applied in characterizing the extremal process.

As we have done in preceding chapters, we begin with a basic (unperturbed) space-time diffusion process $\{X_t, P_{(s,x)}, (s, x) \in [a, b] \times \mathbf{R}^d\}$ determined by a diffusion equation

$$(5.5) \qquad\qquad L p = 0,$$

where L is a parabolic differential operator

$$(5.6) \qquad\qquad L = L(t) = \frac{\partial}{\partial t} + \frac{1}{2}\Delta + b(t, x) \cdot \nabla,$$

with a gauge condition $\operatorname{div} b = 0$. Moreover, we assume that a measurable function $c(t, x)$ and a pair of probability densities $\mu_a(x)$ and $\mu_b(x)$ are given at the initial and terminal times, respectively.

In terms of the given quantities $\{P_{(s,x)}, c(t, x), \mu_a(x), \mu_b(x)\}$ we shall formulate a variational problem, which claims the existence of the extremal diffusion process $\{X_t, Q\}$ in the p-representation such that formula (5.1) holds with the transition density $p(s, x; t, y)$ which is a (weak sense) fundamental solution of the diffusion equation (5.2) and with a pair $\{\hat{\phi}_a(x), \phi_b(x)\}$ of functions solving Schrödinger's problem (5.3).

Remark 5.2. In quantum theory probability densities μ_t are factorized in terms of a wave function ψ_t and its complex conjugate $\overline{\psi}_t$ as

$$\mu_t = \psi_t \overline{\psi_t}$$

under a condition that ψ_t evolves in time with unitary operators

$$\psi_t = U_{t,s} \psi_s, \quad for \ \ t \geq s.$$

If we have a wave function of the form $\psi_t(x) = e^{R(t,x) + i S(t,x)}$ beforehand, then a pair of real-valued functions can be defined by

$$\phi_t(x) = e^{R(t,x) + S(t,x)},$$

$$\widehat{\phi}_t(x) = e^{R(t,x) - S(t,x)}.$$

If this is the case, a remaining problem is to find out appropriate $p(s,x;t,y)$ with which equations (5.4) hold. This is another type of problem setting and will be treated in the next chapter.

Our variational problem is based on Csiszar's projection theorem, which will be explained in the next section.

5.2. Csiszar's Projection Theorem

Let $\{\Omega, \mathcal{B}\}$ be a measurable space,

(5.7) $\mathbf{M}_1(\Omega)$: *the space of probability measures on Ω*,

and the relative entropy $H(P|\overline{P})$ of P relative to $\overline{P}$ be defined for $P, \overline{P} \in \mathbf{M}_1(\Omega)$ by

(5.8) $$H(P \mid \overline{P}) = \int (\log \frac{dP}{d\overline{P}}) \, dP, \ if \ P << \overline{P},$$

$$= \infty, \ otherwise.$$

It should be remarked that the relative entropy is not symmetric in P and $\overline{P}$, and $H(P|\overline{P}) < \infty$ does not imply $H(\overline{P}|P) < \infty$ in general.

Then, Csiszar's projection theorem, which is an analogue of Riesz's projection theorem in Hilbert spaces, states as follows.

Theorem 5.1. (Csiszar (1975)) *Let* $\overline{P} \in M_1(\Omega)$ *be fixed. If a subset* **A** *of* $M_1(\Omega)$ *is convex and variation closed, and the subset* **A** *contains at least one element* P *with* $H(P|\overline{P}) < \infty$, *then there exists a unique* **Csiszar's projection** $Q \in$ **A** *(on the subset* **A***) of* $\overline{P}$ *such that*

$$(5.9) \qquad \inf_{P \in \mathbf{A}} H(P|\overline{P}) = H(Q|\overline{P}),$$

and it satisfies Csiszar's inequality

$$(5.10) \qquad H(P|\overline{P}) \geq H(P|Q) + H(Q|\overline{P}), \quad for \ \forall P \in \mathbf{A}.$$

A proof of the theorem will be given in Chapter 10.

5.3. Reference Processes

In the following let us denote by Ω the space of all $\mathbf{R}^d$-valued continuous functions defined on $[a, b]$, $-\infty < a < b < \infty$;

$$\Omega = C([a, b], \mathbf{R}^d),$$

$$X_t(\omega) = X(t, \omega) = \omega(t), \quad for \ \omega \in \Omega.$$

For a pair of prescribed probability measures $\{\mu_a, \mu_b\}$ on $\mathbf{R}^d$, we define a subset $\mathbf{A}_{a,b}$ of $M_1(\Omega)$ by

$$(5.11) \qquad \mathbf{A}_{a,b} = \{P \in M_1(\Omega) : P \circ X_r^{-1} = \mu_r, \ for \ r = a, b\},$$

where $\mu_r(dx) = \mu_r(x)dx$. The subset $\mathbf{A}_{a,b}$ is *the class of all continuous stochastic processes having the prescribed initial and final distributions.*

Therefore, the diffusion process $\{X_t, Q\}$ we are looking for must be in the subset $\mathbf{A}_{a,b}$. Let us take an element $\overline{P}$ from $M_1(\Omega)$ which does not belong to the subset $\mathbf{A}_{a,b}$, and call it a **reference process**. Then, we *project* the reference process $\overline{P}$ on the subset $\mathbf{A}_{a,b}$ to get a projection $Q \in \mathbf{A}_{a,b}$: To define the projection we measure the distance between P and $\overline{P}$ by the relative entropy $H(P|\overline{P})$, $P \in \mathbf{A}_{a,b}$.

In the following we shall fix a reference process $\overline{P} \in \mathbf{M}_1(\Omega)$ such that the entropy "distance"[3] of $\overline{P}$ from the subset $\mathbf{A}_{a,b}$ is finite; namely, we assume

(5.12)   *the set $\mathbf{A}_{a,b}$ contains at least one element P with finite*
          relative entropy: $\mathrm{H}(P \mid \overline{P}) < \infty$.

When condition (5.12) is satisfied, the pair $\{\mu_a, \mu_b\}$ of probability distributions will be called **admissible** for the reference process $\overline{P}$.

According to Theorem 5.1 the projection $Q \in \mathbf{A}_{a,b}$ must be the Csiszar projection, cf. (5.9).

Lemma 5.1. (i) *The subset $\mathbf{A}_{a,b}$ defined in (5.11) is convex and variation closed.*

(ii) *If a pair $\{\mu_a, \mu_b\}$ of probability distributions on $\mathbf{R}^d$ is admissible for the reference process $\overline{P}$, then there exists the Csiszar projection $Q \in \mathbf{A}_{a,b}$ (on the subset $\mathbf{A}_{a,b}$) of the reference process $\overline{P}$.*

The first assertion is almost clear from the definition of the subset $\mathbf{A}_{a,b}$ (cf. Lemma 8.2) and the second assertion is an application of Theorem 5.1.

Thus our problem of finding out the right $Q \in \mathbf{A}_{a,b}$ is reduced to the problem of constructing an appropriate reference process $\overline{P} \in \mathbf{M}_1(\Omega)$.

For the process $Q \in \mathbf{A}_{a,b}$ to be the right one, it is required that

(5.13)
$$\widehat{\phi}_t = \widehat{\phi}_s \mathrm{P}^c_{t-s} \quad for \ \ s \leq t,$$

$$\phi_s = \mathrm{P}^c_{t-s} \phi_t, \quad for \ \ s \leq t,$$

hold with Kac's semi-group $\{\mathrm{P}^c_{t-s}\}$, where the pair of functions $\widehat{\phi}_t$ and ϕ_t must provide Schrödinger's factorization of the distribution density μ_t of X_t with respect to $Q \in \mathbf{A}_{a,b}$ as

(5.14) $\mu_t = \widehat{\phi}_t \phi_t$.

Generalizing the method applied in the case of bounded creation and killing $c(t,x)$ discussed in Chapter 2 to singular ones, we will construct a

---

[3] The relative entropy is not a "distance". But, because of the inequality (cf. Chapter 10) $\| Q - P \|^2_{\mathrm{Var}} \leq 2\mathrm{H}(Q \mid P)$, we can regard $\sqrt{2\mathrm{H}(Q \mid P)}$ something like a "distance"

reference process $\overline{P}$ to be the *renormalization of the measure with (singular) creation and killing* $c(t, x)$.

As mentioned already, we assume the existence of a space-time diffusion process $\{(t, X_t), P_{(s,x)}; (s,x) \in [a, b] \times \mathbf{R}^d\}$ determined by the diffusion equation (5.5).

In order to cover a class of diffusion processes (Schrödinger processes) $\{X_t, Q\}$ wide enough for applications in quantum mechanics, we consider singular creation and killing with

$c(s, x)$: *a measurable function on* $[a, b] \times \mathbf{R}^d$ *(may be singular)*,

$$c(s, x) = c^+(s, x) - c^-(s, x),$$

where c^+ and c^- denote the positive and negative parts of c, respectively.

In terms of the **killing** part $c^-(s, x)$ we define

$$(5.15) \qquad \overline{T}_s = \inf\{t > s: \int_s^t c^-(r, X_r)dr = \infty\}$$

$$= \infty, \quad \textit{if there is no such } t.$$

In order to avoid overkill, we adopt as our **state space** a subset

$$(5.16) \qquad D = \{(s, x): P_{(s,x)}[b < \overline{T}_s] > 0\} \subset [a, b] \times \mathbf{R}^d.$$

This implies that if we start from a point (s, x) in the space-time state space D, we can survive until terminal time b with positive probability.

Let us define, first of all, the **measure** $P^-_{(s,x)}$ **with killing** $c^-(s, x)$ by

$$(5.17) \qquad P^-_{(s,x)}[F] = P_{(s,x)}[\exp(-\int_s^b c^-(r, X_r)dr)F(\cdot)1_{\{b < \overline{T}_s\}}].$$

Because of (5.16) we have

$$(5.18) \qquad P^-_{(s,x)}[1] > 0, \quad \textit{for } \forall (s,x) \in D.$$

We require an **integrability condition** of the **creation** part $c^+(s,x)$ with respect to the measure $P^-_{(s,x)}$ with killing $c^-(s,x)$

(5.19) $$P^-_{(s,x)}[\exp(\int_s^b c^+(r, X_r) dr)] < \infty, \quad for \; (s,x) \in D.$$

It should be remarked that, as a matter of fact, condition (5.19) is *necessary and sufficient* for the existence of extremal diffusion processes in the framework of the variational problems based on Theorem 5.1, since we need condition (5.19) for the renormalization. The integrability condition (5.19) will play a crucial role in this chapter. For this condition see also Chapter 6.

Now we define the **measure** $P^c_{(s,x)}$ **with creation and killing** $c(s,x)$ $= c^+(s,x) - c^-(s,x)$ by

(5.20) $$P^c_{(s,x)}[F] = P^-_{(s,x)}[\exp(\int_s^b c^+(r, X_r) dr)F(\cdot)]$$

$$= P_{(s,x)}[\exp(\int_s^b c(r, X_r) dr)F(\cdot)1_{\{b < \bar{T}_s\}}],$$

where the second line containing c is defined by the one with c^+ on the first line, since the integral $\int_s^b c(r, X_r)dr$ is not well-defined in general.

As already remarked in Chapter 2 the system $\{P^c_{(s,x)}\}$ of measures does not define either a Markov process or Kac's semi-group.

Let us set

(5.21) $$\xi(s,x) = P^c_{(s,x)}[1]$$

$$= P_{(s,x)}[\exp(\int_s^b c(r, X_r) dr)1_{\{b < \bar{T}_s\}}], \; for \; \forall \, (s,x) \in D.$$

Then, property (5.18) of the state space D and the integrability condition given in (5.19) imply

(5.22) $$0 < \xi(s,x) < \infty, \quad for \; \forall \, (s,x) \in D.$$

The Markov property of $P_{(s,x)}$ yields

(5.23) $$\xi(s,x) = P_{(s,x)}[\exp(\int_s^t c(r,X_r)dr)\xi(t,X_t)1_{\{t < \bar{T}_s\}}].$$

As will be shown below, $\xi(s,x)$ satisfies an integral equation and hence is the sum of a space-time harmonic function and the difference of two potentials. Therefore, $\xi(s,X_s)$ is right-continuous in s (cf., e.g. Theorem 5.7 in Blumenthal-Getoor (1968)).

It is easy to see that $\xi(s,x)$ satisfies an integral equation

$$\xi(s,x) = P_{(s,x)}[\xi(t,X_t)1_{\{t < \bar{T}_s\}}] + \int_s^t dr P_{(s,x)}[c(r,X_r)\xi(r,X_r)1_{\{r < \bar{T}_s\}}],$$

where $t \in (a, b]$ is fixed. In fact (5.23) implies

$$\xi(s,x) = P_{(s,x)}[\xi(t,X_t)1_{\{t < \bar{T}_s\}}]$$

$$+ P_{(s,x)}[\sum_{k=1}^\infty \frac{1}{k!}(\int_s^t c(r,X_r)dr)^k\xi(t,X_t)1_{\{t < \bar{T}_s\}}].$$

where the second term of the right-hand side is equal to

$$P_{(s,x)}[\int_s^t c(r,X_r)dr \sum_{k=1}^\infty \frac{1}{(k-1)!}(\int_r^t c(u,X_u)du)^{k-1}\xi(t,X_t)1_{\{t < \bar{T}_s\}}]$$

$$= \int_s^t dr P_{(s,x)}[c(r,X_r)\exp(\int_r^t c(u,X_u)du)\xi(t,X_t)1_{\{t < \bar{T}_s\}}]$$

substituting (5.21) and applying the Markov property

$$= \int_s^t dr P_{(s,x)}[c(r,X_r)P_{(r,X_r)}[\exp(\int_r^b c(u,X_u)du)1_{\{b < \bar{T}_r\}}]1_{\{r < \bar{T}_s\}}]$$

$$= \int_s^t dr P_{(s,x)}[c(r,X_r)\xi(r,X_r)1_{\{r < \bar{T}_s\}}].$$

Hence $\xi(s,x)$ satisfies the integral equation.

Let us define a multiplicative functional $\overline{N}_s^t$ by

(5.24)
$$\overline{N}_s^t = \exp\left(\int_s^t c(r, X_r)dr\right)\frac{\xi(t, X_t)}{\xi(s, X_s)}1_{\{t < \overline{T}_s\}}.$$

This is the renormalization of a modified Kac multiplicative functional in the case of singular creation and killing $c(t, x)$. Because of (5.23) it satisfies the normality condition

(5.25)
$$P_{(s, x)}[\overline{N}_s^t] = 1, \quad for \;\; \forall (s, x) \in D,$$

and hence it is a continuous[4] multiplicative functional.

Using the multiplicative functional $\overline{N}_s^t$ let us define a system of new transformed probability measures

(5.26)
$$\overline{P}_{(s, x)}[F] = P_{(s, x)}[\overline{N}_s^b F].$$

We will call the transformed diffusion process $\{(t, X_t), \overline{P}_{(s, x)}, (s, x) \in D\}$ the **renormalized process**.

Then, we have

Theorem 5.2. *The renormalized process $\{(t, X_t), \overline{P}_{(s, x)}, (s, x) \in D\}$ is a diffusion process with a semi-group*

(5.27)
$$\overline{P}_{t-s}f((s, x)) = \frac{1}{\xi}P_{t-s}^c(f\xi)(s, x),$$

where P_{t-s}^c is Kac's one defined by

(5.28)
$$P_{t-s}^c f((s, x)) = P_{(s, x)}[m_s^t f(t, X_t)],$$

in terms of the modified Kac multiplicative functional

(5.29)
$$m_s^t = \exp\left(\int_s^t c(r, X_r)dr\right)1_{\{t < \overline{T}_s\}}.$$

The renormalized process $\{(t, X_t), \overline{P}_{(s, x)}, (s, x) \in D\}$ has an additional drift coefficient

---

[4] See Clark (1970), cf., e.g. Liptser-Shiryayev (1977)

(5.30)
$$\bar{a}(s,x) = \sigma^T \sigma \nabla \log \xi(s,x),$$

where $\sigma^T \sigma(x)$ is the diffusion coefficient of the diffusion process $\{(t, X_t), P_{(s,x)}; (s,x) \in [a,b] \times \mathbf{R}^d\}$, and $\xi(s,x) \in C^{1,2}(D)$ is assumed.

Proof. The transformation in terms of a multiplicative functional with the normality condition has been treated in Theorem 2.1. For the renormalisation of (modified) Kac's semi-group see Theorem 2.2. For a proof of (5.30) we apply (3.68) with ξ in place of ϕ ($\xi \in H^{1,2}(D)$ is sufficient, cf. Aebi (1989, 93)).

A **renormalized measure** $\bar{P}$ of $P^c_{(s,x)}$ is defined finally through

(5.31)
$$\bar{P}[F] = \int_{D_a} k(dx)\bar{P}_{(a,x)}[F]$$

$$= \int_{D_a} k(dx) \frac{1}{\xi(a,x)} P^c_{(a,x)}[F],$$

with $k(dx) = k(x)dx$, $k(x) > 0$, a probability measure on $D_a = \{x: (a,x) \in D\}$, such that $\log(\xi(a,x)/k(x)) \in L^1(\mu_a)$, where μ_a is a probability measure which will be specified later on (cf. Lemma 5.2).

We adopt the renormalized measure $\bar{P}$ as our reference process.

Remark 5.3. If the rate $c(t,x)$ of creation and killing is singular, then it induces *intensive creation or killing of particles* near the singular points under the measure $P^c_{(s,x)}$ defined in (5.20). On the other hand, if we look at the process under the renormalized process $\{X_t, \bar{P}_{(s,x)}, (s,x) \in D\}$ we get *strong attractive or repulsive drift* $\bar{a}(s,x) = \sigma^T \sigma \nabla \log \xi(s,x)$ *near the singular points* on ∂D. One can interpret this as follows: We look at the process with time reversed, performing *backward prediction*, namely, we are tracing backward all created particles under the condition of an event "survived until the terminal time b".

5.4. Diffusion Processes in Schrödinger's Representation

We adopt the renormalized process $\bar{P}$ defined in (5.31) as the reference process $\bar{P}$ in Theorem 5.1. Then, we get the Csiszar projection Q (on the set $\mathbf{A}_{a,b}$) of the renormalized process $\bar{P}$, if a prescribed pair $\{\mu_a, \mu_b\}$ of probability distributions is admissible for the renormalized process $\bar{P}$.

In terms of the Csiszar projection Q we define a **diffusion process** $\{X_t, Q\}$ (we will call it *Schrödinger process* for short)[5] **determined by a triplet** $\{P^c_{(s,x)}, \mu_a, \mu_b\}$, the Markov property of which will be shown in Theorem 5.3. The Schrödinger process $\{X_t, Q\}$ has clearly the prescribed marginal distributions $\{\mu_a, \mu_b\}$ at the initial and terminal times $t = a, b$, since the measure Q is an element of the subset $\mathbf{A}_{a,b} \subset \mathbf{M}_1(\Omega)$ defined in (5.11). Moreover, Schrödinger's problem is solved, namely there exists a pair of functions $\{\hat{\phi}_a, \phi_b\}$ with which the finite dimensional distributions of the Schrödinger process $\{X_t, Q\}$ are represented as in (5.1). We shall first prove a lemma, which is a generalization of a theorem settled by Schrödinger (1931) and discussed by Bernstein (1932), Fortet (1940), Beurling (1960), and Jamison (1974,a;75) for positive kernel functions.

Lemma 5.2. (Föllmer (1988), Nagasawa (1990))   *Assume the integrability condition* (5.19) *on the creation and killing* $c(s,x)$. *If a pair* $\{\mu_a, \mu_b\}$ *of probability measures on* $\mathbf{R}^d$ *is admissible for the renormalized process* $\bar{P}$ *of the measure* $P^c_{(s,x)}$ *with creation and killing, then there exists a solution* $\{\hat{\phi}_a, \phi_b\}$ *for Schrödinger's problem*:

(5.32)
$$\mu_a(A) = \int_{D_a} \hat{\phi}_a(x) 1_A(x) dx P^c_{(a,x)}[\phi_b(X_b)],$$

$$\mu_b(B) = \int_{D_a} \hat{\phi}_a(x) dx P^c_{(a,x)}[\phi_b(X_b) 1_B(X_b)],$$

where $\log \hat{\phi}_a \in L^1(\mu_a)$ *and* $\log \phi_b \in L^1(\mu_b)$.

Proof. Let us define a probability measure $\bar{p}$ on $\mathbf{R}^d \times \mathbf{R}^d$ through

$$\bar{p}(A \times B) = \int_{D_a} dx \frac{k(x)}{\xi(a,x)} 1_{\mu_a}(x) 1_A(x) P^c_{(a,x)}[1_{\mu_b}(X_b) 1_B(X_b)]$$

with a positive $k(x)$ such that $\bar{p}$ is a probability measure[6] and $\log(\xi(a,x)/k(x)) \in L^1(\mu_a)$, where 1_μ denotes the indicator function of the

---

[5] The diffusion process was called Schrödinger process, but abandoned to avoid confusion in Chapter 3. But, we will again call it "Schrödinger process" for short

[6] $\bar{p}$ is the marginal distribution of $\bar{P}$ normalized on the support of $\mu_a \times \mu_b$

support of a measure μ. Denote by $\mathcal{E}_{a,b}$ the set of marginal distributions on $\mathbf{R}^d \times \mathbf{R}^d$ at $t = a, b$ of $\forall P \in \mathbf{A}_{a,b}$. Since the set $\mathcal{E}_{a,b}$ is convex and variation closed, under the admissibility condition (5.12) we can apply Theorem 5.1, which yields the unique Csiszar's projection $q(dxdy)$ on the set $\mathcal{E}_{a,b}$ of $\bar{p}(dxdy)$ such that

$$\inf_{p \in \mathcal{E}_{a,b}} H(p \mid \bar{p}) = H(q \mid \bar{p}).$$

Then, applying Theorem 10.5 (or 10.4) on marginal distributions, we get a pair of functions $\{\widehat{\phi}_a, \phi_b\}$ such that

$$\frac{dq}{d\bar{p}} = \frac{\xi(a,x)}{k(x)} \widehat{\phi}_a(x)\phi_b(y),$$

where $\widehat{\phi}_a(x)\phi_b(y) = 0$ on the subset $\{(x, y): \mu_a(x)\mu_b(y) = 0\}$. This completes the proof.

In terms of the density $p(s, x; t, y)$ which will be given in Section 5.5 the Schrödinger problem (5.32) can be expressed as

$$\mu_a(A) = \int_{D_a} \widehat{\phi}_a(x)1_A(x)\,dx \int_{D_b} p(a, x; b, y)dy\phi_b(y),$$

(5.32')

$$\mu_b(B) = \int_{D_a} \widehat{\phi}_a(x)dx \int_{D_b} p(a, x; b, y)dy1_B(x)\phi_b(y).$$

Let us call the pair $\{\widehat{\phi}_a, \phi_b\}$ of functions obtained in Lemma 5.2 **Schrödinger's entrance-exit law**.[7]

Remark 5.4. The pair $\{\widehat{\phi}_a, \phi_b\}$ in Lemma 5.2 is determined modulo c_i and c_i^{-1} depending on regions separated by the zero set of the solution. The state space D is decomposed as $D = \cup_i D^{(i)}$ with the zero set of $\phi(t, x)$, and the diffusion process stays in one component for ever, since the zero set is inaccessible, as will be seen. Therefore, we can choose any positive or negative constant c_i multiplied with $\widehat{\phi}_a$, and c_i^{-1} with ϕ_b on each $D^{(i)}$. Accordingly, $\widehat{\phi}_a$ and ϕ_b may take positive and negative values.[8] However, it does not affect the diffusion process Q (Csiszar's projection). As a matter of fact we have

---

[7] Cf. Getoor-Glover (1984), Getoor-Sharpe (1984) for entrance-exit laws

[8] See Section 4.8 for the importance of this fact in connection with superposition principle of diffusion processes

(5.33) $\mu_a(A) = \sum_i \int_{(D^{(i)})_a} \hat{\phi}_a(x) 1_A(x) dx P^c_{(a,x)}[1_{(D^{(i)})_b}(X_b)\phi_b(X_b)],$

the integrand of which is always non-negative, independent of any choice of constants c_i. This argument applies also to $\mu_b(B)$ in (5.32). Both functions $\hat{\phi}(t,x)$ and $\phi(t,x)$, which will be defined later, will take the same positive (or negative) sign in each $D^{(i)}$ at the same time, and hence the product $\hat{\phi}(t,x)\phi(t,x)$ is always non-negative.

We can now formulate the main existence theorem of diffusion processes in the p-representation:

Theorem 5.3. (Nagasawa (1990)) *Assume the integrability condition* (5.19) *and let $\overline{P}$ be the renormalized measure of $P^c_{(s,x)}$ defined in* (5.31). *Let a pair $\{\mu_a, \mu_b\}$ of probability distributions be admissible for $\overline{P}$.*

Then:

(i) *There exists a diffusion (Schrödinger) process $\{X_t, Q\}$ in the p-representation determined by the given triplet $\{P^c_{(s,x)}, \mu_a, \mu_b\}$, which is expressed in terms of Schrödinger's entrance-exit law $\{\hat{\phi}_a, \phi_b\}$ and the measure $P^c_{(s,x)}$ with creation and killing as*

(5.34) $Q[F] = \int_{D_a} \hat{\phi}_a(x) dx P^c_{(a,x)}[F(\cdot)\phi_b(X_b)].$

(ii) *There exists a pair of functions $\hat{\phi}_t(x)$ and $\phi_t(x)$, with which Schrödinger's factorization* (5.14) *holds.*

(iii) *The diffusion process does not hit the zero set of $\phi_t(x)$, namely the boundary ∂D is inaccessible.*

Proof. Let Q be Csiszar's projection on the set $\mathbf{A}_{a,b}$ of the measure $\overline{P}$ which is the renormalized process (cf. Lemma 5.1), and denote by $\tilde{Q}$ the measure defined on the right-hand side of (5.34). It is clear that finite dimensional distributions of $\tilde{Q}$ coincide with the right hand side of (5.1).[9] Then, with the help of this fact, we can show easily

---

[9] See Section 5.5 for the existence of $p(s,x;t,y)$

$$H(q|\bar{p}) = H(\widetilde{Q}|\bar{P}),$$

where q is the Csiszar projection on the set $\boldsymbol{E}_{a,b}$ given in the proof of Lemma 5.2. On the other hand, Jensen's inequality implies

$$H(q|\bar{p}) \leq H(Q|\bar{P}).$$

In fact, with a convex function $f(x) = x \log x$, $x \geq 0$,

$$H(Q|\bar{P}) = \bar{P}[f(\frac{dQ}{dP})] = \bar{P}[\bar{P}[f(\frac{dQ}{dP})|\sigma(X_a, X_b)]]$$

$$\geq \bar{P}[f(\bar{P}[(\frac{dQ}{dP})|\sigma(X_a, X_b)])]$$

$$= \bar{P}[f(\frac{dq}{dp})] = \bar{p}[f(\frac{dq}{dp})]$$

$$= H(q|\bar{p}).$$

Therefore, the measure $\widetilde{Q}$ must be the Csiszar projection $Q \in \boldsymbol{A}_{a,b}$, and hence we have formula (5.34).

Let us define

$$\phi_t(x) = P^c_{(t,x)}[\phi_b(X_b)].$$

Then, $\phi_t(x)$ satisfies the second equation of (5.13). Denote the zero set of $\phi_t(x)$ by $N = \{(t,x): \phi_t(x) = 0\}$. Then formula (5.34) implies

$$Q[1_N(t, X_t)] = \int_{D_a} \hat{\phi}_a(x) dx P^c_{(a,x)}[1_N(t, X_t)\phi_t(X_t)] = 0,$$

which, combined with the right continuity of $\phi_t(X_t)$, implies

(5.35) $Q[\textit{the process } (t, X_t) \textit{ does not hit the zero set } N \textit{ of } \phi_t(x)] = 1.$

Therefore, the process cannot cross over the zero set N, namely, it is segregated by the zero set N of $\phi(t, x)$.

The measure Q inherits the (strong) Markov property of $P_{(s,x)}$ because of (5.34). In fact, let $G \in \mathcal{F}_a^s$, $G \geq 0$ and f be a non-negative measurable function on $[a, b] \times \mathbf{R}^d$. Then, the (strong) Markov property of $P_{(s,x)}$ yields

(5.36) $Q[G \cdot f(t, X_t)]$

$$= \int_{D_a} \hat{\phi}_a(x) dx P_{(a,x)}[\exp(\int_a^s c(r, X_r) dr) G(\cdot) \phi_s(X_s) q(s, X_s; t, f) 1_{\{s < \bar{T}_a\}}]$$

$$= Q[G \cdot q(s, X_s; t, f)], \quad \text{for } s < t,$$

with the transition probability $q(s, x; t, f)$ defined by

(5.37) $\quad q(s, x; t, f) = P_{(s,x)}[\phi_s(X_s)^{-1} \exp(\int_s^t c(r, X_r) dr) \phi_t(X_t) f(t, X_t) 1_{\{t < \bar{T}_s\}}],$

where we notice (5.35). Thus, the process Q has the (strong) Markov property (replace s by stopping times for the strong Markov property).

It should be remarked that formula (5.36) implies the q-representation of the process Q with the (weak) fundamental solution $q(s, x; t, y)$ (not $p(s, x; t, y)$!), while formula (5.34) gives the p-representation (cf. Theorem 3.6).

It is clear that

$$\hat{\phi}_t(B) = \int_{D_a} \hat{\phi}_a(x) dx P_{(a,x)}^c[1_B(X_t)]$$

satisfies a generalized form of the first equation of (5.13). Formula (5.34) together with the Markov property of the process $\{X_t, P_{(s,x)}\}$ yields

$$Q[f(X_t)] = \int \hat{\phi}_t(dx) \phi_t(x) f(x),$$

which is a generalized form of Schrödinger's factorization (5.14). As a matter of fact there is a density function $\hat{\phi}_t(x)$ of $\hat{\phi}_t(B)$ and (5.14) holds, since we can apply the backward description, as will be shown in the next section, and there exists a (weak) fundamental solution $p(s, x; t, y)$ for the diffusion equation (5.2) with creation and killing. The proof is completed.

5.5. Weak Fundamental Solutions

Let $g^o(s, x; t, y)$, $a \le s < t \le b$, be defined by

$$(5.38) \qquad g^o(s, x; t, y) = g(s, x; t, y) - P_{(s,x)}[g(\overline{T}_s, X_{\overline{T}_s}; t, y); \overline{T}_s < t],$$

in terms of the fundamental solution $g(s, x; t, y)$ of (5.5) and the diffusion process $\{X_t, P_{(s,x)}\}$, where $\overline{T}_s$ is defined in (5.15).

We apply now the backward description, cf. Nagasawa (1989, a; 90). Changing the direction of time, we define the backward (space-time) diffusion process $\{X_s^*, P_{(t,y)}^*\}$, with the time variable s ($<t$) running backward from t to a, in terms of the same fundamental solution $g(s, x; t, y)$. Then, $g^o(s, x; t, y)$ defined in (5.38) can be given in terms of the backward process as

$$(5.39) \qquad g^o(s, x; t, y) = g(s, x; t, y) - P_{(t,y)}^*[g(s, x; \overline{T}_t^*, X_{\overline{T}_t^*}^*); s < \overline{T}_t^*].$$

Therefore, it is differentiable once in s and twice in x (resp. in (t, y)). We can adopt $g^o(s, x; t, y)$ as a transition density function, namely,

$$P_{(s,x)}[f(t, X_t); t < \overline{T}_s] = \int g^o(s, x; t, y) f(t, y) dy.$$

In other words, $g^o(s, x; t, y)$ is the weak fundamental solution of (5.5) with the absorbing boundary condition at the boundary ∂D of the state space D defined in (5.16).

If we define $p(s, x; t, y)$ and $p^*(s, x; t, y)$ by

$$\int p(s, x; t, y) f(t, y) \phi(t, y) dy = P_{(s,x)}\Big[\exp\Big(\int_s^t c(r, X_r) dr\Big) \phi f(t, X_t); t < \overline{T}_s\Big],$$

$$\int dx \hat{\phi}(s, x) f(s, x) p^*(s, x; t, y) = P_{(t,y)}^*\Big[\exp\Big(\int_s^t c(r, X_r^*) dr\Big) \hat{\phi} f(s, X_s^*); \overline{T}_t^* < s\Big],$$

respectively, for any bounded continuous function f on D, then

$$p(s, x; t, y) = p^*(s, x; t, y), \quad \hat{\phi}(s, x) \phi(t, y) dx dy - a.e..$$

Thus we can have a regular version of density function $p(s, x; t, y)$, which is the (weak) fundamental solution of (5.2) with the killing boundary condition at the boundary ∂D. Set

$$p(s, x) = \int p(s, x; b, y) dy f(y);$$

then it satisfies an integral equation with a kernel function $g^o(s, x; t, y)$: for $a \le s < t \le b$,

$$(5.40) \qquad p(s, x) = \int g^o(s, x; t, y) dy p(t, y)$$

$$+ \int_s^t dr \int g^o(s, x; r, y) dy c(r, y) p(r, y).$$

Moreover, we have

$$\phi(s, x) = \int p(s, x; t, y) dy \phi(t, y),$$

$$(5.41)$$

$$\widehat{\phi}(t, y) = \int \widehat{\phi}(s, x) dx p(s, x; t, y).$$

This shows that ϕ and $\widehat{\phi}$ are p-harmonic and p-coharmonic, respectively.

5.6. An Entropy Characterization of the Markov Property

An application of Theorem 5.3 yields an entropy characterization of the Markov property; namely, among stochastic processes with a prescribed flow of probability distributions we can find a Markov process uniquely under an entropy requirement.

Theorem 5.4.[10]   *Let* $\{\mu_t : t \in [a, b]\}$ *be a flow of probability distributions, and define a subset* $\mathbf{A} \subset \mathbf{M}_1(\Omega)$ *by*

$$(5.42) \qquad \mathbf{A} = \{\mathbf{R} \in \mathbf{M}_1(\Omega) : \mathbf{R} \circ X_t^{-1} = \mu_t, \text{ for } \forall t \in [a, b]\},$$

where Ω *denotes the space of right-continuous (resp. continuous) paths.*

[10] Compare with Corollary 6.2

Assume the set $\boldsymbol{A}$ is admissible for a reference Markov process $\mathbf{Q} \in \mathbf{M}_1(\Omega)$ which has transition probability densities; namely there exists at least one $\mathbf{R} \in \boldsymbol{A}$ such that $H(\mathbf{R}|\mathbf{Q}) < \infty$. Then, there exists uniquely a Markov (resp. diffusion) process $\mathbf{P}^0 \in \boldsymbol{A}$ such that

$$(5.43) \qquad \inf_{\mathbf{R} \in \boldsymbol{A}} H(\mathbf{R}|\mathbf{Q}) = H(\mathbf{P}^0|\mathbf{Q}).$$

Proof. The subset $\boldsymbol{A}$ is convex and variation closed. Therefore, there exists a unique Csiszar's projection $\mathbf{P}^0$ of $\mathbf{Q}$ on the set $\boldsymbol{A}$ such that

$$H(\mathbf{P}^0 | \mathbf{Q}) = H(\boldsymbol{A}|\mathbf{Q}).$$

Let us prove $\mathbf{P}^0$ is Markovian. Taking $\mathbf{P} \in \boldsymbol{A}$ with $H(\mathbf{P} | \mathbf{Q}) < \infty$, we define a sequence $\boldsymbol{A}^{(m)}$ of sets of probability measures on the path space by

$$(5.44) \qquad \boldsymbol{A}^{(m)} = \{\mathbf{R} : \mathbf{R} \circ X_t^{-1} = \mathbf{P} \circ X_t^{-1}, \text{ for } t = t_j^{(m)}, \forall j = 0, 1, 2, \dots \},$$

where $t_j^{(m)} = (a + \frac{j}{2^m}) \wedge b$, for $m \in N$. Then, the sets $\boldsymbol{A}^{(m)}$ are convex and variation closed, and satisfy

$$(5.45) \qquad \boldsymbol{A}^{(m)} \supset \boldsymbol{A}^{(m+1)} \supset \boldsymbol{A}, \quad and \quad \bigcap_m \boldsymbol{A}^{(m)} = \boldsymbol{A},$$

because of the right-continuity of paths. Consequently,

$$(5.46) \qquad H(\boldsymbol{A}|\mathbf{Q}) \geq H(\boldsymbol{A}^{(m+1)} | \mathbf{Q}) \geq H(\boldsymbol{A}^{(m)} | \mathbf{Q}),$$

and, furthermore, there exist Csiszar's projections $\mathbf{P}^{(m)}$ such that

$$(5.47) \qquad H(\mathbf{P}^{(m)} | \mathbf{Q}) = H(\boldsymbol{A}^{(m)} | \mathbf{Q}),$$

(cf. Chapter 10). It is clear that $\mathbf{P}^0$ and $\mathbf{P}^{(m)}$ are absolutely continuous with respect to $\mathbf{Q}$ because of the assumption of the theorem combined with

$$(5.48) \qquad H(\mathbf{P}^{(m)}|\mathbf{Q}) \leq H(\mathbf{P}^0|\mathbf{Q}) = H(\boldsymbol{A}|\mathbf{Q}) \leq H(\mathbf{P}|\mathbf{Q}) < \infty.$$

Then the lower semi-continuity of the relative entropy, together with (5.45), (5.46) and (5.47), yields

(5.49) $$\lim_{m \to \infty} H(\mathbf{P}^{(m)} \mid \mathbf{Q}) = H(\mathbf{P}^0 \mid \mathbf{Q}).$$

Because of Csiszar's inequality (10.12), we have

$$H(\mathbf{P}^0 \mid \mathbf{Q}) - H(\mathbf{P}^{(m)} \mid \mathbf{Q}) \geq H(\mathbf{P}^0 \mid \mathbf{P}^{(m)}),$$

since $\mathbf{P}^0 \in \boldsymbol{H}^{(m)}$. Therefore,

(5.50) $$\lim_{m \to \infty} H(\mathbf{P}^0 \mid \mathbf{P}^{(m)}) = 0,$$

i.e., $\mathbf{P}^{(m)}$ converges to $\mathbf{P}^0$ in entropy and hence in variation as m tends to infinity. The Csiszar projection $\mathbf{P}^{(m)}$ is the Schrödinger process determined by $\{\mathbf{Q}, \mu_r(x), r = t_j^{(m)}, t_{j+1}^{(m)}\}$ on each subinterval $[t_j^{(m)}, t_{j+1}^{(m)}]$, $j = 0, 1, 2, \dots$, where $\mu_r(x)$ denotes the given flow (the distribution density of X_r according to $\mathbf{P}$). Therefore, $\mathbf{P}^{(m)}$ is Markovian by Theorem 5.3 (cf. (5.36) and also Theorem 3.6). Consequently, as will be shown, the process $\mathbf{P}^0$ is Markovian as the variation limit of the sequence of the Markovian processes $\mathbf{P}^{(m)}$:

Lemma 5.3. *If a sequence* $\{\mathbf{P}^{(m)}\}$ *of Markov processes converges to* $\mathbf{P}^0$ *in variation, then the limit* $\mathbf{P}^0$ *is also Markovian, namely, for any bounded* $\mathcal{F}_a^t$- *(resp.* $\mathcal{F}_t^b$-*) measurable G and F*

$$\mathbf{P}^0[GF] = \mathbf{P}^0\big[G\,\mathbf{P}^0[F \mid \sigma(X_t)]\big].$$

Proof. We fix a version of $\mathbf{P}^{(m)}[F \mid \sigma(X_t)]$. Then, for any bounded $\sigma(X_t)$-measurable K and $\mathcal{F}_t^b$-measurable F we have

$$\big|\mathbf{P}^0\big[K\,\mathbf{P}^0[F \mid \sigma(X_t)]\big] - \mathbf{P}^0\big[K\,\mathbf{P}^{(m)}[F \mid \sigma(X_t)]\big]\big|$$

$$\leq \big|\mathbf{P}^0[KF] - \mathbf{P}^{(m)}[KF]\big|$$

$$+ \big|\mathbf{P}^{(m)}\big[K\,\mathbf{P}^{(m)}[F \mid \sigma(X_t)]\big] - \mathbf{P}^0\big[K\,\mathbf{P}^{(m)}[F \mid \sigma(X_t)]\big]\big|$$

$$\leq const. \, \|\mathbf{P}^{(m)} - \mathbf{P}^0\|_{Var},$$

which vanishes as $m \to \infty$, namely

(5.51) $\mathbf{P}^{(m)}[F \mid \sigma(X_t)]$ *converges to* $\mathbf{P}^0[F \mid \sigma(X_t)]$ *in* $L^1(\mathbf{P}^0)$.

Moreover, for any bounded $\mathcal{F}_a^t$-measurable G, we have

$$| \mathbf{P}^0[GF] - \mathbf{P}^0[G \, \mathbf{P}^0[F \mid \sigma(X_t)]] | \leq | \mathbf{P}^0[GF] - \mathbf{P}^{(m)}[GF] |$$

$$+ | \mathbf{P}^{(m)}[G \, \mathbf{P}^{(m)}[F \mid \sigma(X_t)]] - \mathbf{P}^0[G \, \mathbf{P}^{(m)}[F \mid \sigma(X_t)]] |$$

$$+ | \mathbf{P}^0[G \, \mathbf{P}^{(m)}[F \mid \sigma(X_t)]] - \mathbf{P}^0[G \, \mathbf{P}^0[F \mid \sigma(X_t)]] |,$$

where the first and second terms are bounded by *const.* $\| \mathbf{P}^{(m)} - \mathbf{P}^0 \|_{Var}$ which vanishes as $m \to \infty$, and the third one also vanishes because of (5.51). This completes the proof.

Remark 5.5. In Lemma 5.3 the convergence in variation of $\mathbf{P}^{(m)}$ cannot be replaced by the weak convergence. It is well-known that the weak convergence of Markov processes $\mathbf{P}^{(m)}$ does not preserve the Markov property in general.[11]

5.7. Remarks

(i) The variational principle (or the least action principle) for diffusion processes in connection with quantum mechanics was first discussed by Fényes (1952) (cf. Appendix a.1), and then in the context of "stochastic mechanics" (see Appendix a.2) by Yasue (1981), who deduced Nelson's "Newtonian equation" from his stochastic variation principle. After Yasue's paper many publications followed, including Guerra-Morato (1983), Nelson (1985), Zao (1986), Zambrini (1985, 86), Blanchard-Combe-Zheng (1987), Guerra-Pavon (1988), Föllmer (1988), Nagasawa (1989, b,c; 90), Wakolbinger (1989), Dawson-Gorostiza-Wakolbinger (1990), Blaquière (1991,92) among others.

(ii) In the traditional variational principle (action principle) we proceed as follows: e.g. in Nagasawa (1989, b,c) an action functional $I(Y)$ of semi-martingale[12] Y_t with drift $b(t, Y_t) + a(t, \cdot)$ is defined by

[11] Consider a uniform motion starting (with probability 1/2) from $1 + 1/m$ and $-1 - 1/m$ with velocities -1 and $+1$, and reflected at $1/m$ and $-1/m$, respectively. Then this Markov process $\mathbf{P}^{(m)}$ converges weakly to a process $\mathbf{P}^0$, but the limit $\mathbf{P}^0$ is not Markovian

[12] The drift coefficient a is allowed to be singular. Therefore, the existence of such singular semi-martingales is not evident. Cf. Liptser-Shiryayev (1977), Föllmer (1988)

(5.52) $$I(Y) = Q[\int_a^b \{\tfrac{1}{2}\, a(s, \cdot)^2 - c(s, Y_s)\} ds],$$

where $c(s, x)$ denotes the creation and killing induced by $\phi(s, x)$. Then it is shown that

(5.53) $$I(Y) = Q[\tfrac{1}{2}\int_a^b \{a(s, \cdot) - \nabla \log \phi(s, Y_s)\}^2 ds]$$

$$+ \int \hat{\phi}(b, x)\phi(b, x) \log\phi(b, x)\, dx - \int \hat{\phi}(a, x)\phi(a, x) \log\phi(a, x)\, dx,$$

where we have assumed $\sigma^{ij} = \delta^{ij}$ and $\phi \in C^{1,2}(D)$ for simplicity.

Therefore, we see immediately from (5.53) that the extremal process X_t which minimizes the action functional must have an additional drift coefficient $\nabla \log \phi(s, x)$ and

$$I(X) = \inf_{Y \in \mathcal{A}} I(Y)$$

$$= \int \hat{\phi}(b, x)\phi(b, x) \log\phi(b, x)\, dx - \int \hat{\phi}(a, x)\phi(a, x) \log\phi(a, x)\, dx,$$

where $\mathcal{A} = \{Y$ *with admissible additional drift coefficient* $a\}$ (cf. Theorem 2 in Nagasawa (1989, b,c)). Therefore, the extremal process must be the Schrödinger process.

(iii) The formulation employed in this chapter for variational principle looks different from the usual one at a glance, but they agree with each other. In fact, applying the Maruyama-Girsanov theorem (cf. Maruyama (1954), Girsanov (1960)), we can compute the relative entropy explicitly, and show that it coincides with the action functional defined in (5.52): namely

(5.54) $$H(P^a \mid \bar{P}) = I(Y) + const.,$$

where P^a denotes the probability measure induced by the semi-martingale Y_t with drift $b(t, Y_t) + a(t, \cdot)$. However, it should be emphasized that the two methods are essentially different. The traditional variational principle explained in (ii) above characterizes the extremal process, but it *does not* imply the existence of the extremal process which is in general a singular diffusion process. On the other hand, our variational principle based on Csiszar's projection theorem claims the existence of the extremal process. This is a definitive advantage of our method, as explained in this chapter.

(iv) According to Föllmer (1985, 86), there exist the forward derivative DY_t and the backward derivative $D*Y_t$ for semi-martingales Y_t, which were defined by Nelson (1966):

$$DY_t = \lim_{h \downarrow 0} \frac{1}{h} (P[Y_{t+h}|Y_t] - Y_t),$$

$$-D*Y_t = \lim_{h \downarrow 0} \frac{1}{h} (P[Y_{t-h}|Y_t] - Y_t).$$

Since

$$DY_t = a(t, \cdot) + b(t, Y_t)$$

$$-D*Y_t = \hat{a}(t, \cdot) - b(t, Y_t), \quad P\text{-a.e.},$$

we have

$$a(t, \cdot)^2 = (DY_t - b(t, Y_t))^2,$$

$$\hat{a}(t, \cdot)^2 = (D*Y_t - b(t, Y_t))^2.$$

Therefore, the integrand of the action functional $I(Y)$ defined in (5.52) is a Lagrangian

$$(5.55) \qquad L(Y_t) = Q[\tfrac{1}{2}\{DY_t - b(t, Y_t)\}^2 - c(t, Y_t)].^{13}$$

A Lagrangian

$$(5.56) \quad \tilde{L}(Y_t) = Q[\tfrac{1}{2}\{\tfrac{1}{2}(DY_t - b(t, Y_t))^2 + \tfrac{1}{2}(D*Y_t - b(t, Y_t))^2\} + V(t, Y_t)]$$

was adopted by Yasue (1981), in which it was not intended to consider its relation to the Schrödinger processes. Instead, he proved that the extremal process satisfies Nelson's "stochastic mechanics" (cf. Appendix). First of all, in $\tilde{L}(Y_t)$ of (5.56) the Lagrangian of the time-reversed process is added, in order to deduce Nelson's "stochastic mechanics" (as a matter of fact it is added to take in account of the duality relation). However, for the

---

[13] Cf. Appendix a.1

Schrödinger process, it is enough to consider a Lagrangian

(5.57) $L_V(Y_t) = Q[\frac{1}{2}\{DY_t - b(t, Y_t)\}^2 + V(t, Y_t)],$

instead of (5.56). Then, the Lagrangians in (5.55) and (5.57) look almost the same, but we must pay attention to a crucial difference between them. In (5.55) we assume that the creation and killing $c(t, x)$ is given, and hence the extremal process is the Schrödinger process as shown above. In contrast to (5.55), we assume in (5.57) that a potential function $V(t, x)$ which should appear in the Schrödinger equation is given instead of the creation and killing $c(t, x)$. Then the Lagrangian $L_V(Y_t)$ in (5.57) turns out to be nonlinear, as is discussed in Section 4.3, since we must substitute

$$V(t, x) = -c(t, x, \phi) - 2\frac{\partial S}{\partial t}(t, x) - (\sigma\nabla S)^2(t, x) - 2b \cdot \nabla S(t, x)$$

in (5.57). Therefore, the results established in this chapter cannot be applied to the Lagrangian (5.57). The existence of the Schrödinger process in this case will be discussed in Chapter 6.

(v) We have shown, applying Csiszar's projection theorem, the existence of diffusion processes $\{X_t, Q\}$ (Schrödinger process) for a given triplet $\{P^c_{(s,x)}, \mu_a, \mu_b\}$ under integrability condition (5.19) and admissibility condition (5.12). The method employed is a generalization of the one for Brownian motions in Föllmer (1988) to the case with singular creation and killing $c(t, x)$. A crucial point in the generalization is the renormalization of $P^c_{(s,x)}$ in the definition (5.31) under integrability condition (5.19). This generalizes the results in Wakolbinger (1989), in which the case of bounded c is treated. As is already seen in Theorem 4.1, Theorem 5.3 provides a way of getting *probabilistic* solutions of Schrödinger equations *not* through solving the equations themselves (cf. section 6.2). This is a significant advantage of the method of variation based on Theorem 5.3.

On the other hand, there are many cases in applications in which $\phi_t(x)$ is given beforehand. In these cases we can construct diffusion processes using a transformation of Markov processes, which will be discussed in the next chapter.

Chapter VI
Diffusion Processes in q-Representation

In this chapter we will assume a (non-negative) measurable function $\phi(t, x)$ is given (but *not* $\hat{\phi}(t, x)$!), and concentrate on the problem of constructing diffusion processes starting from an arbitrary point (s, x) with an additional drift coefficient $\sigma^T \sigma \nabla(\log \phi(t, x))$; namely we will not fix an initial distribution. It is clear, then, that we cannot handle our problem in the framework of the variational method (in the p-representation) formulated in the preceding chapter based on Csiszar's projection theorem for lack of a fixed initial distribution, or in other words, lack of the function $\hat{\phi}(t, x)$. If a transition density $p(s, x; t, x)$ with creation and killing is given and if $\phi(t, x)$ is p-harmonic and *positive*, then the harmonic transformation induces an additional drift term $\sigma^T \sigma \nabla(\log \phi(t, x))$ as we have discussed in Chapter 2. A mathematical problem encountered with $\phi(t, x)$ which vanishes on a subset will be reduced to defining a multiplicative functional properly (cf. (6.5)) and to verifying the normality condition under a mild integrability condition (cf. (6.4)).

6.1. A Multiplicative Functional

For a given $\phi(t, x)$, which will be fixed in the following, we adopt the subset

$$D = N^c, \ where \ N = \{(t, x): \phi(t, x) = 0\},$$

as a **state space**.

Let us assume for simplicity that $\phi(t, x)$ is a (non-negative) bounded continuous function on $[a, b] \times \mathbf{R}^d$ and $\phi \in C^{1,2}(D)$ (for the case of $\phi \in H^{1,2}(D)$ see Aebi (1989, 93)).

First of all we define the rate of **creation and killing** $c(t,x)$ **induced by** the given function $\phi(t,x)$:

$$(6.1) \qquad c(t,x) = - \frac{L\,\phi(t,x)}{\phi(t,x)}, \quad for \ (t,x) \in D,^1$$

where $L = L(t)$ is a parabolic differential operator

$$L = \frac{\partial}{\partial t} + \frac{1}{2}\Delta + b(t,x)\cdot\nabla.$$

Remark 6.1. Since time reversal (duality) will not be our concern in this chapter, we can take

$$\tilde{L} = \frac{\partial}{\partial t} + \frac{1}{2}\,\sigma\sigma^T(t,x)^{ij}\,\frac{\partial^2}{\partial x^i\partial x^j} + \tilde{b}(t,x)^i\frac{\partial}{\partial x^i},$$

instead of L, where the diffusion coefficient $\sigma^T\sigma(t,x)$ may degenerate, assuming the existence of a diffusion process $\{X_t, P_{(s,x)}, (s,x) \in [a,b]\times\mathbf{R}^d\}$ determined by the operator $\tilde{L}$. However, to avoid confusion, we will state results in terms of L given above, unless otherwise stated.

Remark 6.2. The function $\phi(t,x)$ is not necessarily non-negative and may take negative values. Its sign does not affect the definition of creation and killing $c(t,x)$ given in (6.1). However, if we allow $\phi(t,x)$ to take negative values, we must be concerned with the ergodic decomposition of the state space D by the zero set of $\phi(t,x)$. Therefore, to avoid complication of statement we assume $\phi(t,x)$ to be non-negative in this chapter.

Let $\{(t,X_t), P_{(s,x)}; (s,x) \in [a,b]\times\mathbf{R}^d\}$ be the space-time diffusion process determined by the elliptic operator L, and let T_s be the first hitting time to the zero set $N = \{(t,x): \phi(t,x) = 0\}$ of ϕ defined by

$$T_s = \inf\,\{t > s : \phi(t,X_t) = 0\}, \quad if\ such\ t\ exists,$$

$$= \infty, \quad otherwise.$$

With this T_s in place of $\overline{T}_s$ we define a measure $P^-_{(s,x)}$ as we have done in (5.17) :

---

[1] This has appeared already in (3.40)

(6.2) $P^-_{(s,x)}[F] = P_{(s,x)}\big[\exp(-\int_s^b c^-(r,X_r)dr)F(\cdot)1_{\{b<T_s\}}\big],$

where we use the same notation $P^-_{(s,x)}$.

We assume

(6.3) $P^-_{(s,x)}[1] > 0, \quad for \ \forall \ (s,x) \in D.$

This condition restricts the shape of the space-time domain $D = N^c$, namely, paths starting at $(s,x) \in D$ must stay in D up to the terminal time b with positive probability. We require condition (6.3) because of the same reasoning as we adopted for the state space given in (5.16).

We impose an **integrability condition** as in (5.19):

(6.4) $P^-_{(s,x)}\big[\exp(\int_s^b c^+(r,X_r)\,dr)\big]$

$$= P_{(s,x)}\big[\exp(\int_s^b c\,(r,X_r)dr)1_{\{b<T_s\}}\big] < \infty,$$

for $\forall \ (s,x) \in D$, where $c(t,x)$ is defined in (6.1).

The integrability condition (6.4) is local, while condition (5.19) is global. Condition (6.4) will be applied to the uniqueness assertion in Lemma 6.1 so that we can conclude the first claim of Theorem 6.1 below. Examples will be given in Section 6.6 and also in Section 7.9.

Moreover, let us define a **multiplicative functional**[2] by

(6.5) $N^t_s = \exp(-\int_s^t \frac{L\phi}{\phi}(r,X_r)dr)\,\frac{\phi(t,X_t)}{\phi(s,X_s)}1_{\{t<T_s\}},$[3]

in terms of a prescribed function $\phi(s,x)$. Then we have

----

[2] See Chapter 2 for multiplicative functionals

[3] For time-symmetric cases, cf. Donsker-Varadhan (1975), Fukushima-Takeda (1984). Cf. also Oshima (1992)

Theorem 6.1. (Nagasawa $(1989, \text{a})$[4])   *Assume the integrability conditions given in* (6.3) *and* (6.4). *Then:*

(i) *The multiplicative functional* N_s^t *defined in* (6.5) *satisfies the normality condition*

$$(6.6) \qquad\qquad \mathrm{P}_{(s,x)}[N_s^t] \; = 1, \quad for \; \forall \, (s,x) \in D.$$

(ii) *The transformed diffusion process* $\mathrm{Q}_{(s,x)} = N_s^b \mathrm{P}_{(s,x)}$, $\forall \, (s,x) \in D$, *has an additional drift term*

$$a(t,x) = \sigma^T \sigma \nabla \log \, \phi(t,x).$$

(iii) *The space-time diffusion process* $\{(t, X_t), \mathrm{Q}_{(s,x)}; (s,x) \in D\}$ *does not hit the zero set* N *of* ϕ.

This theorem can be regarded as the existence claim of diffusion (Schrödinger) processes in the q-representation. The transformation by the multiplicative functional N_s^t is a kind of generalization of the so-called "ground state representation" to the case of arbitrary states (especially excited states).

Proof.   We can assume that the basic diffusion process $\{X_t, \mathrm{P}_{(s,x)}\}$ satisfies the stochastic differential equation

$$X_t = X_a + \int_a^t \sigma(X_r)dB_r + \int_a^t \widetilde{b}(r, X_r)\, dr. \text{[5]}$$

Define a sequence of stopping times

$$(6.7) \qquad\qquad \tau_n = \inf\{t > s : \int_s^t \|\sigma^T \sigma \nabla \log \, \phi(r, X_r)\|^2 dr \ge n\},$$

and denote $\alpha = \alpha_n = \tau_n \wedge T_s$. An application of Itô's formula to $\log \phi$ yields

$$(6.8) \qquad \exp(\int_s^{t \wedge \alpha} c(r, X_r)dr) \frac{\phi(t \wedge \alpha, X_{t \wedge \alpha})}{\phi(s,x)} = M_s^{t \wedge \alpha}, \quad \mathrm{P}_{(s,x)}\text{-}a.e.,$$

---

[4] It was presented in 1987 to commemorate the centenary of E. Schrödinger's birth
[5] See (2.14) and what follows in Chapter 2

where $M_s^{t \wedge \alpha}$ is the exponential martingale of Maruyama-Girsanov for the drift $h = \sigma^T \sigma \nabla(\log \phi)$

$$M_s^{t \wedge \alpha} = \exp(\int_s^{t \wedge \alpha} h(r, X_r) \cdot dB_r - \frac{1}{2} \int_s^{t \wedge \alpha} \| h(r, X_r) \|^2 dr),$$

the expectation of which is equal to one. Multiplying $1_{\{t \wedge \alpha < T_s\}}$ to both sides of (6.8), we get

(6.9) $\qquad N_s^{t \wedge \alpha} = M_s^{t \wedge \alpha} 1_{\{t \wedge \alpha < T_s\}} = M_s^{t \wedge \tau_n} 1_{\{t \wedge \alpha < T_s\}} \leq M_s^{t \wedge \tau_n},$

where $N_s^{t \wedge \alpha} = N_s^{t \wedge \tau_n}$.

Then, by the Fatou lemma,

$$P_{(s,x)}[N_s^t] = P_{(s,x)}[\liminf_{n \to \infty} N_s^{t \wedge \tau_n}] \leq \liminf_{n \to \infty} P_{(s,x)}[N_s^{t \wedge \tau_n}]$$

$$\leq \liminf_{n \to \infty} P_{(s,x)}[M_s^{t \wedge \tau_n}] = 1,$$

for $\forall (s,x) \in D$. Thus we have an inequality $P_{(s,x)}[N_s^t] \leq 1$, namely

(6.10) $\qquad P_{(s,x)}^-\Big[\exp(\int_s^t c^+(r, X_r) dr) \phi(t, X_t)\Big] \leq \phi(s, x) < \infty.$

We shall prove that under integrability condition (6.4), equality (6.6) holds. Let T^n be defined by

$$T^n = \inf \{r \in (s, b] : |\phi(r, X_r)| \leq \frac{1}{n}\},$$

$$= \infty, \quad \text{if there is no such } r.$$

Then $T^n < T_s$ and $T^n \to T_s$ as n tends to infinity, if $T_s < \infty$. An application of Itô's formula yields, since $-L\phi = c\phi$ by definition (6.1),

$$\phi(s, x) = P_{(s,x)}[\phi(t \wedge T^n, X_{t \wedge T^n})]$$

$$+ P_{(s,x)}\Big[\int_s^{t \wedge T^n} dr\, c(r, X_r) \phi(r, X_r) : t \wedge T^n < T_s\Big].$$

Applying Lemma 6.2 which will be given in Section 6.5 with slight modification, we can show that $\phi(s,x)$ satisfies

$$\phi(s,x) = P^-_{(s,x)}[\phi(t \wedge T^n, X_{t \wedge T^n})] + P^-_{(s,x)}[\int_s^{t \wedge T^n} dr\, c^+(r,X_r)\phi(r,X_r)],$$

where $P^-_{(s,x)}$ is defined in (6.2), from which it follows that $\phi(s,x)$ is a solution of an integral equation

(6.11) $u(s,x) = P^-_{(s,x)}[\phi(t,X_t)] + \int_s^t dr P^-_{(s,x)}[c^+(r,X_r)u(r,X_r)],$

for $s \in [a,t]$, where $t \in (a,b]$ is fixed.

On the other hand, define $u^o(s,x)$ by

(6.12) $u^o(s,x) = P_{(s,x)}[\exp(\int_s^t c(r,X_r)dr)\phi(t,X_t)1_{\{t < T_s\}}]$

$$= P^-_{(s,x)}[\exp(\int_s^t c^+(r,X_r)dr)\phi(t,X_t)],$$

which is dominated by $\phi(s,x)$ as is shown in (6.10) and hence well-defined. It is easy to see that $u^o(s,x)$ also satisfies the integral equation (6.11). We will prove that $\phi(s,x)$ coincides with $u^o(s,x)$, namely;

Lemma 6.1. *The uniqueness holds for bounded solutions of integral equation* (6.11) *under integrability condition* (6.4).

Proof. Let u_1 and u_2 be any bounded solutions of the integral equation (6.11). Then, it is easy to verify by induction that $k = |u_1 - u_2|$ satisfies an inequality

$$k(s,x) \leq \int_s^t dr P^-_{(s,x)}[c^+(r,X_r)k(r,X_r)]$$

$$\leq \int_s^t dr P^-_{(s,x)}[\frac{1}{(n-1)!}(\int_s^r c^+(u,X_u)du)^{n-1}c^+(r,X_r)k(r,X_r)],$$

which vanishes as n tends to infinity under integrability condition (6.4). Therefore, we have the uniqueness for bounded solutions of the integral equation (6.11) under integrability condition (6.4). This completes the proof of the lemma.

Therefore, $\phi(s, x)$ and $u^o(s, x)$ must coincide, and hence we have

$$\phi(s, x) = P_{(s,x)}\left[\exp\left(\int_s^t c(r, X_r)dr\right)\phi(t, X_t)1_{\{t < T_s\}}\right],$$

which is nothing but (6.6). Moreover, N_s^t is a continuous multiplicative functional. For this we can adopt the same arguments applied to $\overline{N}_s^t$ defined in (5.24). The second assertion (ii) follows from formula (3.68). The third assertion (iii) is clear because of (6.5) and (6.6). This completes the proof of the theorem.

Remark 6.3.   In Lemma 6.1, the boundedness assumption on $u(t, x)$ can be relaxed and replaced by the following integrability condition:

$$\int_s^b dr P_{(s,x)}^-\left[\exp\left(\int_s^r c^+(\xi, X_\xi)d\xi\right)u(r, X_r)c^+(r, X_r)\right] < \infty,$$

which reduces to condition (6.4) if $u(s, x)$ is bounded. In later applications we must handle unbounded solution in general (cf. examples in Section 7.9).

By the generalized Maruyama-Girsanov theorem (cf. e.g., Liptser-Shiryayev (1977)) the process $\{X_t, Q_{(s,x)} = N_s^b \cdot P_{(s,x)}; (s, x) \in D\}$ obtained through the transformation is a semi-martingale of the form, for $t \in [s, b]$,

$$(6.13)\quad X_t = X_s + \int_s^t \sigma(X_r)dW_r + \int_s^t \tilde{b}(r, X_r)dr + \int_s^t \sigma^T\sigma \nabla\log\phi(r, X_r)dr,$$

where $\tilde{b} = b + \frac{1}{2}\frac{1}{\sqrt{\sigma_2}}\nabla(\sigma^T\sigma\sqrt{\sigma_2})$ and W_t is a d-dimensional Brownian motion with respect to $Q_{(s,x)}$ (*not* $P_{(s,x)}$!) defined by

$$W_t = \int_s^t \sigma^{-1}(X_r)dX_r - \int_s^t \left\{\sigma^{-1}\tilde{b}(r, X_r) + \sigma\frac{\nabla\phi}{\phi}(r, X_r)\right\}dr, \quad s \le t \le b.[6]$$

---

[6] Cf. Aebi (1989, 93) for general cases

Remark 6.4. In some articles the stochastic differential equation such as (6.13) in terms of a given d-dimensional Brownian motion B_t in place of $\{W_t, Q_{(s,x)}\}$ is employed, because it gives an "intuitive picture". However, if we do so, the stochastic differential equation is not well-defined because of the singularity of the drift coefficient in the third integral on the right-hand side of (6.13). Therefore, to make sense we must show the inaccessibility of the space-time process (t, X_t) to the zero set N of $\phi(t, x)$ beforehand, and then apply approximation arguments. This makes things complicated, and hence we avoid it.

The method of transformation in terms of a multiplicative functional N_s^t has an advantage in applications, since it is not necessary to specify an initial distribution beforehand and one can start from any point $(s, x) \in D$. In this respect, the diffusion process $\{X_t, Q_{(s,x)}; (s, x) \in D\}$ in Theorem 6.1 is finer than diffusion processes constructed in the preceding chapter. After having constructed the diffusion process $\{X_t, Q_{(s,x)}; (s, x) \in D\}$, we can choose an arbitrary non-negative function $\hat{\phi}_a$ and let the processes start with the initial distribution density $\mu_a = \hat{\phi}_a \phi_a$, in order to define the diffusion (Schrödinger) process $\{X_t, Q\}$ in q-representation. It is easy to see that Schrödinger's factorization $\mu_t = \hat{\phi}_t \phi_t$ holds for $t \in [a, b]$.

6.2. Flows of Distribution Densities

As corollaries of Theorem 6.1 we can formulate a couple of assertions on the existence of diffusion processes for given flows of distribution densities. These corollaries are useful in applications, and they have been already applied in Chapter 4, especially in constructing a superposition of diffusion processes.

Corollary 6.1. *Let* $\{\mu_t(x), t \in [a, b]\}$ *be a flow of non-negative distribution densities and set*

(6.14)
$$\phi_t(x) = \sqrt{\mu_t(x)} \ e^{S(t,x)},$$

$$\hat{\phi}_t(x) = \sqrt{\mu_t(x)} \ e^{-S(t,x)}.$$

Require $\phi_t \in C^{1,2}(D)$ *(resp.* $H^{1,2}(D)$*), where* $D = \{(t, x): \mu(t, x) \neq 0\}$, *and assume* (6.3) *and the integrability condition* (6.4) *for the creation and killing* $c = -L\phi_t/\phi_t$ *induced by* ϕ_t *(cf.* (6.1)*) with a prescribed* L. *Then, there exists a space-time diffusion process* $\{(t, X_t), Q_{(s,x)}; (s, x) \in D\}$ *with an additional drift coefficient* $a(t, x) = \sigma^T \sigma \nabla \log \phi_t(x)$.

The same argument applies to $\widehat{\phi}_t$ with $\widehat{c} = -\widehat{L}\widehat{\phi}_t/\widehat{\phi}_t$, which implies the existence of a space-time diffusion process $\{(s, X_s), \widehat{Q}_{(t,x)}; (t,x) \in \widehat{D}\}$ (in reversed time) with an additional drift term $\widehat{a}(s, x) = \sigma^T \sigma \nabla \log \widehat{\phi}_s(x)$.

In Corollary 6.1 diffusion processes are not determined uniquely by a flow $\mu_t, t \in [a, b]$, because of an ambiguity depending on a choice of a (phase) function $S(t, x)$. Moreover the distribution densities of the constructed processes do not coincide with the given flow. To specify the distribution of the process $\{X_t, Q_{\mu_a}\}$ (resp. $\{X_t, \widehat{Q}_{\mu_b}\}$) we must require an additional condition, as will be explained in the following corollaries.

Corollary 6.2.[7] *Keep the notations and assumptions of Corollary 6.1. The distribution density of the process $\{(t, X_t), Q_{\mu_a}\}$, where*

$$(6.15) \qquad Q_{\mu_a}[\,\cdot\,] = \int \mu_a(x) dx \, Q_{(a,x)}[\,\cdot\,],$$

coincides with the given flow $\mu_t, t \in [a, b]$, if and only if

$$(6.16) \qquad \frac{\partial R}{\partial t} + \frac{1}{2}\Delta S + (\sigma \nabla S) \cdot (\sigma \nabla R) + \boldsymbol{b} \cdot \nabla R = 0,$$

where $\log \mu_t = 2R$.

Under condition (6.16), the diffusion processes with the additional drift coefficients $\boldsymbol{a} = \sigma^T \sigma \nabla \log \phi_t$ and $\widehat{a} = \sigma^T \sigma \nabla \log \widehat{\phi}_t$, respectively, are time reversal of each other (in duality with respect to the given μ_t), namely, Q_{μ_a} has the q-representation and the $\widehat{q}$-representation

$$Q_{\mu_a} = [\,\mu_a q \gg \,=\, \ll \widehat{q} \mu_b\,],$$

which can be given also in the p-representation (cf. Theorem 3.6)

$$Q_{\mu_a} = [\,\widehat{\phi}_a p \gg \ll p \, \phi_b\,].$$

Proof. The two diffusion processes are in duality with respect to $\mu_t(x)$, if and only if the Fokker-Planck equations $B^\circ \mu = 0$ and $(\widehat{B})^\circ \mu = 0$ hold at the same time and hence (3.35'), which is our requirement (6.16).

[7] Compare with Theorem 5.4

Corollary 6.3. *Keep the assumptions of Corollary 6.1. If the given flow $\mu_t, t \in [a,b]$, satisfies the Fokker-Planck equation $B^\circ \mu = 0$, then the distribution density of the process $\{(t, X_t), Q_{\mu_a}\}$, where Q_{μ_a} is defined by (6.15), coincides with the given flow.*

Proof.   In this corollary the parabolic operator B is prescribed beforehand, namely $a(t,x)$ is given. The function $\phi_t(x)$ can be determined through the formula $a(t,x) = \sigma^T \sigma \nabla \log \phi_t(x)$, and then the function $S(t,x)$ is uniquely prescribed by the first equation of (6.14).

Remark 6.5.   Since we do not consider time reversal (or duality) in Corollary 6.3, we can begin with

$$\widetilde{L} = \frac{\partial}{\partial t} + \frac{1}{2}\sigma\sigma^T(t,x)^{ij}\frac{\partial^2}{\partial x^i \partial x^j} + \widetilde{b}(t,x)^i\frac{\partial}{\partial x^i},$$

so that

$$B = \frac{\partial}{\partial t} + \frac{1}{2}\sigma^T\sigma(t,x)^{ij}\frac{\partial^2}{\partial x^i \partial x^j} + \{\widetilde{b}(t,x)^i + a(t,x)^i\}\frac{\partial}{\partial x^i},$$

where the diffusion coefficient $\sigma^T\sigma(t,x)$ may degenerate.

Corollary 6.4.   *Let $\psi_t(x) = e^{R(t,x) + iS(t,x)}$, $t \in [a,b]$, be a flow of complex valued functions with*

$$\int |\psi_t(x)|^2 dx = 1,$$

and set

(6.17)
$$\phi_t(x) = e^{R(t,x) + S(t,x)},$$

$$\widehat{\phi}_t(x) = e^{R(t,x) - S(t,x)}.$$

Under the same assumptions as in Corollary 6.1 (resp. 6.2) the same conclusions hold with the pair of functions ϕ_t and $\widehat{\phi}_t$ defined above.

Notice that in Chapter 5 the entropy condition (5.12) played a crucial role, but the condition is not required in Theorem 6.1 nor in the above corollaries.

Theorem 6.1 clarifies a typical case of inadmissibility of a pair $\{\mu_a, \mu_b\}$ of distributions:

Corollary 6.5. *Let $\phi(t,x)$ be given and $c(t,x)$ be the creation and killing induced by $\phi(t,x)$ (respectively, $c(t,x)$ be given and $\phi(t,x)$ a (weak) solution of $L\phi + c\phi = 0$). Assume the process $\{(t,X_t),\ Q_{(s,x)};\ (s,x) \in D\}$ is segregated into subsets D_1 and D_2 by the zero set of $\phi(t,x)$. Choose $\widehat{\phi}_a$ so that the distribution $\mu_a = \mu_a(x)\,dx$ charges on D_1 but not on D_2, and take another μ_b which charges on D_2 but not on D_1. Then the pair $\{\mu_a, \mu_b\}$ is not admissible for the renormalized process $\overline{P}$ of the measure $P^c_{(s,x)}$ with creation and killing $c(t,x)$.*

6.3. Discussions on the q-Representation

Time reversal disappeared in this chapter. The reason is this: To discuss time reversal of Markov processes, we must fix an initial distribution, since the transition probability of the time reversed process depends on the initial distribution (cf. Theorem 3.2). Therefore, it is clear that we cannot apply time reversal, since we avoid fixing initial distributions in this chapter. The crucial fact in "time reversal" is that it establishes the duality relation of drift coefficients which contain the most important quantity for the duality relation, the *phase factor* $S(t,x)$ (beside $R(t,x)$ which is directly determined by the distribution of the process X_t, in other words by the initial distribution of the process).

In the modern theory of diffusion (Markov) processes, we abandon to some extent the beautiful symmetry of time reversal to gain generality. We consider, instead of fixing an initial distribution, a family of processes which start from arbitrary points of a given state space.[8] Thus "time reversal" fades out somehow into the background.

However, there is a quantity with which we can recover contents of time reversal and all connected. It is the function $\phi(t,x)$. This function contains an important quantity, the phase factor $S(t,x)$ *implicitly* (to make it explicit we need the adjoint quantity $\widehat{\phi}(t,x)$ for which we must fix an initial distribution, but we do not in this chapter).

Now about the function $\phi(t,x)$. It has played a central role in this chapter. Let L be a parabolic differential operator defined in (5.5) and consider the basic unperturbed diffusion process defined by the operator L. The function $\phi(t,x)$ determines "perturbation" through the additional drift coefficient

---

[8] This is the reason why time reversal is somewhat a strange subject for those who were educated by the modern theory of Markov processes

$$a(t,x) = \sigma^T \sigma(x) \nabla \log \phi(t,x)$$

on the one hand, and in terms of the creation and killing

$$c(t,x) = -\frac{L\phi(t,x)}{\phi(t,x)}$$

on the other hand (notice both quantities are defined in terms of $\phi(t,x)$). The relation between the two quantities $a(t,x)$ and $c(t,x)$ is the key of the whole story.

We know these functions have already appeared in (3.37) and (3.40). As we have seen, both quantities induce severe "perturbation". Namely, the $a(t,x)$ represents "drift" from unperturbed motion, while the $c(t,x)$ gives another kind of "modification" of the unperturbed motion in terms of the rate of "creation and annihilation".

The multiplicative functional defined in (6.5)

$$N_s^t = \exp\left(\int_s^t c(r,X_r)dr\right) \frac{\phi(t,X_t)}{\phi(s,X_s)} 1_{\{t < T_s\}}$$

provides a "bridge" connecting two pictures drawn by the drift coefficient $a(t,x)$ and the rate of creation and killing $c(t,x)$ which look completely different at a glance. This connects implicitly the diffusion equation with creation and killing $c(t,x)$ to the diffusion equation with the drift coefficient $a(t,x)$. We have shown that an application of the transformation in terms of the multiplicative functional N_s^t yields immediately the process with the drift coefficient $a(t,x)$. According to Theorem 6.1, the multiplicative functional N_s^t is in fact a renormalization of the modified Kac functional

$$m_s^t = \exp\left(\int_s^t c(r,X_r)dr\right)1_{\{t < T_s\}} \ ,$$

by means of a given $(L + c)$-harmonic function $\phi(s,x)$.

In connection with this, one can define a "transition function" $p(s,x;t,y)$ with creation and killing through the Feynman-Kac formula

$$\int p(s,x;t,y)f(y)dy = P_{(s,x)}[\exp(\int_s^t c(r,X_r)dr)1_{\{t<T_s\}}f(X_t)],$$

and hence one might expect the possibility of constructing a process with creation and killing by means of the transition density $p(s,x;t,y)$. However, as already remarked, it is not possible to construct a "process with creation and killing" (killing is o.k. but creation!). One possible way is to define a probability measure Q by (3.49) in p-representation. But for this we need an additional function $\hat{\phi}$, which we avoid in this chapter.

Another way is offered in (3.44), namely, we define

(6.18) $$q(s,x;t,y) = \frac{1}{\phi(s,x)}p(s,x;t,y)\phi(t,y).$$

This is a transition *probability* density by definition, if $\phi(t,x)$ is space-time p-harmonic. This is what we have done in this chapter in terms of the q-representation. In fact the transition probability density $q(s,x;t,y)$ of the transformed process satisfies

$$\int q(s,x;t,y)f(y)dy = P_{(s,x)}[\exp(\int_s^t c(r,X_r)dr)\frac{\phi(t,X_t)}{\phi(s,X_s)}1_{\{t<T_s\}}f(X_t)],$$

where c is the creation and killing induced by ϕ. We now have a diffusion process $\{X_t, Q_{(s,x)}; (s,x) \in D\}$, as stated in Theorem 6.1, which can start from any point $(s,x) \in D$ in the space-time state space. Thus we have obtained a family of processes as is usual in the general theory of Markov processes.

Remark 6.6. If $\phi(t,x)$ and $\hat{\phi}(t,x)$ are not known beforehand in the first statement of Theorem 4.1, Theorem 5.3 in terms of variation can be applied to the existence claim of diffusion processes. If we begin with a wave function $\psi = e^{R+iS}$ as in the converse assertion (ii) of Theorem 4.1, we can apply Theorem 6.1, since we get $\phi(t,x)$ (and $\hat{\phi}(t,x)$) automatically.

Remark 6.7. In the *fictitious* description of the p-representation the time symmetry is achieved at the cost of abandoning the Markov property, and the formula $[\hat{\phi}_a \, p >><< p \, \phi_b]$ clearly shows Eddington-Schrödinger's (Markov field type) prediction. By contrast the *real* q-representation recovers the Markov property through "conditioning". Namely, the transition probability $q(s,x;t,y)$ is defined by the so-called harmonic transformation (6.18), through which we control the transition continuously in time so that the controlled process will be distributed at $t = b$ according to a prescribed distribution μ_b. Therefore, information from the future is built-in also in the q-representation through the function

$$\phi(s, x) = \int p(s, x; b, y) dy \phi_b(y),$$

which is determined by the data $\phi_b(y)$ at $t = b$.

6.4. What is the Feynman Integral ?

In terms of the the complex-valued fundamental solution $\psi(s, z; t, x)$, $a \leq s < t \leq b$,[9] of the Schrödinger equation

$$i \frac{\partial \psi}{\partial t} + \frac{1}{2} \Delta \psi + i b(t, x) \cdot \nabla \psi - V(t, x) \psi = 0,$$

the so-called "Feynman integral" (cf. Feynman (1948)) is defined through

$(6.19) \quad \mathcal{F}[f(X_a, X_{t_1}, \dots, X_{t_{n-1}}, X_b)]$

$$= \int dx_0 \psi(a, x_0) \psi(a, x_0; t_1, x_1) dx_1 \psi(t_1, x_1; t_2, x_2) dx_2 \cdots$$

$$\cdots \psi(t_{n-1}, x_{n-1}; b, x_n) dx_n f(x_0, x_1, \dots, x_n),$$

where $a < t_1 < t_2 < \cdots < t_{n-1} < b$.

It is well-known that this does not define a measure on the path space, since it oscillates as $n \to \infty$ and does not converge in general. This fact has caused problems mathematically.

There have been various interesting attempts to justify the "Feynman integral" mathematically in the last decades.[10] Nonetheless, people have never felt satisfied completely with these justifications, because they are always somehow artificial. We know that Itô's stochastic integral does not define a measure too, but it is a well-developed integration theory, which is

---

[9] We can require $\psi(t, x; r, y)$, $a \leq t < r \leq b$, is the fundamental solution of the time reversed equation, namely the complex conjugate of the Schrödinger equation

[10] For extensive literature cf., e.g. Albeverio- Høegh-Krohn (1976). Cf. also Itô (1961, 67)

indispensable in the theory of stochastic processes. The practical calculus which we enjoy with the theory of Itô's stochastic integral is more or less missing in these justifications of the Feynman integral.

In many cases of applications, the Feynman integral is applied together with the so-called "imaginary time argument $t \rightarrow it$", through which one can get a measure on the path space. This measure is useful in computing various quantities because it is really *a measure*, but the original Feynman integral then plays little role or none at all. This means that we are actually avoiding the Feynman integral. Then we cannot stop asking: What is the Feynman integral? and what do we need it for? People thought first that the Feynman integral provides a way of "quantization" which is different from Heisenberg-Born-Jordan's, and moreover that Feynman's way gives an intuitive picture, *paths*! However, since we do not need "quantization" to get the Schrödinger equation in diffusion theory, as we have shown already in Chapter 4, the first point can be put more or less aside. Secondly Feynman's picture of paths is rather obscure, since the Feynman integral does not give a measure on the path space, as we have remarked, and moreover it is complex-valued, even though it would be well-defined. Therefore, I wonder what Feynman imagined with his integral, even though he spoke about "quantum paths".[11] I suspect that he was probably always thinking of paths of diffusion processes by himself but speaking of *quantum paths* guided by his physical intuition. Let us assume so. Then, we should and can consider diffusion processes consistently in Schrödinger's and Kolmogoroff's representations instead of Feynman's integral. Remember that Schrödinger and diffusion equations are equivalent.

Namely, we can adopt and begin with the p-representation (3.49):

$$Q[f(X_a, X_{t_1}, \ldots, X_{t_{n-1}}, X_b)]$$

$$= \int dx_0 \widehat{\phi}(a, x_0) p(a, x_0; t_1, x_1) dx_1 p(t_1, x_1; t_2, x_2) dx_2 \cdots$$

$$\cdots p(t_{n-1}, x_{n-1}; b, x_n) \phi(b, x_n) dx_n f(x_0, x_1, \ldots, x_n),$$

although we have constructed the measure Q not through the above formula but through our variational method in terms of Csiszar's projection (cf. Chapter 5), where $p(s, x; t, y)$ is the weak fundamental solution of the diffusion equation

---

[11] He speaks of "probability amplitude for a path", but the probability amplitude is not a well-defined mathematical object in probability theory

$$\frac{\partial \phi}{\partial t} + \frac{1}{2} \Delta \phi + b(t,x) \cdot \nabla \phi + c(t,x)\phi = 0,$$

instead of the complex-valued kernel $\psi(s,x;t,y)$ (or wave functions) of the Schrödinger equation

$$i \frac{\partial \psi}{\partial t} + \frac{1}{2} \Delta \psi + i b(t,x) \cdot \nabla \psi - V(t,x)\psi = 0;$$

or the q-representation (3.47), i.e.,

$$Q[f(X_a, X_{t_1}, \ldots, X_{t_{n-1}}, X_b)]$$

$$= \int dx_0 \widehat{\phi}_a(x_0)\phi_a(x_0)q(a,x_0;t_1,x_1)dx_1 q(t_1,x_1;t_2,x_2)dx_2 \cdots$$

$$\cdots q(t_{n-1},x_{n-1};b,x_n)dx_n f(x_0,x_1,\ldots,x_n)$$

$$= \int f(x_0, x_1, \ldots, x_n)dx_0 \widehat{q}(a,x_0;t_1,x_1)dx_1 \widehat{q}(t_1,x_1;t_2,x_2) \cdots$$

$$\cdots dx_{n-1}\widehat{q}(t_{n-1},x_{n-1};b,x_n)\widehat{\phi}(b,x_n)\phi(b,x_n)dx_n \, ,$$

which we have constructed through a transformation by a multiplicative functional in this chapter.

Moreover, we have, for the distribution density of the process, the factorization formula

$$\mu_t(x) = \overline{\psi_t}(x)\psi_t(x) = \widehat{\phi}_t(x)\phi_t(x).$$

Thus we have a nice probability measure Q (diffusion process) defined on the path space for quantum mechanics. In this way we can adopt diffusion processes instead of the "Feynman integral". Furthermore, the fact that the diffusion process is not with imaginary-time but with *real time* should be emphasized.

Remark 6.8. The so-called "imaginary time argument $t \to it$" should be avoided, because it is hardly justifiable and moreover cannot be applied to the case with vector potentials as we have remarked already (notice there is no such problem with our q (or p)-representation).[12] Nonetheless, the results obtained through the imaginary time argument can be appropriately

---

[12] In this connection cf. Doss (1980)

reinterpreted. Namely, the diffusion equation deduced through the imaginary-time argument turns out to be correct with an appropriate modification of potential functions (cf. Chapter 4), if no vector potential is involved, and results obtained in terms of the diffusion equation remain valid according to the equivalence theorem (cf. Theorem 4.1). The key point is this: one needs *no* analytic continuation, since what one treats is *not imaginary-time but real correct time*. If one needs to write down results in terms of wave functions, what one must do is not "analytic continuation" but the transformation

$$\phi(t, x) = e^{R(t,x) + S(t,x)} \implies \psi(t, x) = e^{R(t,x) + iS(t,x)}.$$

Moreover, one must and can proceed further to the statistical mechanical picture in order to give consistent meanings to the distribution density $\mu_t = \overline{\psi}_t \psi_t = \hat{\phi}_t \phi_t$, as will be seen in the following chapters.

Remark 6.9. One might expect an analogy of the p-representation would rescue Fynman's idea. Namely, we modify (6.19) as

(6.20) $\quad \widetilde{F}[f(X_a, X_{t_1}, \dots, X_{t_{n-1}}, X_b)]$

$$= \int dx_0 \psi(a, x_0) \psi(a, x_0; t_1, x_1) dx_1 \psi(t_1, x_1; t_2, x_2) dx_2 \cdots$$

$$\cdots \psi(t_{n-1}, x_{n-1}; b, x_n) \overline{\psi}(b, x_n) dx_n f(x_0, x_1, \dots, x_n),$$

where $\overline{\psi}(b, y)$ is the complex conjugate of

$$\psi(b, y) = \int \psi(a, x_0) dx_0 \psi(a, x_0; b, y).$$

As we have transformed the p-representation $[\hat{\phi}_a\, p \gg \ll p\, \phi_b]$ into the q-representation $[\hat{\phi}_a \phi_b\, q \gg$, we can rewrite (6.20) in terms of $\overline{\psi}$-transformed kernel function

(6.21) $\qquad \widetilde{\psi}(s, x; t, y) = \dfrac{1}{\overline{\psi}(s, x)} \psi(s, x; t, y) \overline{\psi}(t, y).$

In fact, formula (6.20) can be reformed with

(6.22) $\qquad \widetilde{\psi}(s, x; t, y) = \int d\xi\, \dfrac{\overline{\psi}(s, \xi)}{\overline{\psi}(s, x)}\, \overline{\psi}(s, \xi; t, y) \psi(s, x; t, y).$

as

(6.23) $\widetilde{F}[f(X_a, X_{t_1}, \ldots, X_{t_{n-1}}, X_b)]$

$$= \int dx_0 \psi(a, x_0) \overline{\psi}(a, x_0) \, \widetilde{\psi}(a, x_0; t_1, x_1) dx_1 \, \widetilde{\psi}(t_1, x_1; t_2, x_2) dx_2 \cdots$$

$$\cdots \widetilde{\psi}(t_{n-1}, x_{n-1}; b, x_n) dx_n f(x_0, x_1, \ldots, x_n),$$

However, it seems to me that both (6.20) and (6.23) do not help us.

6.5. A Remark on Kac's Semi-Group

In the proof of Theorem 6.1 we have applied the uniqueness of solutions of the integral equation (6.11), which is defined in terms of the measure $P^-_{(s,x)}$ and $c^+(t, x)$. Then a question arises: Does the uniqueness also hold for an integral equation defined in terms of $P_{(s,x)}$ and $c(t, x)$?

Let us denote for $s \leq u \leq t$

$$g_s^u f(s, x) = P_{(s,x)}[f(u, X_u) \, 1_{\{u < T_s\}}],$$

(6.24) $\overline{g}_s^u f(s, x) = P_{(s,x)}[e^{-c^-[s,u]} f(u, X_u) \, 1_{\{u < T_s\}}],$

$$c^-[s, u] = \int_s^u c^-(r, X_r) dr,$$

with $c = c^+ - c^-$, and consider non-negative solutions of the pair of integral equations:

(6.25) $p(s) = \overline{g}_s^t f + \int_s^t du \, \overline{g}_s^u (c^+ p),$

(6.26) $p(s) = g_s^t f + \int_s^t du \, g_s^u (cp),$

where t is fixed.

We have shown in Lemma 6.1 that uniqueness holds for bounded solutions of integral equation (6.25). Does the uniqueness hold also for integral equation (6.26) ?

In the following we consider non-negative solutions and formulate our problem in the following form. In order to keep our notation unchanged, let $c^-(s, x)$ and $c^+(s, x)$ denote *arbitrary* non-negative measurable function (therefore, $c^+(s, x) c^-(s, x) \neq 0$, in general), and set

$$c(s, x) = c^+(s, x) - c^-(s, x).$$

Lemma 6.2. *Under the above-mentioned convention, let p be a non-negative solution of the integral equation (6.26). Then, it satisfies integral equation (6.25).*

Proof.[13] We first assume that $c^-(s, x)$ is bounded. The boundedness requirement on $c^-(s, x)$ will be removed later on. Let p be a non-negative solution of (6.26) and set

(6.27) $$y(u) = \overline{g}_s^{t-u} p = I + II,$$

with

$$I = \overline{g}_s^{t-u} (g_{t-u}^t f) = P_{(s, x)}[e^{-c^-[s, t-u]} f(t, X_t) : t < T_s],$$

and

$$II = \int_{t-r}^{t} dr \, P_{(s, x)}[e^{-c^-[s, t-u]} c(r, X_r) p(r, X_r) : r < T_s],$$

where we have applied the Markov property of X_t. Therefore, we have

$$\frac{dI}{du} = P_{(s, x)}[c^-(t-u, X_{t-u}) e^{-c^-[s, t-u]} f(t, X_t) : t < T_s],$$

$$\frac{dII}{du} = P_{(s, x)}[e^{-c^-[s, t-u]} c(t-u, X_{t-u}) p(t-u, X_{t-u}) : t-u < T_s]$$

$$+ \int_{t-r}^{t} dr \, P_{(s, x)}[c^-(t-u, X_{t-u}) e^{-c^-[s, t-u]} c(r, X_r) p(r, X_r) : r < T_s]$$

$$= \left(\frac{dII}{du}\right)_1 + \left(\frac{dII}{du}\right)_2 ;$$

and hence

---

[13] Due to K. Uchiyama, cf. Nagasawa (1990)

$$\frac{dI}{du} + \left(\frac{dII}{du}\right)_2 = P_{(s,x)}\left[c^-(t-u,X_{t-u})e^{-c^-[s,t-u]}\left\{P_{(t-u,X_{t-u})}[f(t,X_t):t<T_{t-u}]\right.\right.$$

$$\left.\left. + \int_{t-u}^t dr\, P_{(t-u,X_{t-u})}[c(r,X_r)p(r,X_r):r<T_{t-u}]\right\}:t-u<T_s\right]$$

$$= P_{(s,x)}\left[c^-(t-u,X_{t-u})e^{-c^-[s,t-u]}p(t-u,X_{t-u}):t-u<T_s\right],$$

which implies

$$\frac{dy}{du} = \left(\frac{dII}{du}\right)_1 + \left\{\frac{dI}{du} + \left(\frac{dII}{du}\right)_2\right\}$$

$$= P_{(s,x)}\left[c^+(t-u,X_{t-u})e^{-c^-[s,t-u]}p(t-u,X_{t-u}):t-u<T_s\right].$$

Consequently, because $p(t) = f(t)$,

$$p(s) - \overline{g}_s^t p = \int_0^{t-s} du\, \frac{dy}{du}$$

$$= \int_s^t du\, P_{(s,x)}\left[e^{-c^-[s,u]}c^+(u,X_u)p(u,X_u):u<T_s\right]$$

$$= \int_s^t du\, \overline{g}_s^u(c^+p),$$

which proves that p satisfies (6.25).

Now removing the boundedness assumption on c^-, we define a sequence of operators $\overline{g}(n)_s^u$ by (6.24) with $c_n^- = c^- \wedge n$ in place of c^-. Let p be a non-negative solution of (6.26). Then, applying the above arguments with $c_n^- = c^- \wedge n$, we get

$$(6.28) \quad p(s) = \overline{g}(n)_s^t p + \int_s^t du\, \overline{g}(n)_s^u(c^+p) - \int_s^t du\, \overline{g}(n)_s^u\{(c^- - c_n^-)p\}.$$

Since $c^- - c_n^- \downarrow 0$ as $n \to \infty$ and also $\overline{g}(n)_s^u f \downarrow \overline{g}_s^u f$ for non-negative f, the third term of the right-hand side of (6.28) vanishes as $n \to \infty$ and the first and second terms converge to the ones without n. This proves that a non-negative solution of (6.26) satisfies (6.25) without the boundedness condition on c^-. This completes the proof of the lemma.

Lemma 6.3. *The uniqueness holds for bounded* [14] *non-negative solutions of the integral equation* (6.26) *under the integrability condition* (6.4).

Proof. Combine Lemma 6.1 and Lemma 6.2.

6.6. A Typical Case

Let a pair $\{\phi_t, \hat{\phi}_t\}$ of functions be given by (4.15) in terms of a solution of the Schrödinger equation in (4.1) in the exponential form $\psi = e^{R + iS}$. Then, we should take the function $c(t, x, \phi)$ of creation and killing given at (4.16), i.e.,

$$c(t, x, \phi) = - V(t, x) - 2\frac{\partial S}{\partial t}(t, x) - (\sigma \nabla S)^2(t, x) - 2\,b\cdot\nabla S(t, x).$$

Therefore, the integrability condition (6.4) turns out to be

$$(6.29) \qquad P_{(s,x)}\Big[\exp\Big(\int_s^b \{V^- + 2\Big(\frac{\partial S}{\partial t}\Big)^- + 2\,(b\,\nabla S)^-\}(t, X_t)dt\Big)\Big] < \infty.$$

To give a sufficient condition to the integrability requirement (6.4), especially (6.29), let us consider functions in the Kato class which was discussed in Aizenman-Simon:[15]

We say that a function $c(s, x)$ is in the *Kato class* (for space-time processes), if it is measurable and

$$(6.30) \qquad \lim_{r \to 0} \sup_{(s,x) \in D} P^-_{(s,x)}\Big[\int_s^{(s+r)\wedge b} c^+(t, X_t)dt\Big] = 0.$$

Namely, we first kill the space-time process with the negative part c^- to get $P^-_{(s, x)}$ and then control the positive part c^+ by means of $P^-_{(s, x)}$. If $c(s, x)$ is in the Kato class, then we have

---

[14] Cf. Remark 6.1 after Lemma 6.2

[15] Cf. Aizenman-Simon (1982), Stummer (1990) and also Sturm (1989). For Schrödinger operators see Chung-Rao (1981)

(6.31) $$\sup_{(s,x)\in D} P^-_{(s,x)}[\exp(\int_s^b c^+(t, X_t)dt)] < \infty,$$

by Hasiminsky's lemma (1959).[16]

As a typical case let us assume in (4.1) that there is no vector potential and $V(x)$ is a potential function in the Kato class, i.e., we consider a Schrödinger equation

$$i\frac{\partial\psi}{\partial t} + \frac{1}{2}\Delta\psi - V(x)\psi = 0.$$

Then, by Carmona (1985) there exists a "weak" solution $\psi = e^{R+iS}$ of the Schrödinger equation such that $\partial S/\partial t$ and ∇S are well-defined.

Therefore, if

$$\lim_{r\to 0} \sup_{(s,x)\in D} P^-_{(s,x)}[\int_s^{(s+r)\wedge b} (\frac{\partial S}{\partial t})^-(t, X_t)dt] = 0,$$

namely, $-\partial S/\partial t$ is in the Kato class, then the integrability condition (6.29) is fulfilled uniformly in $(s,x)\in D$, and hence the corresponding diffusion processes exist by Theorem 5.3 or Corollary 6.2.

6.7.   Hydrogen Atom

As a typical example[17] let us consider a hydrogen atom. According to the conventional interpretation of quantum mechanics, the Schrödinger equation describes states of a single electron in the hydrogen atom, and therefore, a single diffusion particle. In an excited state, as we have shown, the corresponding diffusion process cannot move across over the zero set of its distribution density, and is trapped in one of the regions surrounded by the zero set of the solution. This means that there must be various kinds of different states of the hydrogen atom in the excited state depending on the localization of the electron which is trapped in one of the closed regions. This is improbable. A more definitive fact is, as I have remarked already, that a single electron (diffusion particle) *cannot* reproduce the distribution density $|\psi|^2$, but just a part of it. Therefore, we must leave the single-

---

[16] Cf. Stummer (1990)

[17] Non-typical examples with severely singular potentials see Section 7.9

electron picture; namely we must abandon the conventional interpretation of quantum mechanics.

This suggests that we rather adopt the quantum field picture instead of the single-electron picture, which means that we give a statistical mechanical interpretation to solutions of Schrödinger equations. By this I mean, we consider virtual photons and virtual electrons in a hydrogen atom. Because of the creation and annihilation of virtual photons and virtual electrons, we can not specify the location of an electron in the hydrogen atom. Assuming that the distributions of virtual electrons and photons, respectively, are proportional, let us consider virtual photons.[18] This means that we adopt an interpretation that the distribution of any state of a hydrogen atom determined by the Schrödinger equation gives a distribution of virtual photons in a hydrogen atom. One possibility is to proceed as follows: We apply diffusion approximation to, say, a lattice model of the field of virtual photons (or electrons). Then we get a system of infinitely many interacting diffusion processes confined in a hydrogen atom. Then, the probability that we find an electron somewhere in the hydrogen atom must be proportional to the spatial distribution of virtual photons, that is, it agrees with the distribution given in terms of a solution of the Schrödinger equation of a hydrogen atom. This is, in a sense, a revival of the chemists' favorite picture (originally due to Schrödinger), namely, *cloud of charge* in a hydrogen atom, but now what we have is a *cloud of virtual photons (or electrons)*, instead of charge.

To justify the interpretation explained above, in other words to make it not just an "interpretation" but a concrete "model", we must build a reasonable statistical theory and prove a limit theorem, say, the propagation of chaos. This will be discussed in the following chapters.

6.8. A Remark on $\{\mu_a, \mu_b\}$

As shown in Chapter 5, we can construct a Schrödinger process $\{X_t, Q\}$ for a creation and killing $c(t, x)$ and a pair of probability distributions μ_a and μ_b under admissibility condition (5.12), and moreover, the Schrödinger equation determined by the Schrödinger process $\{X_t, Q\}$ is non-linear as shown in Theorem 4.1 (cf. (4.19)).

Solving an initial value problem of the Schrödinger equation (4.1) for a given initial value ψ_a we obtain a solution $\psi_t = e^{R + iS}$. Let us define the creation and killing $c = -L\phi/\phi$ induced by $\phi = e^{R + S}$. Then we construct a

---

[18] We take virtual photons to avoid theoretical complication of fermions

Schrödinger process for arbitrary μ_a and μ_b under the admissibility condition. This Schrödinger process provides a pair of functions $\phi' = e^{R' + S'}$ and $\widehat{\phi}' = e^{R' - S'}$. The corresponding wave function $\psi_t' = e^{R' + iS'}$ satisfies in general another Schrödinger equation with an additional non-linear potential function determined by S'. This Schrödinger equation coincides with the given one (4.1) if and only if we take $\mu_a = \psi_a \overline{\psi}_a$ and $\mu_b = \psi_b \overline{\psi}_b$.

Chapter VII
Segregation of a Population

7.1. Introduction

Let φ be an eigenfunction of

$$-\frac{1}{2}\Delta\varphi + V(x)\varphi = E\varphi,$$

where $V(x)$ is in the Kato class. We have shown in the preceding chapters there exists a diffusion process $\{X_t, Q\}$ with the probability density $\mu = \varphi^2$ (in general $|\sum \alpha_i \varphi_i|^2$). In this chapter we regard it as the spatial distribution density of a population. The diffusion process $\{X_t, Q\}$, therefore, describes the movement of *a typical particle* in the population when the size of the population becomes sufficiently large.

This means that a system of interacting n-diffusion processes should be constructed whose distribution converges to the infinite (independent) copies of the process $\{X_t, Q\}$ as n tend to infinity. If we denote the system by $P^{(n)}$, then we can express it symbolically

$$P^{(n)} \to Q^\infty, \quad as \ n \to \infty,$$

the precise meaning of which will be explained later on. This is a general assertion in the context of statistical mechanics.

This will be done in terms of the propagation of chaos of a system of interacting diffusion processes, which is due to McKean (1966, 67). In this chapter we shall show such statistical models are mathematically possible for simple cases in one dimension. General cases in higher dimensions will be discussed in Chapter 8.

Based on this mathematical facts, several applications were discussed in Nagasawa (1980, 81) and Nagasawa-Yasue (1982), which will be explained in Chapter 9, namely, on the segregation of monkey populations, septation of *Escherichia coli*, and the mass spectrum of mesons.

7.2. Harmonic Oscillator

Let us consider as an example the distribution density of the first excited state of the one-dimensional harmonic oscillator :

$$(7.1) \qquad \mu(x) = \varphi^2(x) = \beta^{-1} x^2 e^{-x^2},$$

where β is a normalizing constant. Following McKean (1966, 67), we consider a system of n-interacting diffusion particles on the line, and let n tend to infinity. Then as a limit we obtain, under some regularity conditions on interactions, a "Boltzmann equation" (of Kac caricature, McKean says) for the distribution density $u(t, x)$ of particles:

$$(7.2) \qquad \frac{\partial u}{\partial t} = \frac{1}{2} \frac{\partial^2 u}{\partial x^2} - \frac{\partial}{\partial x}(b[x, u]u),$$

where the drift coefficient is given by

$$(7.3) \qquad b[x, u] = \int b(x, y) u(t, y) dy,$$

with a pair interaction $b(x, y)$. Equation (7.2) is often called in the contexts of the propagation of chaos the "McKean-Vlasov equation".

Let us assume in (7.2) that $u(t, x)$ is a stationary state and it coincides with the distribution density $\mu(x)$. This means that the drift coefficient $b[x, \mu]$ satisfies the Kolmogoroff's duality relation (3.1), i.e.,

$$b[x, \mu] = \frac{1}{2} \frac{d}{dx}(\log \mu(x)),$$

and (7.3) at the same time and hence, identifying them, we have

$$(7.4) \qquad \frac{1}{2} \frac{d}{dx}(\log \mu(x)) = \int b(x, y) \mu(y) dy.$$

For the distribution density $\mu(x)$ given in (7.1), equation (7.4) has no non-trivial solution $b(x, y)$, if we will not introduce two segregated (coloured) groups separated by the zero of the distribution $\mu(x)$. If one allows this segregation, one can get a solution $b(x, y) = b(x - y)$ (= a pair interaction) due to Föllmer-Nagasawa (cf. Nagasawa-Tanaka (1985)):

$$(7.5) \qquad b(x) = \beta\left(\frac{3}{x^4} + \frac{2}{x^2}\right) + b_0(x),$$

$$b_0(x) = O\left(\frac{1}{x^2}\right), \quad as \ x \downarrow 0.$$

The theorem of McKean (1967) on the propagation of chaos cannot be applied to the case of interactions such as given in (7.5), because of its strong singularity. In Nagasawa-Tanaka (1986), a system of stochastic differential equations of two types of (coloured) particles is considered:
For $i, j = 1, 2, \ldots, n$,

$$X_i(t) = X_i(0) + B_i^+(t) + \int_0^t ds\left\{\frac{1}{n}\sum_{j=1}^n h(X_i(s) - Y_j(s)) + h_0(X_i(s))\right\} + \xi_i(t),$$

$$(7.6)$$

$$Y_j(t) = Y_j(0) + B_j^-(t) + \int_0^t ds\left\{-\frac{1}{n}\sum_{i=1}^n h(X_i(s) - Y_j(s)) + h_0(Y_j(s))\right\} - \eta_j(t),$$

where $\{(B_i^+(t), B_j^-(t)): i, j = 1, 2, \ldots, n\}$ is a family of mutually independent Brownian motions, $X_i(0)$ (resp. $Y_j(0)$) has a common distribution on $[0, \infty)$ (resp. on $(-\infty, 0]$), $h(x)$ is a non-increasing continuous function on $(0, \infty)$, which is bounded below but may diverge at the origin like the $b(x)$ given in (7.5), and $h_0(x)$ is an odd function which is continuous and non-increasing in $\mathbf{R} - \{0\}$ satisfying a growth condition (cf. (7.70) in Section 7.8). The processes $\xi_i(t)$ and $\eta_j(t)$ are local times which make the origin a reflecting boundary for $X_i(t)$ and $Y_j(t)$, respectively. Then we have

Theorem 7.1. *The propagation of chaos holds for the system* (7.6) *of interacting coloured diffusion particles; that is, the distribution of the process* $\{(Y_i(t), X_i(t)): i = 1, 2, \ldots, n\}$ *converges weakly to the infinite independent copies of a diffusion process* $(Y(t), X(t))$, *as* $n \to \infty$, *and the distribution of the limiting process is governed by a non-linear equation like* (7.2),[1] *but one half of the distribution* $u(t, x)$ *on* $(-\infty, 0]$ *is the distribution of* $Y(t)$ *and another half on* $[0, \infty)$ *is of* $X(t)$.

---

[1] For the precise meaning see below and Theorem 7.7

Let us assume there is a unique solution $\{(Y_i(t), X_i(t)): i = 1, 2, \ldots, n\}$ of (7.6) and $\{Y_i(t), X_i(t)\}$ are asymptotically independent and identically distributed as the total number n increases. Because of the law of large numbers, the empirical distributions converge to non-random distributions

$$\frac{1}{n}\sum_{i=1}^{n} \delta_{X_i(s)} \to u_X(s, \cdot), \quad and \quad \frac{1}{n}\sum_{i=1}^{n} \delta_{Y_i(s)} \to u_Y(s, \cdot),$$

respectively. Tending n to infinity formally in (7.6), we obtain a pair of equations

$$X(t) = X(0) + B^+(t) + \int_0^t ds\{\int_{(-\infty, 0]} h(X(s) - y)u_Y(s, dy) + h_0(X(s))\} + \varphi(t),$$

$$Y(t) = Y(0) + B^-(t) + \int_0^t ds\{-\int_{[0, \infty)} h(x - Y(s))u_X(s, dy) + h_0(Y(s))\} - \psi(t).$$

This is the assertion of the propagation of chaos in Theorem 7.1, a proof of which will be given later in Section 7.8, after preparing the pathwise existence and uniqueness of solutions to a system of equations and a limit theorem in Sections 7.6 and 7.7, respectively.

Thus for the first excited state of a one-dimensional harmonic oscillator, there is a microscopic model which realizes the distribution $\mu(x) = \varphi^2(x) = \beta^{-1}x^2e^{-x^2}$, when the population size n tends to infinity.

In the model treated above the segregating point between two groups is fixed at the origin. If we allow moving segregating points, then the problem of the propagation of chaos becomes much more complicated and difficult to handle, as will be seen in the following sections (cf. Nagasawa-Tanaka (1987,a,b)).

7.3. Segregation of a Finite-System of Particles

Suppose we are watching a football game in *one*-dimension played between two teams "*red caps*" and "*blue caps*". We watch the movement of two players with the red and blue who are nearest among the others. Suppose, if the two players meet and pass over, they exchange their caps quickly so that we cannot see it. Assume we can follow the movement of players and distinguish them only by the colours of their caps. Then we see no mixing occurs between the two teams, i.e., the two teams will stay always completely segregated. However, the segregating point between the two teams moves, i.e., it is a time-dependent random variable. Let us formulate the football game as a system of stochastic differential equations with reflection.

Let $\{B_i^-(t), B_j^+(t); 1 \le i \le m, 1 \le j \le n\}$ be independent one dimensional Brownian motions starting at 0. Under a constraint

(7.7) $$\max_{1 \le i \le m} X_i(t) \le \min_{1 \le j \le n} Y_j(t), \quad for \ \forall \, t \ge 0,$$

and with local times

(7.8) $\Phi_{ij}(t)$ *is continuous, monotone non-decreasing,*

$$\Phi_{ij}(0) = 0, \ and \ \ \mathrm{supp}(d\Phi_{ij}) \subset \{t \ge 0 : X_i(t) = Y_j(t)\},$$

we consider a system of stochastic differential equations

$$X_i(t) = X_i(0) + B_i^-(t) + \frac{1}{m} \sum_{k=1}^{m} \int_0^t b_{11}(X_i(s), X_k(s))ds$$

$$+ \frac{1}{n} \sum_{k=1}^{n} \int_0^t b_{12}(X_i(s), Y_k(s))ds - \sum_{k=1}^{n} \Phi_{ik}(t), \quad 1 \le i \le m,$$

(7.9)

$$Y_j(t) = Y_j(0) + B_j^+(t) + \frac{1}{m} \sum_{k=1}^{m} \int_0^t b_{21}(Y_j(s), X_k(s))ds$$

$$+ \frac{1}{n} \sum_{k=1}^{n} \int_0^t b_{22}(Y_j(s), Y_k(s))ds + \sum_{k=1}^{m} \Phi_{kj}(t), \quad 1 \le j \le n,$$

where the initial values are independent of the Brownian motions. System (7.9) is a special case of Tanaka's equations in convex regions.[2]

This is a stochastic model of the one-dimensional football game with exchange of caps.

Theorem 7.2. (Nagasawa-Tanaka (1987,a)) *If the interactions $b_{ij}(x,y)$ satisfy a Lipschitz condition*

---

[2] Cf. Tanaka (1979) and also Lions-Sznitman (1984)

(7.10) $\displaystyle\sum_{i=1}^{2}\sum_{j=1}^{2} |b_{ij}(x,y) - b_{ij}(x',y')|^2 \le c\{|x - x'|^2 + |y - y'|^2\},$

then there exists a (pathwise) unique solution $\{(X_1(t), \ldots , X_m(t), Y_1(t),$ *... , $Y_n(t)$), $\Phi_{ij}(t)$, $1 \le i \le m$, $1 \le j \le n\}$ of system (7.9) of stochastic differential equations with (7.7) and (7.8).*

In a proof of Nagasawa-Tanaka (1987, a) the idea "to exchange caps of players" provides the right way of treating the problem.   In fact, the exchange of caps produces appropriate local times $\Phi_{ij}(t)$. To see this let us consider the simplest game between two one-player-teams.

Let $x_1(t)$ and $x_2(t)$ be solutions of a stochastic differential equation

$$x(t) = x(0) + B(t) + \int_0^t b(s,x(s))ds,$$

with independent Brownian motions $B_1(t)$ and $B_2(t)$, respectively. If the players $x_1(t)$ and $x_2(t)$ exchange their caps, when they meet and pass over, then what we see through the movement of caps is not $x_1(t)$ and $x_2(t)$ but

$$\tilde{x}_1(t) = x_1(t) \wedge x_2(t) \quad and \quad \tilde{x}_2(t) = x_1(t) \vee x_2(t).$$

We apply Itô's formula (through approximation) to the pair of functions $x_1 \wedge x_2$ and $x_1 \vee x_2$. Choose a non-negative continuous even function g on $\mathbf{R}^1$ of compact support such that

$$\int g(x)dx = 1,$$

and put

$$f_N(x) = \int_0^x dy \int_0^y Ng(Nz)dz.$$

Then, $f_N \in C^2(\mathbf{R}^1)$ and $f_N(x) \to \frac{1}{2}|x|$ uniformly as $N \to \infty$, and hence

$$f_N(x_1 - x_2) + \frac{1}{2}(x_1 + x_2) \to x_1 \vee x_2.$$

An application of Itô's formula yields

$$f_N(x_1(t) - x_2(t)) + \frac{1}{2}(x_1(t) + x_2(t)) - f_N(x_1(0) - x_2(0)) - \frac{1}{2}(x_1(0) + x_2(0))$$

$$= \int_0^t \{\frac{1}{2} + f_N'(x_1(s) - x_2(s))\}\{dB_1(s) + b(s, x_1(s))ds\}$$

$$+ \int_0^t \{\frac{1}{2} - f_N'(x_1(s) - x_2(s))\}\{dB_2(s) + b(s, x_2(s))ds\}$$

$$+ \int_0^t Ng(N(x_1(s) - x_2(s)))ds$$

from which follows, as $N \to \infty$,

$$\tilde{x}_2(t) - \tilde{x}_2(0) = B^+(t) + \int_0^t b(s, \tilde{x}_2(s))ds + \Phi(t).$$

The formula for $\tilde{x}_1(t)$ can be obtained similarly. Hence, we see that the process $(\tilde{x}_1(t), \tilde{x}_2(t))$ satisfies a system of stochastic differential equations

$$X(t) = X(0) + B^-(t) + \int_0^t b(s, X(s))ds - \Phi(t),$$

$$Y(t) = Y(0) + B^+(t) + \int_0^t b(s, Y(s))ds + \Phi(t),$$

with the initial value $(\tilde{x}_1(0), \tilde{x}_2(0))$. The new Brownian motions $B^-(t)$, $B^+(t)$, and a local time $\Phi(t)$ are given explicitly by

$$B^-(t) = \int_0^t 1_{\{\tilde{x}_1(s) = x_1(s)\}}dB_1(s) + \int_0^t 1_{\{\tilde{x}_1(s) = x_2(s)\}}dB_2(s),$$

$$B^+(t) = \int_0^t 1_{\{\tilde{x}_2(s) = x_1(s)\}}dB_1(s) + \int_0^t 1_{\{\tilde{x}_2(s) = x_2(s)\}}dB_2(s),$$

and

$$\Phi(t) = \frac{1}{2}\int_0^t \delta_{\{x_1(s) = x_2(s)\}}ds = \frac{1}{2}\int_0^t \delta_{\{\tilde{x}_1(s) = \tilde{x}_2(s)\}}ds.$$

It is straightforward to extend this to a play between an m-players-team and an n-players-team (for detail cf. Nagasawa-Tanaka (1987, a)).

The propagation of chaos for the system (7.9) of interacting coloured diffusion processes will be discussed in the following sections.

7.4. A Formulation of the Propagation of Chaos

For a given bounded continuous functions $\{b_{ij}(x,y); i,j = 1,2\}$ on $\mathbf{R}^2$, let us define

$$(7.11) \quad b(x,y,u) = b_{11}(x,y)\frac{1}{\theta}1_{(-\infty,\gamma(u)]}(y) + b_{12}(x,y)\frac{1}{1-\theta}1_{(\gamma(u),\infty)}(y),$$
$$\text{for } x \le \gamma(u),$$

$$= b_{21}(x,y)\frac{1}{\theta}1_{(-\infty,\gamma(u)]}(y) + b_{22}(x,y)\frac{1}{1-\theta}1_{(\gamma(u),\infty)}(y),$$
$$\text{for } x > \gamma(u),$$

where $0 < \theta < 1$ is a constant, which will be specified in the following, and $\gamma(u)$ denotes the segregating front of a probability distribution u on $\mathbf{R}^1$ defined by

$$(7.12) \qquad \gamma(u) = \gamma(u, \theta) = \min \{x: u((-\infty,x)) \le \theta \le u((-\infty,x])\}.[3]$$

In (7.11) $b_{11}(x,y)$ stands for the interaction between reds at x and y, $b_{12}(x,y)$ is the influence of a blue at y on a red at x, and so on.

Moreover, we set

$$(7.13) \qquad\qquad b[x,u] = \int b(x,y,u)u(dy).$$

This $b[x,u]$ defined in terms of $b(x,y,u)$ in (7.11) is the interaction between two groups divided by the segregating front $\gamma(u)$ depending on the distribution u.

Let us set $\theta = n/(n+m)$ in (7.11) and (7.12) and consider a system of stochastic differential equations for $(X_1(t), X_2(t), \ldots , X_{n+m}(t))$ with the interaction $b[x,u]$:

---

[3] γ is the θ-quantile of the probability distribution u

$$(7.14) \quad X_i(t) = X_i(0) + B_i(t) + \int_0^t b[X_i(s), U^{(n+m)}(s)]ds, \quad i = 1, 2, \dots, n+m,$$

where

$$U^{(n+m)}(t) = \frac{1}{n+m} \sum_{i=1}^{n+m} \delta_{X_i(t)}$$

is the empirical distribution of $(X_1(t), X_2(t), \dots, X_{n+m}(t))$.

Since the system of equations (7.9) contains reflections but system (7.14) does not, they look completely different except for the total number of involved particles. However, if we apply the order statistics, namely the change of caps (cf. section 2 of Nagasawa-Tanaka (1987, a)) to the solution $(X_1(t), X_2(t), \dots, X_{n+m}(t))$ of (7.14), then we can obtain another system $(X_1(t), \dots, X_m(t), Y_1(t), \dots, Y_n(t))$ of two segregated groups which satisfies (7.7) and (7.9). Therefore, it is enough to show the propagation of chaos for the solution of system (7.14) without reflecting boundary conditions.

7.5. The Propagation of Chaos

We will treat the problem of the propagation of chaos in a general formulation in $\mathbf{R}^d$. Let $b^{(n)}[x, u]$ be $\mathbf{R}^d$-valued measurable functions on $\mathbf{R}^d \times \mathbf{M}_1(\mathbf{R}^d)$, where $\mathbf{M}_1(\mathbf{R}^d)$ denotes the space of probability distributions on $\mathbf{R}^d$. We consider a system of interacting diffusion processes on $\mathbf{R}^d$ described by

$$(7.15) \quad X_i^{(n)}(t) = X_i^{(n)}(0) + B_i(t) + \int_0^t b^{(n)}[X_i^{(n)}(s), U^{(n)}(s)]ds, \quad i = 1, 2, \dots, n.$$

In order to show the propagation of chaos for system (7.15) of stochastic differential equations, we require the following three conditions on the boundedness, continuity, and uniqueness.

In the following $<u, f>$ denotes the bilinear form defined by

$$<u, f> = \int u(dx) f(x), \quad u \in \mathbf{M}_1(\mathbf{R}^d).$$

Conditions:

(B) $b[x,u]$ and $b^{(n)}[x,u]$ are $\mathbf{R}^d$-valued measurable functions on $\mathbf{R}^d \times \mathbf{M}_1(\mathbf{R}^d)$, which are bounded uniformly in n.

(C) If $u_n \in \mathbf{M}_1(\mathbf{R}^d)$ converges weakly to $u \in \mathbf{M}_1(\mathbf{R}^d)$ which has a strictly positive density (almost everywhere) with respect to the Lebesgue measure of $\mathbf{R}^d$, then for $\forall f \in C_b(\mathbf{R}^d)$

$$<u_n, b^{(n)}[\,\cdot\,, u_n]f> \quad converges \ to \quad <u, b[\,\cdot\,, u]f> \quad as \ n \to \infty.$$

(U) The uniqueness holds for $\mathbf{M}_1(\mathbf{R}^d)$-valued solutions of the initial value problem

(7.16) $$\frac{d}{dt}<u(t), f> = <u(t), A_{u(t)}f>, \quad for \ f \in C_K^\infty(\mathbf{R}^d),$$

$$u(0) = u \in \mathbf{M}_1(\mathbf{R}^d),$$

where

(7.17) $$A_u f = \tfrac{1}{2}\Delta + b[\,\cdot\,, u]\cdot\nabla f.$$

The boundedness condition (B) is not desired (but technically required) and must be removed, if possible. The continuity condition (C) covers the case of our interest. For the uniqueness condition (U) we have an analytic sufficient condition given in the following:

Lemma 7.1. (due to H. Amann) *Define a non-linear operator* $G(u)$ *from* $(C_0)^*$ *into* $(C_0^1)^*$ *by* [4]

(7.18) $$<G(u), f> = <u, b[\,\cdot\,, u]\cdot\nabla f>, \quad for \ f \in C_0^1.$$

If $G(u)$ *defined above satisfies*

(7.19) $$\| G(u) - G(w) \|_1^* \le const. \| u - w \|_0^*, \quad for \ u, w \in (C_0)^*,$$

then the uniqueness holds for the initial value problem (7.16).

Proof. See Appendix of Nagasawa-Tanaka (1987, a).

---

[4] $(C_0)^*$ and $(C_0^1)^*$ denote the dual Banach spaces of C_0 and C_0^1, respectively. C_0 denotes the space of continuous functions vanishing at infinity, and $C_0^1 = \{f \in C_0: \partial_x f \in C_0\}$.

Let $u \in \mathbf{M}_1(\mathbf{R}^d)$. A sequence $\{u_n \in \mathbf{M}_1((\mathbf{R}^d)^n): n = 1, 2, \ldots\}$ is u-**chaotic**, if

$$(7.20) \qquad \lim_{n \to \infty} <u_n, f_1 \otimes f_2 \otimes \cdots \otimes f_m \otimes 1 \otimes \cdots \otimes 1> = \prod_{k=1}^{m} <u, f_k>,$$

for fixed but arbitrary m and $f_1, \ldots, f_m \in C_b(\mathbf{R}^d)$.

Equation (7.20) is equivalent to

$$(7.21) \qquad u_n\left[\left\{(x_1, \cdots, x_n) : \left|\frac{1}{n}\sum_{i=1}^{n} f(x_i) - <u, f>\right| > \varepsilon\right\}\right] \to 0,$$

as $n \to \infty$ for any $\varepsilon > 0$ and $f \in C_b(\mathbf{R}^d)$ (cf. Nagasawa-Tanaka (1987, b)).

Theorem 7.3. (Nagasawa-Tanaka (1987, b))   *Let $\{u_n \in \mathbf{M}_1((\mathbf{R}^d)^n): n = 1, 2, \ldots\}$ be u-chaotic and $(X_1^{(n)}(t), X_2^{(n)}(t), \ldots, X_n^{(n)}(t))$ be a solution of system (7.15) of stochastic differential equations with an initial value distributed according to the given u_n. Under the conditions (B), (C), and (U) stated above, the following assertions hold:*

(i) *The empirical distribution $U^{(n)}(t)$ converges in probability to some (non-random) limit $u(t) \in \mathbf{M}_1(\mathbf{R}^d)$ which is a solution of the initial value problem (7.16).*

(ii) *For each (fixed) $m \leq n$ the process $(X_1^{(n)}(t), X_2^{(n)}(t), \ldots, X_m^{(n)}(t))$ converges in law as $n \to \infty$ to $(X_1(t), X_2(t), \ldots, X_m(t))$, which are mutually independent, and each $X_i(t)$, $1 \leq i \leq m$ is a copy of the solution $X(t)$ of McKean-Vlasov's stochastic differential equation*

$$(7.22) \qquad X(t) = X(0) + B(t) + \int_0^t b[X(s), u(s)]ds,$$

where $X(0)$ is u-distributed and $u(s)$ denotes the probability distribution of $X(s)$.

The existence claim of a (unique) solution of (7.16) (resp. (7.22)) is part of the claim of the theorem.   For a proof we refer to Nagasawa-Tanaka (1987, a, b).

Since it is easy to see that the $b[x, u]$ defined in (7.13) satisfies the conditions (B), (C), and (U), Theorem 7.3 implies that the propagation of chaos holds for system (7.14) and hence for system (7.9) of interacting coloured diffusion particles.

System (7.8) which we have treated is nothing but a simple one-dimensional toy-model. Nonetheless, the proof given in Nagasawa-Tanaka (1987, a, b) is rather involved. I can imagine, therefore, what kind of difficulties might be encountered, if one will attempt to extend it to higher dimensions with singular interactions (one will need probably a new idea). I have given up this program for the moment because of this, and I will try to approach the problem through another bypass in the next chapter.

Remark 7.1. For the propagation of chaos see: McKean (1966, 67), Tanaka (1984), Kusuoka-Tamura (1984), Shiga-Tanaka (1958), Dawson-Gärtner (1984, 89), Ölschläger (1989), Sznitman (1989), and references given in these articles.

Remark 7.2. To handle non-linear diffusion equations there is another probabilistic method based on the branching property. Though it does not belong to the present contexts, there must be an interesting interrelation between the non-linearity which appeared in McKean-Vlasov's equation (7.2) or (7.22) and non-linearity (4.20). It is known (Cf., e.g. Nagasawa (1968, 72, 77)) that we can construct generalized branching Markov processes (on an enlarged state spaces) which provide stochastic solutions of non-linear equations such as

$$(7.23) \qquad \frac{\partial u}{\partial t} = \frac{1}{2} \frac{\partial^2 u}{\partial x^2} + \sum_{n,m} c_{n,m} u^n \left(\frac{\partial u}{\partial x} \right)^m.$$

For non-linearity induced by the (generalized) branching property, see Chapter 12.

7.6. Skorokhod's Problem with Singular Drift

In the following sections we denote $\Omega = C([0, \infty), \mathbf{R})$, $\Omega^+ = C([0, \infty), \mathbf{R}^+)$, and Φ_ξ with $\xi \in \Omega^+$ the class of all functions φ satisfying

(7.24) φ *is continuous, non-decreasing with* $\varphi(0) = 0$, *and constant on each connected component of* $\{t > 0: \xi(t) > 0\}$.

In other words

(7.24') $\text{supp } d\varphi \subset \{t \geq 0 : \xi(t) = 0\}$,

or equivalently

(7.24") $\varphi(t) = \displaystyle\int_0^t 1_{\{0\}}(\xi(s)) d\varphi(s)$.

We consider Skorokhod's problem:[5] Given $\omega \in \Omega$ with $\omega(0) \geq 0$ and a function $b(t, x)$ on $\mathbf{R}^+ \times \mathbf{R}^+$,[6] find a pair $\{\xi(t), \varphi(t)\}$, $\xi \in \Omega^+$, $\varphi \in \Phi_\xi$, satisfying

(7.25) $\xi(t) = \omega(t) + \displaystyle\int_0^t b(s, \xi(s)) ds + \varphi(t)$.

Such a pair $\{\xi(t), \varphi(t)\}$, $\xi \in \Omega^+$, $\varphi \in \Phi_\xi$, will be called simply a solution of equation (7.25).

We shall often apply the following formula which represents a solution $\{\xi(t), \varphi(t)\}$ of (7.25) as

(7.26) $\xi(t) = \max_{0 \leq s \leq t}\{\widetilde{\omega}(t) - \widetilde{\omega}(s)\} \vee \widetilde{\omega}(t)$

$$= \begin{cases} \widetilde{\omega}(t), & \text{for } 0 \leq t \leq T_0, \\ \widetilde{\omega}(t) + \max_{T_0 \leq s \leq t}\{-\widetilde{\omega}(s)\}, & \text{for } t > T_0, \end{cases}$$

and

(7.27) $\varphi(t) = \begin{cases} 0, & \text{for } 0 \leq t \leq T_0, \\ \max_{T_0 \leq s \leq t}\{-\widetilde{\omega}(s)\}, & \text{for } t > T_0, \end{cases}$

where

$$\widetilde{\omega}(t) = \omega(t) + \int_0^t b(s, \xi(s)) ds$$

and

$$T_0 = \inf\{t > 0 : \widetilde{\omega}(t) < 0\}.$$

We prove the existence and uniqueness of solutions of the Skorokhod problem (7.25) under the following condition on $b(t, x)$

---

[5] Cf. Skorokhod (1961)

[6] $b(t, 0)$ may be $+\infty$, if the integral of $b(s, \xi(s))$ is absolutely convergent

(7.28)   *$b(t,x)$ is continuous in $(t,x) \in [0,\infty) \times (0,\infty)$, and decreasing*
          in x for each $t \geq 0$.

It should be noticed that the limit $b(t,0) = \lim_{x \downarrow 0} b(t,x)$ may diverge.[7]

The main theorem in this section claims the existence and comparison of
solutions of (7.25) :

Theorem 7.4. (Nagasawa-Tanaka (1985))

(i) *If coefficient $b(t,x)$ in (7.25) satisfies condition (7.28), then there
exists a unique solution of equation (7.25).*

(ii) *If coefficients $b_1(t,x)$ and $b_2(t,x)$ satisfy the condition (7.28),
and if*

(7.29) $$b_1(t,x) \leq b_2(t,x),$$

then

(7.30) $$\xi_1(t) \leq \xi_2(t),$$

(7.31) $$d\varphi_1(t) \geq d\varphi_2(t),$$

*where $\{\xi_i, \varphi_i\}$, $i = 1, 2$, are solutions of equation (7.25) with the
coefficient $b_i(t,x)$ in place of $b(t,x)$.*

For a proof of the theorem we prepare some lemmas.

Lemma 7.2. *Let $\xi_i, \eta_i \in \Omega^+$ and $\varphi_i \in \Phi_{\xi_i}$ satisfy*

$$\xi_i(t) = \omega(t) + \int_0^t b_i(s, \eta_i(s))ds + \varphi_i(t),$$

for $i = 1, 2$. Then

(7.32) $$|\xi_1(t) - \xi_2(t)|^2 \leq 2 \int_0^t \{\xi_1(s) - \xi_2(s)\}\{b_1(s, \eta_1(s)) - b_2(s, \eta_2(s))\}ds.$$

Proof. Setting

---

[7] An example is given at (7.5). Cf. Section 7.9 for further examples of such functions

$$\psi_i(t) = \int_0^t b_i(s, \eta_i(s))ds + \varphi_i(t), \quad i = 1, 2,$$

and

$$\psi(t) = \psi_1(t) - \psi_2(t),$$

we have

$$|\xi_1(t) - \xi_2(t)|^2 = |\psi(t)|^2 = 2\int_{\{0 < r < s < t\}} d\psi(r)d\psi(s)$$

$$= 2\int_0^t \{\psi_1(s) - \psi_2(s)\}\{d\psi_1(s) - d\psi_2(s)\}$$

$$= 2\int_0^t \{\xi_1(s) - \xi_2(s)\}\{b_1(s, \eta_1(s)) - b_2(s, \eta_2(s))\}ds$$

$$+ 2\int_0^t \{\xi_1(s) - \xi_2(s)\}d\varphi_1(s) + 2\int_0^t \{\xi_2(s) - \xi_1(s)\}d\varphi_2(s).$$

where the last two integrals are non-positive, since φ_i, $i = 1, 2$, satisfy (7.24'). Thus we have inequality (7.32).

Lemma 7.3. *Let $b(t, x)$ be continuous in $(t, x) \in \mathbf{R}^+ \times \mathbf{R}^+$, and satisfy*

(7.33) $\qquad |b(t, x) - b(t, y)| \le K|x - y|, \quad \text{for } \forall t \ge 0, \, \forall x, y \ge 0,$

with a positive constant K. Then there exists a unique solution of equation (7.25).

Proof. The uniqueness is immediate. In fact, if $\{\xi_i, \varphi_i\}$, $i = 1, 2$, are solutions of (7.25), then Lemma 7.2 together with (7.33) yields

$$|\xi_1(t) - \xi_2(t)|^2 \le 2K\int_0^t |\xi_1(s) - \xi_2(s)|^2 ds,$$

and hence they coincide by Gronwall's lemma.

Define $\{\xi_0, \varphi_0\}$ using (7.26) and (7.27) with ω in place of $\widetilde{\omega}$; namely, we begin with a solution of

$$\xi_0(t) = \omega(t) + \varphi_0(t).$$

For $n \geq 1$ define $\{\xi_n, \varphi_n\}$ successively to be a solution of

$$\xi_n(t) = \omega(t) + \int_0^t b(s, \xi_{n-1}(s))ds + \varphi_n(t).$$

The solution $\{\xi_n, \varphi_n\}$ can be given explicitly using (7.26) and (7.27). An application of Lemma 7.2 yields

$$|\xi_{n+1}(t) - \xi_n(t)|^2 \leq 2\int_0^t \{\xi_{n+1}(s) - \xi_n(s)\}\{b_1(s, \xi_n(s)) - b_2(s, \xi_{n-1}(s))\}ds$$

$$\leq K\int_0^t 2|\xi_{n+1}(s) - \xi_n(s)||\xi_n(s) - \xi_{n-1}(s)|ds$$

$$\leq K\int_0^t |\xi_{n+1}(s) - \xi_n(s)|^2ds + K\int_0^t |\xi_n(s) - \xi_{n-1}(s)|^2ds.$$

Therefore, by Gronwall's lemma[8] we have

$$|\xi_{n+1}(t) - \xi_n(t)|^2 \leq Ke^{Kt}\int_0^t |\xi_n(s) - \xi_{n-1}(s)|^2ds,$$

from which follows

$$|\xi_{n+1}(t) - \xi_n(t)|^2 \leq c\frac{(Ke^{KT}T)^{n-1}}{(n-1)!}$$

in a finite interval $[0, T]$, where c is a constant depending on the interval, and hence

$$\sum_{n=1}^{\infty} |\xi_{n+1}(t) - \xi_n(t)|^2 < \infty.$$

Therefore ξ_n converges uniformly on $[0, T] \times \Omega$. It is easy to see that the

[8] See Chapter 2

pair of the limits

$$\xi(t) = \lim_{n \to \infty} \xi_n(t),$$

$$\varphi(t) = \lim_{n \to \infty} \varphi_n(t)$$

gives a solution of (7.25). This completes the proof.

Lemma 7.4. *Let* $b_i(t,x)$, $i = 1, 2$, *satisfy the condition in* Lemma 7.3 *and assume in addition*

(7.34) $$b_1(t,x) \le b_2(t,x), \quad for \ \forall t \ge 0, \ \forall x \ge 0.$$

(i) *Let* $0 \le x_1 \le x_2$, *and* $\{\xi_i, \varphi_i\}$, $i = 1, 2$, *be solutions of*

$$\xi_i(t) = x_i + \omega(t) + \int_0^t b_i(s, \xi_i(s))ds + \varphi_i(t).$$

Then

(7.35) $$d\varphi_1(t) \ge d\varphi_2(t).$$

(ii) *Let* $0 \le r$, $0 \le x_1 \le x_2$ *and* $r \le x_2$, *and* (7.34) *hold for all* $x \ge r$. *Let* $\{\xi_1, \varphi_1\}$ *be the same as in the assertion* (i), *and* $\{\xi_2, \varphi_2\}$ *be the solution of the Skorokhod problem on* $[r, \infty)$

$$\xi_2(t) = x_2 + \omega(t) + \int_0^t b_2(s, \xi_2(s))ds + \varphi_2(t).$$

Then

(7.36) $$\xi_1(t) \le \xi_2(t).$$

Proof. Let us prove assertion (ii) first. Take $a > 0$ and let $\{\xi_2^a, \varphi_2^a\}$ be the solution of a Skorokhod problem on $[r, \infty)$

$$\xi_2^a(t) = a + x_2 + \omega(t) + \int_0^t \{b_2(s, \xi_2^a(s)) + a\}ds + \varphi_2^a(t).$$

We will prove

(7.37) $$\xi_1(t) \le \xi_2^a(t), \quad for \ all \ t \ge 0,$$

from which (7.36) follows through $a \downarrow 0$. Define

$$T = \inf \{t > 0: \xi_1(t) > \xi_2^a(t)\}$$

$$= \infty, \quad \textit{if there is no such } t.$$

Suppose $T < \infty$. Then by the definition of T

(7.38) $\xi_1(T) = \xi_2^a(T) \geq r,$

and moreover we have

(7.39)   *there exists a sequence $\{t_n > 0\}$ tending to zero such that*

$$\xi_1(T + t_n) > \xi_2^a(T + t_n).$$

For brevity let us denote

$$\eta_1(t) = \xi_1(T + t), \qquad \psi_1(t) = \varphi_1(T + t) - \varphi_1(t),$$

$$\eta_2(t) = \xi_2^a(T + t), \qquad \psi_2(t) = \varphi_2^a(T + t) - \varphi_2^a(t).$$

First of all, let us consider the case $r > 0$ in (7.38). It is clear that the pair $\{\eta_1, \psi_1\}$ solves a Skorohod problem on $[0, \infty)$:

(7.40) $\eta_1(t) = \xi_1(T) + \omega(T + t) - \omega(T) + \displaystyle\int_0^t b_1(T + s, \eta_1(s))ds + \psi_1(t),$

while the pair $\{\eta_2, \psi_2\}$ does a Skorohod problem on $[r, \infty)$:

(7.41) $\eta_2(t) = \xi_1(T) + \omega(T + t) - \omega(T) + \displaystyle\int_0^t \{b_2(T + s, \eta_2(s)) + a\}ds + \psi_2(t).$

Because of (7.38) we have $\eta_1(0) \geq r > 0$ and hence $\psi_1(t) = 0$ for all sufficiently small $t > 0$. Therefore, a comparison of (7.40) with (7.41) shows $\eta_1(t) \leq \eta_2(t)$ for all sufficiently small $t > 0$. But this contradicts (7.39). Hence $T = \infty$, namely (7.37) holds.

Secondly, the case $r = 0$ can be treated as follows: In this case, equation (7.41) turns out to be a Skorohod problem on $[0, \infty)$. Suppose $\xi_1(T) > 0$. Then $\psi_1(t) = \psi_2(t) = 0$ for all sufficiently small $t > 0$, and hence a comparison of (7.40) and (7.41) yields that $\eta_1(t) \leq \eta_2(t)$ for all sufficiently small $t > 0$. This contradicts (7.39). Suppose $\xi_1(T) = 0$. Then, applying formula (7.26), we obtain

$$\eta_1(t) = \max_{0 \leq s \leq t} \left\{ \omega(T+t) - \omega(T+s) + \int_s^t b_1(T+r, \eta_1(r))dr \right\}$$

$$\eta_2(t) = \max_{0 \leq s \leq t} \left\{ \omega(T+t) - \omega(T+s) + \int_s^t (b_2(T+r, \eta_1(r)) + a)dr \right\},$$

a comparison of which shows that $\eta_1(t) \leq \eta_2(t)$ for all sufficiently small $t > 0$. This contradicts (7.39), and hence the inequality (7.37) holds.

Now we prove assertion (i). It is enough to show the inequality (7.35) assuming that $b_1(t, x)$ is strictly smaller than $b_2(t, x)$, since we can consider $b_2(t, x) + \varepsilon$ and then let $\varepsilon \downarrow 0$. Let us show

$$\varphi_1(t+s) - \varphi_1(t) \geq \varphi_2(t+s) - \varphi_2(t)$$

for all sufficiently small $s > 0$ and for any $t \geq 0$ such that $\xi_2(t) = 0$ (and hence $\xi_1(t) = 0$).

Because of the assumption $b_1(t, x) < b_2(t, x)$, we have

(7.42) $$b_1(t+s, \xi_1(t+s)) < b_2(t+s, \xi_2(t+s))$$

for all sufficiently small $s > 0$. Denote for $i = 1, 2$,

$$\widetilde{\omega}_i(s) = \omega(t+s) - \omega(t) + \int_0^s b_i(t+r, \xi_i(t+r))dr ,$$

and apply formula (7.27). Then

$$\varphi_2(t+s) - \varphi_2(t) = \max_{0 \leq r \leq s} \{ -\widetilde{\omega}_2(r) \}$$

$$= \max_{0 \leq r \leq s} \left\{ -\omega(t+r) + \omega(t) - \int_0^r b_2(t+\tau, \xi_2(t+\tau))d\tau \right\}$$

$$\leq \max_{0 \leq r \leq s} \left\{ -\omega(t+r) + \omega(t) - \int_0^r b_1(t+\tau, \xi_1(t+\tau))d\tau \right\}$$

$$= \varphi_1(t + s) - \varphi_1(t),$$

where we have applied (7.42). This completes the proof.

Lemma 7.5. *Assume that* $b(t, x)$ *satisfies condition* (7.28) *and moreover it is bounded from below with* $-M > -\infty$. *Then there exists a unique solution* $\{\xi^a, \varphi^a\}$ *of a Skorokhod problem on* $[a, \infty)$

$$(7.43) \qquad \xi^a(t) = a + \omega(t) + \int_0^t b(s, \xi^a(s))ds + \varphi^a(t).$$

Proof. We approximate the coefficient $b(t, x)$ with a sequence of uniformly Lipschiz continuous $b_n(t, x)$ in such a way that

$$b_n(t, x) \downarrow b(t, x), \quad for \ x \geq a.$$

For equation (7.43) with $b_n(t, x)$ in place of $b(t, x)$ the existence of the unique solution $\{\xi_n^a, \varphi_n^a\}$ is already shown in lemma 7.3. By Lemma 7.4

$$\xi_n^a(t) \downarrow \xi^a(t) \quad and \quad \varphi_n^a(t) \uparrow \varphi^a(t).$$

Moreover it is easy to see

$$\lim_{n \to \infty} \int_0^t b_n(s, \xi_n^a(s))ds = \int_0^t b(s, \xi^a(s))ds,$$

and hence $\{\xi^a, \varphi^a\}$ is a solution of (7.43).

Applying the same approximation arguments of coefficients we can prove

Lemma 7.6. *Require the same conditions as in* Lemma 7.5 *on the coefficients* $b_i(t, x)$. *Then assertions* (i) *and* (ii) *of* Lemma 7.4 *hold for equation* (7.43).

Proof of Theorem 7.4. We will first prove the theorem under an additional condition

$$b(t, x) \geq -M > -\infty.$$

Let $\{\xi^a, \varphi^a\}$ be a solution of (7.43).   Because of Lemma 7.6 $\xi^a(t)$ is decreasing when $a \downarrow 0$ and hence the limit

$$\xi(t) = \lim_{a \downarrow 0} \xi^a(t)$$

exists. Set

$$\bar{b}(t, x) = b(t, x) + M \geq 0.$$

Then

$$\xi^a(t) + Mt = a + \omega(t) + \int_0^t \bar{b}(s, \xi^a(s))ds + \varphi^a(t),$$

and hence

$$0 \leq \int_0^t \bar{b}(s, \xi^a(s))ds \leq \xi^a(t) + Mt - a - \omega(t),$$

from which we get

$$0 \leq \int_0^t \bar{b}(s, \xi(s))ds \leq \lim_{a \downarrow 0} \int_0^t \bar{b}(s, \xi^a(s))ds < \infty.$$

Therefore the integral

$$\int_0^t b(s, \xi(s))ds$$

is absolutely convergent and equals

$$\lim_{a \downarrow 0} \int_0^t b(s, \xi^a(s))ds < \infty.$$

Consequently, the limit

$$\varphi(t) = \lim_{a \downarrow 0} \varphi^a(t)$$

also exists, and the pair $\{\xi, \varphi\}$ satisfies equation (7.25). It is easy to see that $\xi(t)$ is continuous on the subset $\{t : \xi(t) > 0\}$. If $\xi(t) = 0$, then we can show

$$\lim_{\varepsilon \downarrow 0} \sup_{|t - s| < \varepsilon} \xi(s) = 0.$$

Therefore, ξ is continuous and hence $\xi \in \Omega^+$ and also $\varphi \in \Phi_\xi$, and hence the pair $\{\xi, \varphi\}$ is a solution of (7.25).

We can carry over the proof without the additional boundedness assumption $b(t, x) \geq - M$. In fact, set

$$b_M(t, x) = b(t, x) \vee (-M).$$

Let $\{\xi_M, \varphi_M\}$ be the solution of (7.25) with $b_M(t, x)$. Since

$$b_M(t, x) \downarrow b(t, x), \quad as \ M \to \infty,$$

we get

$$\xi_M(t) \downarrow \xi(t).$$

For fixed t set

$$N = \sup_{0 \leq s \leq t} \xi(s) \leq \sup_{0 \leq s \leq t} \xi_M(s) < \infty.$$

Then for M large enough we have

$$b(s, x) = b_M(s, x), \quad for \ \forall s \leq t \ and \ \forall x \leq N + 1,$$

and hence $\xi(s) = \xi_M(s)$ for all $s \leq t$. Therefore

$$\int_0^t b(s, \xi(s))ds = \int_0^t b_M(s, \xi(s))ds$$

is absolutely convergent. This implies that

$$\varphi(t) = \lim_{M \to \infty} \varphi_M(t) \quad (= \varphi_M(t), \ for \ sufficiently \ large \ M),$$

and the pair $\{\xi, \varphi\}$ is a solution of (7.25). Assertion (ii) of the theorem can be shown similarly with the same approximation methods.

The uniqueness of the solution is a simple consequence of Lemma 7.2. This completes the proof of Theorem 7.4.

7.7. A Limit Theorem

Let $\mu^{(+)}$, $\mu^{(-)}$ be probability measures on Ω and consider a system

$$\xi(t, w) = w(t) + \int_0^t ds \int_\Omega b(\xi(s, w), \eta(s, \tilde{z}))\mu^{(-)}(d\tilde{z}) + \varphi(t, w),$$

(7.44)

$$\eta(t, w) = z(t) + \int_0^t ds \int_\Omega b(\eta(s, z), \xi(s, \tilde{w}))\mu^{(+)}(d\tilde{w}) + \psi(t, z),[9]$$

where $w, z \in \Omega$ have non-negative initial values, and $\varphi \in \Phi_\xi$, $\psi \in \Phi_\eta$.

Moreover we assume

(7.45) $$b(x, y) = h(x + y) + h_0(x),$$

where $h(x)$ is a non-increasing continuous function on $(0, \infty)$ which may diverge at the origin, and $h_0(x)$ is an odd function which is non-increasing and continuous in $\mathbf{R} - \{0\}$.

Let $\{\zeta, \varphi\}$ be the solution of a Skorokhod problem

(7.46) $$\zeta(t, w) = w(t) \vee 1 + (w(t) - w(0)) + ct + \varphi(t, w),$$

where $c \geq h(1) + h_0(1)$. The solution $\{\zeta, \varphi\}$ can be given explicitly by the formulae (7.26) and (7.27). We call $\zeta(t, w)$ a *dominating path* for the solution of (7.44), because

(7.47) $$0 \leq \xi(t, w) \leq \zeta(t, w), \qquad 0 \leq \eta(t, z) \leq \zeta(t, z),$$

which follows from Theorem 7.4, since $c \geq h(x + y) + h_0(x)$ for $x \geq 1$ and $y \geq 0$.

We require, in addition, integrability conditions on $\mu^{(+)}$, $\mu^{(-)}$, h, and h_0

(7.48) $$\int_\Omega |w(t)|\mu(dw) < \infty, \qquad \mu = \mu^{(+)}, \mu^{(-)},$$

---

[9] This is a special case of Tanaka's equations in convex regions. Cf. Tanaka (1979)

$$\int_{\Omega \times \Omega} h^-(\zeta(s,w) + \zeta(s,z))\mu^{(+)}(dw)\mu^{(-)}(dz) \le K_T < \infty,$$

$$\int_{\Omega} h_0^-(\zeta(s,w))\mu(dw) \le K_T < \infty, \qquad \mu = \mu^{(+)}, \mu^{(-)}$$

on each finite time interval $[0, T]$, where h^- denotes the negative part of h.

Theorem 7.5. (Nagasawa-Tanaka (1986)) *Assume conditions (7.45) and (7.48). Then there exists a unique solution* $\{(\xi, \varphi), (\eta, \psi)\}$, ξ, $\eta \in \Omega^+$, $\varphi \in \Phi_\xi$, $\psi \in \Phi_\eta$, *of the system of equation (7.44).*

Proof. Let us define

$$a^{(+)}(s,x) = -\int_{\Omega} b(x, \zeta(s,z))\mu^{(-)}(dz),$$

$$a^{(-)}(s,x) = -\int_{\Omega} b(x, \zeta(s,w))\mu^{(+)}(dw),$$

and let $\{(\xi^{(k)}, \varphi^{(k)}), (\eta^{(k)}, \psi^{(k)})\}$, $k \ge 0$, be the solution of the systems of equations

$$\xi^{(0)}(t,w) = w(t) + \int_0^t ds \, a^{(+)}(s, \xi^{(0)}(s,w)) + \varphi^{(0)}(t,w),$$

(7.49)

$$\eta^{(0)}(t,z) = z(t) + \int_0^t ds \, a^{(-)}(s, \eta^{(0)}(s,z)) + \psi^{(0)}(t,z),$$

and for $k \ge 1$

$$\xi^{(k)}(t,w) = w(t) + \int_{[0,t] \times \Omega} b(\xi^{(k)}(s,w), \eta^{(k-1)}(s,\tilde{z}))ds \, \mu^{(-)}(d\tilde{z}) + \varphi^{(k)}(t,w),$$

(7.50)

$$\eta^{(k)}(t,z) = z(t) + \int_{[0,t] \times \Omega} b(\eta^{(k)}(s,z), \xi^{(k-1)}(s,\tilde{w}))ds \, \mu^{(+)}(d\tilde{w}) + \psi^{(k)}(t,z).$$

Theorem 7.4 implies the existence of unique solutions to the equations (7.49) and (7.50), and moreover that the solutions satisfy a sequence of inequalities

(7.51) $0 \leq \xi^{(0)}(t, w) \leq \xi^{(k)}(t, w) \leq \zeta(t, w),$ for all $k \geq 1,$

and

$$0 \leq \xi^{(0)} \leq \xi^{(2)} \leq \xi^{(4)} \leq \cdots \leq \xi^{(5)} \leq \xi^{(3)} \leq \xi^{(1)} \leq \zeta,$$

(7.52)

$$\varphi^{(1)} \leq \varphi^{(3)} \leq \varphi^{(5)} \leq \cdots \leq \varphi^{(4)} \leq \varphi^{(2)} \leq \varphi^{(0)}.$$

The same inequalities hold for $(\eta^{(k)}, \psi^{(k)})$.

Therefore, there exist monotone limits

$$\xi_-(t, w) = \lim_{k \to \infty} \xi^{(2k)}(t, w), \qquad \varphi_-(t, w) = \lim_{k \to \infty} \varphi^{(2k)}(t, w),$$

$$\overline{\xi}(t, w) = \lim_{k \to \infty} \xi^{(2k+1)}(t, w), \qquad \overline{\varphi}(t, w) = \lim_{k \to \infty} \varphi^{(2k+1)}(t, w),$$

and correspondingly $\eta_-, \psi_-, \overline{\eta}, \overline{\psi}$. They satisfy

$$\xi_-(t, w) \leq \overline{\xi}(t, w), \qquad \varphi_-(t, w) \geq \overline{\varphi}(t, w),$$

(7.53)

$$0 \leq \mu^{(+)}[\xi_-(t)] \leq \mu^{(+)}[\overline{\xi}(t)] \leq \mu^{(+)}[\zeta(t)] < \infty,$$

and

$$\overline{\xi}(t, w) = w(t) + \int_{[0, t] \times \Omega} b(\overline{\xi}(s, w), \eta_-(s, \tilde{z})) ds \, \mu^{(-)}(d\tilde{z}) + \overline{\varphi}(t, w),$$

(7.54)

$$\xi_-(t, w) = w(t) + \int_{[0, t] \times \Omega} b(\xi_-(s, w), \overline{\eta}(s, \tilde{z})) ds \, \mu^{(-)}(d\tilde{z}) + \varphi_-(t, w).$$

Moreover, the corresponding ones hold for $\eta_-, \psi_-, \overline{\eta}, \overline{\psi}$. To get (7.54) from (7.50) we apply

$$b(\xi^{(2k+1)}, \eta_-) \leq b(\xi^{(2k+1)}, \eta^{(2k)}) \leq b(\overline{\xi}, \eta^{(2k)}),$$

$$b(\xi^{(2k+1)}, \eta_-) \uparrow, \quad and \quad b(\overline{\xi}, \eta^{(2k)}) \downarrow, \quad as \ k \to \infty,$$

for the first equation. The same arguments can be applied to the second equation of (7.54).

As a matter of fact we can show

(7.55) $\xi_- = \bar{\xi}, \quad \varphi_- = \bar{\varphi}, \quad \eta_- = \bar{\eta}, \quad \psi_- = \bar{\psi},$

and hence we get a solution to system (7.44).

In fact, because of the integrability conditions (7.48), $\{\xi_-, \bar{\xi}, \varphi_-, \bar{\varphi}\}$ and $\{\eta_-, \bar{\eta}, \psi_-, \bar{\psi}\}$ are integrable with respect to $\mu^{(+)}$ and $\mu^{(-)}$, respectively. Therefore, we have

$$0 \leq \mu^{(+)}[\bar{\xi}(t) - \xi_-(t)] + \mu^{(-)}[\bar{\eta}(t) - \eta_-(t)]$$

$$\leq \int_{[0,\,t]\times\Omega} \{h_0(\bar{\xi}(s,w)) - h_0(\xi_-(s,w))\}\mu^{(+)}(dw) + \mu^{(+)}[\bar{\varphi}(t) - \varphi_-(t)]$$

$$+ \int_{[0,\,t]\times\Omega} \{h_0(\bar{\eta}(s,z)) - h_0(\eta_-(s,w))\}\mu^{(-)}(dz) + \mu^{(-)}[\bar{\psi}(t) - \psi_-(t)]$$

$$\leq \mu^{(+)}[\bar{\varphi}(t) - \varphi_-(t)] + \mu^{(-)}[\bar{\psi}(t) - \psi_-(t)] < \infty.$$

This inequality combined with $\bar{\varphi}(t) - \varphi_-(t) \leq 0$ and $\bar{\psi}(t) - \psi_-(t) \leq 0$ implies (7.55) almost everywhere with respect to $\mu^{(+)}$ (or $\mu^{(-)}$). Now define

$$c(s,x) = \int_\Omega b(x, \bar{\eta}(s,z))\mu(dz),$$

and consider a Skorokhod problem on $[0, \infty)$

$$\xi(t) = w(t) + \int_0^t c(s, \xi(s))ds + \varphi(t).$$

Theorem 7.4 claims that the solution of the equation is uniquely determined and hence we have

$$\xi(t, w) = \bar{\xi}(t, w) = \xi_-(t, w),$$

$$\varphi(t, w) = \bar{\varphi}(t, w) = \varphi_-(t, w), \text{ for all } w \in \Omega.$$

The same thing can be shown for $\{\eta, \psi\}$. Thus we have proven (7.55). The uniqueness of solutions can be shown also with the help of the same monotone arguments. This completes the proof.

We formulate a limit theorem of a Skorokhod's problem for systems of equations, which will be applied to a proof of Theorem 7.1 on the propagation of chaos of coloured particles.

Theorem 7.6. (Nagasawa-Tanaka (1986))   *Besides (7.45) assume that* $h(x)$ *is bounded below. Let* $\mu^{(+)}$, $\mu^{(-)}$, $\mu_n^{(+)}$, $\mu_n^{(-)}$, $n = 1, 2, \ldots$, *b e probability measures on* Ω *satisfying (7.48). Let* $\{\xi, \eta\}$ *and* $\{\xi_n, \eta_n\}$ *be the solutions of (7.44) with* $\{\mu^{(+)}, \mu^{(-)}\}$ *and* $\{\mu_n^{(+)}, \mu_n^{(-)}\}$, *respectively. If* $\mu_n^{(+)}$ *and* $\mu_n^{(-)}$ *converge weakly to* $\mu^{(+)}$ *and* $\mu^{(-)}$, *respectively, then*

(7.56)
$$\lim_{n \to \infty} \xi_n(t, w) = \xi(t, w),$$

$$\lim_{n \to \infty} \eta_n(t, z) = \eta(t, z),$$

uniformly in $(t, w, z) \in [0, T] \times K_1 \times K_2$, *for any compact subsets* K_1 *and* K_2 *of* Ω *and* $T > 0$.[10]

For a proof we prepare some lemmas.

Lemma 7.8. *Let* $b(t, x)$ *satisfy condition (7.28) and define*

(7.57) $b_a(t, x) = b(t, x \vee a)$,   *for* $(t, x) \in [0, \infty) \times (0, \infty)$,

for $a > 0$. *Let* $\{\xi, \varphi\}$ *and* $\{\xi_a, \varphi_a\}$ *be the solutions of the equations (7.25) with* $b(t, x)$ *and* $b_a(t, x)$, *respectively. Then*

$$\xi_a(t) \uparrow \xi(t), \quad as \quad a \downarrow 0.$$

Proof. The comparison assertion of Theorem 7.4 implies $\xi_a(t) \uparrow$ and $\varphi_a(t) \downarrow$ as as $a \downarrow 0$. Therefore, we have

$$\int_0^t ds\, b_a(s, \xi_a(s)) = \xi_a(t) - w(t) - \varphi_a(t) \uparrow \quad as \quad a \downarrow 0,$$

and hence the limit

$$\lim_{a \downarrow 0} \int_0^t ds\, b_a(s, \xi_a(s))$$

exists. Denote

---

[10] We define on Ω a metric of the uniform convergence on each finite interval

$$\widetilde{\xi}(t) = \lim_{a \downarrow 0} \xi_a(t).$$

Since $b_a(s, \xi_a(s))1_{(\varepsilon, \infty)}(\xi_a(s))$ decreases to $b(s, \widetilde{\xi}(s))1_{(\varepsilon, \infty)}(\widetilde{\xi}(s))$ as $a \downarrow 0$, for $\varepsilon > 0$, we have

$$\lim_{a \downarrow 0} \int_0^t ds b_a(s, \xi_a(s))$$

$$= \int_0^t ds b(s, \widetilde{\xi}(s))1_{(\varepsilon, \infty)}(\widetilde{\xi}(s)) + \lim_{a \downarrow 0} \int_0^t ds b_a(s, \xi_a(s))1_{[0, \varepsilon]}(\xi_a(s))$$

$$= \int_0^t ds b(s, \widetilde{\xi}(s))1_{(0, \infty)}(\widetilde{\xi}(s)) + \lim_{\varepsilon \downarrow 0} \lim_{a \downarrow 0} \int_0^t ds b_a(s, \xi_a(s))1_{[0, \varepsilon]}(\xi_a(s)).$$

Therefore, $\widetilde{\xi}(t)$ satisfies

(7.58) $$\widetilde{\xi}(t) = w(t) - ct + \int_0^t ds\{b(s, \widetilde{\xi}(s)) + c\} + \widetilde{\varphi}(t),$$

where a constant c is chosen to be $b(t, 1) + c > 0$, for $\forall\, t \in [0, T]$, and $\widetilde{\varphi}$ is defined by

$$\widetilde{\varphi}(t) = \lim_{\varepsilon \downarrow 0} \lim_{a \downarrow 0} \int_0^t ds\{b_a(s, \xi_a(s)) + c\}1_{[0, \varepsilon]}(\xi_a(s)) + \lim_{a \downarrow 0} \varphi_a(t).$$

It is easy to see that $\widetilde{\xi}(t)$ is continuous in t (cf. the preceding section) and hence $\widetilde{\varphi}(t)$ too. Moreover, $\widetilde{\varphi}(t)$ is monotone non-decreasing and constant on each open interval in which $\widetilde{\xi}(t) > 0$. Notice that equations (7.58) and (7.25) are the same, and the uniqueness of solutions holds. Therefore, $\{\widetilde{\xi}, \widetilde{\varphi}\}$ must coincides with $\{\xi, \varphi\}$. This completes the proof.

Lemma 7.9. *Under condition* (7.28) *the solution* $\xi(t, w)$ *of equation* (7.25) *is continuous in* $(t, w) \in [0, T] \times \Omega$, *for any* $T > 0$.

Proof. If $b(t, x)$ is Lipschitz continuous, then as shown in the proof of Lemma 7.3, $\xi(t, w)$ is continuous in $(t, w) \in [0, T] \times \Omega$ as a uniform limit of $\xi^{(k)}(t, w)$ which is continuous in $(t, w) \in [0, T] \times \Omega$ by the definition.

Let us define a dominating path ζ as the solution of a Skorokhod problem on $[1,\infty)$

$$\zeta(t,w) = w(0) \vee 1 + (w(t) - w(0)) + ct + \psi(t,w),$$

with $c \geq \sup_{t \in [0,T]} b(t,1)$. Let K be an arbitrary compact subset of Ω and denote $M = \sup_{t \in [0,T]} \zeta(t,w) < \infty$. The coefficient $b_a(t,x) = b(t,x \vee a)$ in the preceding Lemma can be approximated from below on $[0,M+1]$ by Lipschitz continuous ones. Therefore $\xi_a(t,w)$ is lower semi-continuous in $(t,w) \in [0,T] \times K$ as the increasing limit of continuous functions. By the same argument we can show that $\xi^a(t,w)$ in Lemma 7.5 is upper semi-continuous in $(t,w) \in [0,T] \times K$ as the decreasing limit of continuous functions. By Lemmas 7.8 and 7.6 we have

$$\xi_a(t,w) \uparrow \xi(t,w), \quad and \quad \xi^a(t,w) \downarrow \xi(t,w), \quad as \quad a \downarrow 0.$$

Therefore, $\xi(t,w)$ is lower and upper semi-continuous in $(t,w) \in [0,T] \times K$ and hence continuous. This completes the proof.

Lemma 7.10. *Besides condition (7.28) require*

$$\lim_{n \to \infty} b_n(t,x) = b(t,x), \quad uniformly \ in \ (t,x) \in [0,T] \times [a, \tfrac{1}{a}],$$

(7.59)

$$\sup_{n \geq 1} \max_{t \in [0,T]} b_n(t,a) < \infty,$$

for $\forall a > 0$ *and* $\forall T > 0$. *Then for any compact subset K of Ω*

(7.60) $\quad \lim_{n \to \infty} \xi_n(t,w) = \xi(t,w), \quad uniformly \ in \ (t,w) \in [0,T] \times K,$

where $\xi(t,w)$ and $\xi_n(t,w)$ denote the solutions of (7.25) with $b(t,x)$ and $b_n(t,x)$, respectively.

Proof. Let us first prove the lemma assuming that $b(t,x)$ is bounded above and $b_n(t,x)$ converges uniformly on $[0,T] \times [0,a^{-1}]$.

With the help of Lemma 7.2 we have

$$|\xi_n(t) - \xi(t)|^2 \leq 2 \int_0^t \{\xi_n(s) - \xi(s)\}\{b_n(s,\xi_n(s)) - b_n(s,\xi(s))\}ds +$$

$$+ 2 \int_0^t \{\xi_n(s) - \xi(s)\} \{b_n(s, \xi(s)) - b(s, \xi(s))\} ds,$$

where the first integral is non-positive and the second one is dominated by

$$\int_0^t |\xi_n(s) - \xi(s)|^2 ds + \int_0^t |b_n(s, \xi(s)) - b(s, \xi(s))|^2 ds,$$

by Schwarz's inequality. Therefore, with the help of Gronwall's lemma,

$$(7.61) \quad |\xi_n(t, w) - \xi(t, w)|^2 \leq e^T \int_0^T |b_n(s, \xi(s, w)) - b(s, \xi(s, w))|^2 ds.$$

Since, for any compact subset K of Ω,

$$\sup_{w \in K} \sup_{s \in [0, T]} \xi(s, w) \leq \sup_{w \in K} \sup_{s \in [0, T]} \zeta(s, w) < \infty,$$

the integral of the right-hand side of (7.61) is bounded and converges to zero uniformly in $(t, w) \in [0, T] \times K$ as $n \to \infty$ because of (7.59). Hence, $\xi_n(t, w)$ converges uniformly to $\xi(t, w)$ on $[0, T] \times K$.

Secondly we apply what we have shown above to the modified coefficients defined in Lemmas 7.8 and 7.5. Then we get

$$(7.62) \qquad \begin{aligned} (\xi_n)_a(t, w) &\to \xi_a(t, w), \\[1mm] (\xi_n)^a(t, w) &\to \xi^a(t, w), \quad \text{as } n \to \infty, \end{aligned}$$

uniformly on $[0, T] \times K$. Since $\xi_a(t, w) \uparrow \xi(t, w)$ and $\xi^a(t, w) \downarrow \xi(t, w)$, there exists $a > 0$ such that, by Dini's theorem,

$$0 \leq \xi^a(t, w) - \xi_a(t, w) < \varepsilon, \quad \textit{uniformly in } (t, w) \in [0, T] \times K.$$

For this $a > 0$ we have, because of (7.62),

$$|(\xi_n)_a(t, w) - \xi_a(t, w)| < \varepsilon,$$

$$|(\xi_n)^a(t, w) - \xi^a(t, w)| < \varepsilon,$$

uniformly on $[0, T] \times K$, for sufficiently large n. Therefore we have

$$|\xi(t, w) - \xi_n(t, w)| < 3\varepsilon, \quad \text{uniformly in} \quad (t, w) \in [0, T] \times K,$$

where use is made of the inequality $(\xi_n)_a \leq \xi_n \leq (\xi_n)^a$. This completes the proof.

Proof of Theorem 7.6. Let $\{\xi^{(k)}, \eta^{(k)}\}$ be defined in (7.49) and (7.50), and define $\{\xi_n^{(k)}, \eta_n^{(k)}\}$ by the same equations with $\mu_n^{(-)}$ and $\mu_n^{(+)}$. We prove first of all $\xi_n^{(k)}(t, w)$ converges to $\xi^{(k)}(t, w)$ as $n \to \infty$ uniformly on $[0, T] \times K$, where K is a compact subset of Ω (we shall state all claims only for ξ's, since the role of ξ's and η's can be interchanged). We shall show this by induction in k.

First of all $\xi_n^{(0)}(t, w)$ converges to $\xi^{(0)}(t, w)$ as $n \to \infty$ uniformly on $[0, T] \times K$, since $a^{(+)}(s, x)$ and $a_n^{(+)}(s, x)$ satisfy the conditions of Lemma 7.10. Assume $\xi_n^{(k)}(t, w)$ converges to $\xi^{(k)}(t, w)$ as $n \to \infty$ uniformly on $[0, T] \times K$ and define

$$b^{(k+1)}(s, x) = \int_\Omega h(x + \eta^{(k)}(s, z)) \mu^{(-)}(dz) + h_0(x),$$

$$b_n^{(k+1)}(s, x) = \int_\Omega h(x + \eta_n^{(k)}(s, z)) \mu_n^{(-)}(dz) + h_0(x).$$

We will prove that $b^{(k+1)}(s, x)$ and $b_n^{(k+1)}(s, x)$ satisfy the conditions of Lemma 7.10, which implies that $\xi_n^{(k+1)}(s, w)$ converges to $\xi^{(k+1)}(s, w)$, as $n \to \infty$, uniformly on $[0, T] \times K$.

Since condition (7.28) is clear, it is enough to prove condition (7.59). Because of the definition we have

(7.63) $\quad |b_n^{(k+1)}(s, x) - b^{(k+1)}(s, x)|$

$$\leq \int_\Omega |h(x + \eta_n^{(k)}(s, z)) - h(x + \eta^{(k)}(s, z))| \mu_n^{(-)}(dz)$$

$$+ |\int_\Omega h(x + \eta^{(k)}(s, z))\{ \mu_n^{(-)}(dz) - \mu^{(-)}(dz)\}|$$

$$= \mathrm{I} + \mathrm{II}.$$

For $a > 0$ we can find a compact subset K_0 of Ω such that

$$\mu_n^{(-)}(K_0^c) \leq \frac{\varepsilon}{2h(a)},$$

since $\mu_n^{(-)}$ converges to $\mu^{(-)}$ weakly by the assumption. Therefore

$$(7.64) \qquad \mathrm{I} \leq \int_{K_0} |h(x + \eta_n^{(k)}(s,z)) - h(x + \eta^{(k)}(s,z))|\,\mu_n^{(-)}(dz) + \varepsilon,$$

where the integral is also smaller than ε for sufficiently large n uniformly in $(s,x) \in [0,T] \times [a, a^{-1}]$, since the integrand converges to zero uniformly in $(s,x,z) \in [0,T] \times [a, a^{-1}] \times K_0$; and for the second integral II

$$(7.65) \qquad \mathrm{II} \leq \left| \int_{K_0} h(x + \eta^{(k)}(s,z))\{ \mu_n^{(-)}(dz) - \mu^{(-)}(dz)\} \right| + \varepsilon,$$

where the integral is also smaller than ε for sufficiently large n, since $h(x + \eta^{(k)}(s,z))$ is uniformly continuous in $(s,x,z) \in [0,T] \times [a, a^{-1}] \times K_0$ and $\mu_n^{(-)}$ converges to $\mu^{(-)}$ weakly by the assumption.[11] Thus we have

$$\sup_{(s,x) \in [0,T] \times [a, a^{-1}]} |b_n^{(k+1)}(s,x) - b^{(k+1)}(s,x)| \leq 4\varepsilon,$$

for sufficiently large n. Hence $b^{(k+1)}(s,x)$ and $b_n^{(k+1)}(s,x)$ satisfy condition (7.59), the second condition in it being trivial.

Now, the proof can be completed as follows. Because of Theorem 7.5, especially the inequalities (7.52), Lemma 7.9, and Dini's theorem we can find k_0 for any $\varepsilon > 0$ such that

$$(7.66) \qquad 0 \leq \xi^{(2k+1)}(t,w) - \xi^{(2k)}(t,w) \leq \varepsilon, \quad \text{for } k \geq k_0,$$

uniformly in $(t,w) \in [0,T] \times K$. On the other hand, as we have shown, there exists n_0 such that for $n \geq n_0$

$$(7.67) \qquad |\xi_n^{(2k)}(t,w) - \xi^{(2k)}(t,w)| \leq \varepsilon,$$

$$|\xi_n^{(2k+1)}(t,w) - \xi^{(2k+1)}(t,w)| \leq \varepsilon,$$

uniformly in $(t,w) \in [0,T] \times K$.

[11] Cf. Theorem 6.8 (p.51) of Parthasarathy (1967)

Therefore, combining (7.66) and (7.67) with

$$\xi^{(2k)}(t, w) \le \xi(t, w) \le \xi^{(2k+1)}(t, w),$$

$$\xi_n^{(2k)}(t, w) \le \xi_n(t, w) \le \xi_n^{(2k+1)}(t, w),$$

we get

$$\sup_{(t, w) \in [0, T] \times K} |\xi(t, w) - \xi_n(t, w)| \le 3\varepsilon, \quad for \;\; n \ge n_0.$$

Thus $\xi_n(t, w)$ converges to $\xi(t, w)$, as $n \to \infty$, uniformly on $[0, T] \times K$ for any $T > 0$ and compact subset K of Ω. This completes the proof of Theorem 7.6.

7.8. A Proof of Theorem 7.1

Theorem 7.6 of the preceding section can be applied in proving Theorem 7.1. For this we shall construct a sequence of solutions for system (7.6) of interacting diffusion processes and show that they satisfy the requirements of Theorem 7.6.

With a pair of probability measure $\{\nu^{(+)}, \nu^{(-)}\}$ on $\mathbf{R}^+$ let us define probability measures on Ω

$$(7.68) \qquad \mu^{(+)} = \int_{\mathbf{R}^+} \nu^{(+)}(dx) P_x, \qquad \mu^{(-)} = \int_{\mathbf{R}^+} \nu^{(-)}(dx) P_x,$$

where $\{P_x, x \in \mathbf{R}\}$ is a one-dimensional diffusion process defined on Ω, and requires for $\{\nu^{(+)}, \nu^{(-)}\}$ the following conditions

$$(7.69) \qquad \int_{\mathbf{R}^+} x \nu^{(+)}(dx) < \infty, \qquad \int_{\mathbf{R}^+} x \nu^{(-)}(dx) < \infty,$$

so that $\mu^{(+)}$ and $\mu^{(-)}$ satisfy the first of conditions (7.48), the second one of which is automatically satisfied, since $h(x)$ is bounded below. We assume the third one. A sufficient condition for this is given as growth conditions:

$$(7.70) \quad h_0(x) \le c(1 + |x|^\alpha), \;\; \textit{for some non-negative constants } c \textit{ and } \alpha \,,$$

(Cf. Proposition 1 of Nagasawa-Tanaka (1985)).

Then Theorem 7.5 claims that there exists a unique solution $\{\xi(t, w),$ $\eta(t, z), \varphi(t, w), \psi(t, z)\}$ of system (7.44) for $(w, z) \in \Omega \times \Omega$. Set

$$\Omega^2 = \Omega \times \Omega, \quad \mu = \mu^{(+)} \otimes \mu^{(-)},$$

and let us define a stochastic process (X, Y) of two particles on the probability space (Ω^2, μ) by

$$(X(t), Y(t)) = (\xi(t), -\eta(t)).$$

Therefore, $X(t)$ moves on $[0, \infty)$, while $Y(t)$ stays on $(-\infty, 0]$. Because of (7.45) it is easy to see that the process (X, Y) satisfies

$$X(t) = X(0) + B^+(t) + \int_0^t ds\{ \int_{(-\infty, 0]} h(X(s) - y)u_Y(s, dy) + h_0(X(s))\} + \varphi(t),$$
(7.71)

$$Y(t) = Y(0) + B^-(t) + \int_0^t ds\{ -\int_{[0, \infty)} h(x - Y(s))u_X(s, dy) + h_0(Y(s))\} - \psi(t),$$

where u_X and u_Y denote the distributions of $X(t)$ and $Y(t)$, respectively, and $B^+(t)$ and $B^-(t)$ are independent one-dimensional Brownian motions, for simplicity assuming that the process P_x stands for a Brownian motion.

Let us set

$$\widetilde{\Omega} = \Omega^2 \times \Omega^2 \times \cdots,$$
(7.72)
$$\mathbf{P} = \mu \otimes \mu \otimes \cdots,$$

where $\mu = \mu^{(+)} \otimes \mu^{(-)}$ and define

(7.73) $\mathbf{X}(t, \widetilde{\omega}) = ((X_1(t, \widetilde{\omega}), Y_1(t, \widetilde{\omega})), (X_2(t, \widetilde{\omega}), Y_2(t, \widetilde{\omega})), \dots),$

for $\widetilde{\omega} = ((w_1, z_1), (w_2, z_2), \dots) \in \widetilde{\Omega}$, where

$$X_i(t, \widetilde{\omega}) = \xi(t, w_i),$$
(7.74)
$$Y_i(t, \widetilde{\omega}) = -\eta(t, z_i), \quad i = 1, 2, \dots .$$

We define the empirical distributions of $(w_1, \dots, w_n)$ and $(z_1, \dots, z_n)$ by

$$(7.75) \qquad \mu_n^{(+)}(\widetilde{\omega}, \cdot\,) = \frac{1}{n} \sum_{i=1}^{n} \delta_{w_i}(\cdot\,), \quad \mu_n^{(-)}(\widetilde{\omega}, \cdot\,) = \frac{1}{n} \sum_{i=1}^{n} \delta_{z_i}(\cdot\,),$$

where $\widetilde{\omega} = ((w_1, z_1), (w_2, z_2), \ldots) \in \widetilde{\Omega}$.

Since the random variables (X_i, Y_i), $i = 1, 2, \ldots$, are independent by definition, the sequences $\{\mu_n^{(+)}(\widetilde{\omega}, \cdot\,)\}$ and $\{\mu_n^{(-)}(\widetilde{\omega}, \cdot\,)\}$ converge weakly to the probability measures $\mu^{(+)}$ and $\mu^{(-)}$, respectively, for $\mathbf{P}$-a.e. $\widetilde{\omega} \in \widetilde{\Omega}$ by the strong law of large numbers.

By Theorem 7.5 there exists for each $(w, z) \in \Omega \times \Omega$ a unique solution $\{\xi_n(t, w, \widetilde{\omega}), \eta_n(t, z, \widetilde{\omega})\}$ of system (7.44) with $\mu_n^{(-)}(\widetilde{\omega}, \cdot\,)$ and $\mu_n^{(+)}(\widetilde{\omega}, \cdot\,)$ in place of $\mu^{(-)}$ and $\mu^{(+)}$, respectively. Now define

$$X_i^{(n)}(t, \widetilde{\omega}) = \xi_n(t, w_i, \widetilde{\omega}),$$

(7.76)

$$Y_i^{(n)}(t, \widetilde{\omega}) = -\eta_n(t, z_i, \widetilde{\omega}),$$

and set

$$(7.77) \qquad \mathbf{X}^{(n)}(t) = ((X_1^{(n)}(t), Y_1^{(n)}(t)), \ldots, (X_n^{(n)}(t), Y_n^{(n)}(t))).$$

Then, it is easy to see that $\{(X_i^{(n)}(t), Y_i^{(n)}(t)): i = 1, 2, \ldots, n\}$ is a system of coloured particles described by (7.6), i.e., for $i, j = 1, 2, \ldots, n$,

$$X_i(t) = X_i(0) + B_i^+(t) + \int_0^t ds \{\frac{1}{n} \sum_{j=1}^{n} h(X_i(s) - Y_j(s)) + h_0(X_i(s))\} + \xi_i(t),$$

$$Y_j(t) = Y_j(0) + B_j^-(t) + \int_0^t ds \{-\frac{1}{n} \sum_{i=1}^{n} h(X_i(s) - Y_j(s)) + h_0(Y_j(s))\} - \eta_j(t),$$

where $\{(B_i^+(t), B_j^-(t)): i, j = 1, 2, \ldots, n\}$ is a family of mutually independent Brownian motions, $X_i(0)$ (resp. $Y_j(0)$) has a common distribution on $[0, \infty)$ (resp. on $(-\infty, 0]$), $h(x)$ is a non-increasing continuous function on $(0, \infty)$, which is bounded below but may diverge at the origin like the $b(x)$ given in (7.5), and $h_0(x)$ is an odd function which is continuous and non-increasing in $\mathbf{R} - \{0\}$ satisfying a growth condition, e.g. (7.70).

Theorem 7.6 implies that for **P**-a.e. $\widetilde{\omega} \in \widetilde{\Omega}$

(7.78) $(\xi_n(t, w, \widetilde{\omega}), \eta_n(t, z, \widetilde{\omega})) \rightarrow (\xi(t, w), \eta(t, z)),$ *as* $n \rightarrow \infty,$

> *uniformly in* $(t, w, z) \in [0, T] \times K_1 \times K_2$, *for* $\forall T > 0$ *and any compact subsets* K_1 *and* K_2 *of* Ω.

Thus we have shown

Theorem 7.7. (Nagasawa-Tanaka (1986)) *Assume conditions* (7.45), (7.69) *and the third one of* (7.48). *Then, the propagation of chaos holds*: *for arbitrary but fixed* $m \leq n$ *and for* **P**-a.e. $\widetilde{\omega} \in \widetilde{\Omega},$

$(X_1^{(n)}(t, \widetilde{\omega}), Y_1^{(n)}(t, \widetilde{\omega}), \dots, X_m^{(n)}(t, \widetilde{\omega}), Y_m^{(n)}(t, \widetilde{\omega}))$ *converges uniformly in* $t \in [0, T]$ *to* $(X_1(t, \widetilde{\omega}), Y_1(t, \widetilde{\omega}), \dots, X_m(t, \widetilde{\omega}), Y_m(t, \widetilde{\omega}))$ *as* $n \rightarrow \infty$.

Namely, the diffusion process on $(\mathbf{R}^2)^n$ *described by the system of equations* (7.6) *converges in law to the infinite direct product of the diffusion process described by the system of equations* (7.71).

This is the substance of Theorem 7.1.

7.9. Schrödinger Equations with Singular Potentials

We begin with the harmonic oscillator in one-dimension

(7.79) $$-\frac{1}{2}\frac{d^2\psi}{dx^2}(x) + \frac{1}{2}x^2\,\psi(x) = \lambda\,\psi(x),$$

with the boundary condition $\lim_{x \rightarrow \pm\infty} \psi(x) = 0$. Then it is easy to see that solutions must be of the form $H(x)\,e^{-\frac{1}{2}x^2}$.

The ground state is given by

$$\psi_0(x) = e^{-\frac{1}{2}x^2},$$

which corresponds to the smallest eigenvalue $\lambda_0 = 1/2$. The diffusion process with the distribution density $|\psi_0(x)|^2$ has the drift coefficient $b_0(x) = d(\log \psi_0(x))/dx = -x$, which is determined by the duality relation (or time reversal). In this case the origin is accessible, and Theorem 7.4 provides the solution $\{\xi(t), \varphi(t)\}$ of equation (7.25).

The first excited state corresponding to the eigenvalue $\lambda_1 = 3/2$ is given by

$$\psi_1(x) = x e^{-\frac{1}{2}x^2}, \quad for \quad x \in \mathbf{R}^1.$$

It vanishes at the origin $N = \{x: \psi_1(x) = 0\} = \{0\}$, and hence the diffusion process with the distribution density $|\psi_1(x)|^2$ has singular drift

$$b_1(x) = \frac{d \log \psi_1(x)}{dx} = \frac{1}{x} - x.$$

The creation and killing induced by $\psi_1(x)$ is

$$c(x) = -L\psi_1(x)/\psi_1(x) = 1 - \frac{1}{2}x^2,$$

which satisfies integrability condition (6.4), and hence the origin is inaccessible by Theorem 6.1.[12] Therefore, we can consider equation (7.25) with singular drift $b(x) = 1/x - x$ on the positive half line $(0, \infty)$. There exists a unique solution $\{\xi(t), \varphi(t)\}$ of equation (7.25) by Theorem 7.4. Since the origin is inaccessible, $\varphi(t) \equiv 0$ for almost all Brownian path $\omega(t)$ in equation (7.25).

We are interested in the influence of a severely singular perturbation on the eigenvalues of the harmonic oscillator.

(i) As the first example of Schrödinger equations with a severely singular perturbation we consider[13]

$$-\frac{1}{2}\frac{d^2\psi}{dx^2}(x) + \left\{\frac{1}{2}x^2 + \frac{1}{2}\frac{\gamma^2}{x^{2(1+\gamma)}} - \left(\frac{\gamma(1+\gamma)}{2}\frac{1}{|x|^2+\gamma} + \frac{\gamma}{|x|^\gamma}\right)\right\}\psi(x) = \lambda\,\psi(x),$$

in $\mathbf{R}^1$, where $\gamma > 0$ and we put the boundary condition $\lim_{x \to \pm\infty}\psi(x) = 0$. Through substitution it is easy to see that

---

[12] The inaccessibility follows also from Feller's test, cf. Section 2.11

[13] For further examples, cf. Ezawa-Klauder-Shepp (1975)

$$\psi(x) = e^{-(\frac{1}{|x|^\gamma} + \frac{1}{2}x^2)}, \quad for \quad x \in \mathbf{R}^1,$$

solves the equation with $\lambda = 1/2$.

It is interesting to observe that the eigenvalue $\lambda = 1/2$ does not depend on $\gamma > 0$, which is caused by the well-balanced cancellation of the positive and negative parts of the additional singular potential.

The function $\psi(x)$ vanishes at the origin and the drift coefficient $b(x)$ $= d(\log \psi(x))/dx$ is given by

$$b(x) = \mathrm{sgn}(x)\frac{\gamma}{|x|^{1+\gamma}} - x,$$

which is singular at the origin. Since the following two functions

$$\psi^+(x) = e^{-(\frac{1}{|x|^\gamma} + \frac{1}{2}x^2)}, \quad for \quad x > 0,$$

$$= 0, \quad\quad\quad\quad for \quad x \le 0,$$

and

$$\psi^-(x) = 0, \quad\quad\quad\quad for \quad x \ge 0,$$

$$= e^{-(\frac{1}{|x|^\gamma} + \frac{1}{2}x^2)}, \quad for \quad x < 0,$$

satisfy the same Schrödinger equation with eigenvalue 1/2, the corresponding state is degenerated, which is caused by the severely singular additional potential, or in other words, the severely singular repulsive drift $b(x) = \mathrm{sgn}(x)\gamma|x|^{-(1+\gamma)} - x$ at the origin. Because of the singular repulsive drift, the origin is inaccessible, and hence $\varphi(t) \equiv 0$ for almost all Brownian paths $\omega(t)$ in equation (7.25).

(ii) As the second example of Schrödinger equations with a severely singular perturbation[14] we consider

(7.80) $$-\frac{1}{2}\frac{d^2\psi}{dx^2}(x) + \left\{ \frac{1}{2}x^2 + V(x) \right\} \psi(x) = \lambda \, \psi(x),$$

with the boundary condition $\lim_{x \to \pm\infty} \psi(x) = 0$, where the additional singular

---

[14] The following results have not been published elsewhere. In connection with the problem cf. Sturm (1992, preprint), and also Baras-Goldstein (1984)

potential is

(7.81) $$V(x) = \frac{\varepsilon}{8}\frac{1}{x^2}, \quad \varepsilon \geq -1.$$

We denote the potential $V(x)$ by $V^{(+)}(x)$ if $\varepsilon \geq 0$, and $V^{(-)}(x)$ if $-1 \leq \varepsilon \leq 0$.

We are interested in the influence of the additional singular potential $V(x)$ on the eigenvalues λ of the one-dimensional harmonic oscillator (7.79). Based on the energy form of the theory of self-adjoint operators, we expect that the additional positive potential $V^{(+)}(x)$ pushes up the eigenvalues of equation (7.79), while the additional negative one $V^{(-)}(x)$ pulls them down.[15]

It is enough to consider equation (7.80) on the positive half-line, since potentials are symmetric.

Eigenvalues and eigenfunctions of equation (7.80) with the positive $V^{(+)}(x)$ (resp. the negative $V^{(-)}(x)$) will be denoted with (+) (resp. (-)).

We look for solutions $\psi \in L^2((0, \infty), dx)$ of the Schrödinger equation (7.80), since $|\psi|^2$ must be an invariant density of a diffusion process. Further conditions such as $\psi \in H^1((0, \infty), dx)$ will not be required. Since solutions should vanish at infinity, we consider functions of the form

(7.82) $$u(x) = x^{(1-A)/2} e^{-x^2/2}.$$

Then, we can easily see that it satisfies

(7.83) $$-\frac{1}{2}\frac{d^2u}{dx^2}(x) + \left\{\frac{1}{2}x^2 + \frac{A^2-1}{8}\frac{1}{x^2}\right\}u(x) = \frac{2-A}{2}u(x).$$

An identification of (7.83) with (7.80) yields

$$A^2 - 1 = \varepsilon, \quad and \quad \lambda = \frac{2-A}{2},$$

from which we get

$$A = \pm\sqrt{1+\varepsilon}.$$

---

[15] For equation (7.80) without the term $1/2x^2$ cf. e.g., Spohn (1991), Baras-Goldstein (1984). Cf. Titchmarsh (1962) for eigenvalue problems

(*Negative* $V^{(-)}(x)$)   Let us first consider $V^{(-)}(x)$ with $-1 \leq \varepsilon \leq 0$. To distinguish this from the case of positive ε, let us set

(7.84) $\kappa = -\varepsilon$ *with* $0 \leq \kappa \leq 1$.

Then, a substitution $A = \sqrt{1 - \kappa}$ in (7.82) shows that

(7.85) $\psi_0^{(-)}(x) = x^{(1 - \sqrt{1 - \kappa})/2} e^{-x^2/2}$, *for* $x \in \mathbf{R}^+$,

solves (7.80) for the first eigenvalue

$$\lambda_0^{(-)} = \frac{2 - \sqrt{1 - \kappa}}{2}.$$

Furthermore, with $A = -\sqrt{1 - \kappa}$, we see that

(7.86) $\psi_1^{(-)}(x) = x^{(1 + \sqrt{1 - \kappa})/2} e^{-x^2/2}$,

solves the same equation for the second eigenvalue

$$\lambda_1^{(-)} = \frac{2 + \sqrt{1 - \kappa}}{2}.$$

It is clear that, if $0 < \kappa < 1$, then

$$\frac{1}{2} < \frac{2 - \sqrt{1 - \kappa}}{2} < 1 < \frac{2 + \sqrt{1 - \kappa}}{2} < \frac{3}{2}.$$

If $\kappa \downarrow 0$, the additional singular potential $V^{(-)}(x)$ vanishes, and

(7.87) $\lambda_0^{(-)} \downarrow \frac{1}{2}$, $\lambda_1^{(-)} \uparrow \frac{3}{2}$,   *as* $\kappa \downarrow 0$.

Here we observe interesting facts:[16]

(7.88)   *The eigenfunctions $\psi_0^{(-)}(x)$ and $\psi_1^{(-)}(x)$ are positive for $x > 0$, but vanish at the origin.*

(7.89)   *The additional negative singular potential $V^{(-)}(x)$ pushes up the eigenvalue $\lambda_0 = \frac{1}{2}$ and pulls down the eigenvalue $\lambda_1 = \frac{3}{2}$.*

---

[16] I would like to thank F. den Hollander for stimulating discussion on this problem

It is surprising that the *negative* singular potential $V^{(-)}(x)$ lifts up the ground state eigenvalue $\lambda_0 = 1/2$ of the one-dimensional harmonic oscillator.

It is clear that $\psi_0^{(-)} \in L^2((0,\infty), dx)$, and it fulfills our requirement. However,

$$\int_0^\infty dx \frac{1}{x^2} \{\psi_0^{(-)}(x)\}^2 = \infty,$$

namely $\psi_0^{(-)} \notin H^1((0,\infty), dx)$, and hence one should abandon $\psi_0^{(-)}$, if one must apply the energy form of the theory of self-adjoint operators.[17] Nonetheless, this function $\psi_0^{(-)}$ is the right one in diffusion theory, as will be seen, and hence we adopt it.

In connection with phenomena (7.89) let us consider the truncated potential $V^{(-)}(x) \vee (-n)$. It pulls down eigenvalues of the one-dimensional harmonic oscillator (7.79). If we let n tend to infinity, then the perturbed ground state eigenvalue disappears to "$-\infty$", and the perturbed first excited state eigenvalue comes down near to the value $\lambda_0 = 1/2$ and becomes the "ground state" eigenvalue $\lambda_0^{(-)} = (2 - \sqrt{1-\kappa})/2$ of equation (7.80) with the additional negative singular potential $V^{(-)}(x) = -8^{-1}\kappa x^{-2}$. If we abandon the eigenvalue $\lambda_0^{(-)} = (2 - \sqrt{1-\kappa})/2$, then the ground state eigenvalue $\lambda_0 = 1/2$ of (7.79) will never be recovered even if we let $\kappa \downarrow 0$.

(*Positive $V^{(+)}(x)$*) Now let us consider equation (7.80) with the positive singular potential $V^{(+)}(x) = 8^{-1}\varepsilon x^{-2}$. A substitution $A = \sqrt{1+\varepsilon}$ in (7.82) shows that

(7.90) $$\psi_0^{(+)}(x) = x^{(1-\sqrt{1+\varepsilon})/2} e^{-x^2/2}, \quad for \ x > 0,$$

solves (7.80) with the first eigenvalue

$$\lambda_0^{(+)} = \frac{2 - \sqrt{1+\varepsilon}}{2}.$$

Secondly, another substitution of $A = -\sqrt{1+\varepsilon}$ in (7.82) yields

(7.91) $$\psi_1^{(+)}(x) = x^{(1+\sqrt{1+\varepsilon})/2} e^{-x^2/2},$$

which solves the same equation with the second eigenvalue

---

[17] See the remark given at the end of this section

$$\lambda_1^{(+)} = \frac{2 + \sqrt{1 + \varepsilon}}{2} .$$

It is clear that the eigenvalues satisfy, if $\varepsilon > 0$,

$$\frac{2 - \sqrt{1 + \varepsilon}}{2} < \frac{1}{2}, \qquad \frac{3}{2} < \frac{2 + \sqrt{1 + \varepsilon}}{2} .$$

If $\varepsilon \downarrow 0$, then the additional singular potential $V^{(+)}(x)$ vanishes, and

(7.92) $\lambda_0^{(+)} \uparrow \frac{1}{2}$, $\lambda_1^{(+)} \downarrow \frac{3}{2}$,   as $\varepsilon \downarrow 0$.

We observe here again interesting fact:

(7.93)   *The eigenfunction $\psi_0^{(+)}(x)$ is positive and diverges at the origin,*
         and another eigenfunction $\psi_1^{(+)}(x)$ is also positive for $x > 0$ but
         vanishes at the origin.

(7.94)   *The additional positive singular potential $V^{(+)}(x)$ pushes down*
         the ground state eigenvalue $\lambda_0 = \frac{1}{2}$ of the harmonic oscillator
         and lifts up the eigenvalue $\lambda_1 = \frac{3}{2}$.

It is again a surprise that the *positive* singular potential $V^{(+)}(x)$ pushes down the ground state eigenvalue $\lambda_0 = 1/2$ of the one-dimensional harmonic oscillator.

It is clear that $\psi_0^{(+)} \in L^2((0, \infty), dx)$ if $\varepsilon < 3$, and hence it satisfies our requirement. However,

$$\int_0^\infty dx \frac{1}{x^2} \{ \psi_0^{(+)}(x) \}^2 = \infty,$$

namely $\psi_0^{(+)} \notin H^1((0, \infty), dx)$. Therefore, the function $\psi_0^{(+)}$ must be carefully investigated in the theory of self-adjoint operators. Nonetheless the function $\psi_0^{(+)}$ is a good one in diffusion theory, as will be seen, and we should not abandon $\psi_0^{(+)}$ even though it is somehow peculiar.

Notice that $\psi_0^{(+)} \notin L^2((0, \infty), dx)$ if $\varepsilon \geq 3$, and it does not satisfy our requirement. As will be seen, $\varepsilon \geq 3$ is prohibited in diffusion theory.

In connection with phenomena (7.94), we can argue as follows: If one applies truncation $V^{(+)}(x) \wedge n$, it pushes upward the eigenvalues of the one-dimensional harmonic oscillator (7.79). If one lets n tend to ∞, then the ground state eigenvalue $\lambda_0 = 1/2$ of the one-dimensional harmonic oscillator is pushed up by the truncated potential, and becomes the "excited state" eigenvalue $\lambda_1^{(+)} = (2 + \sqrt{1 + \varepsilon})/2$, as $n \uparrow \infty$. In fact, the convergence formula in (7.92) shows that it becomes the first excited state eigenvalue $\lambda_1 = 3/2$ of the one-dimensional harmonic oscillator as $\varepsilon \downarrow 0$. However, we cannot find where the ground state eigenvalue $\lambda_0^{(+)} = (2 - \sqrt{1 + \varepsilon})/2$ comes from through the truncation . For this solution see the remark given at the end of this section. If we abandon $\lambda_0^{(+)} = (2 - \sqrt{1 + \varepsilon})/2$, which converges to $1/2$ as $\varepsilon \downarrow 0$, then the eigenvalue $\lambda_0 = 1/2$ will never be recovered.

(*Diffusion theory*) Now we consider diffusion processes with the distribution densities $|\psi_i^{(\pm)}(x)|^2$. The diffusion processes corresponding to the ground states $\psi_0^{(\pm)}(x)$ and the ones corresponding to the first excited states $\psi_1^{(\pm)}(x)$ behave differently near the origin. To see this we look at their drift coefficients.

The drift coefficients $b_i^{(\pm)}(x) = d(\log \psi_i^{(\pm)}(x))/dx$ determined by time reversal (duality) are as follows:

$$b_i^{(\pm)}(x) = \delta_i^{(\pm)} \frac{1}{x} - x,$$

where

$$\delta_0^{(-)} = (1 - \sqrt{1 - \kappa})/2, \qquad \delta_1^{(-)} = (1 + \sqrt{1 - \kappa})/2, \qquad 0 \leq \kappa < 1,$$

$$\delta_0^{(+)} = (1 - \sqrt{1 + \varepsilon})/2, \qquad \delta_1^{(+)} = (1 + \sqrt{1 + \varepsilon})/2, \qquad \varepsilon \geq 0.$$

For the construction of the diffusion processes with the distribution densities $|\psi_i^{(\pm)}(x)|^2$ we consider

(7.95) $$\xi(t) = \omega(t) + \int_0^t b_i^{(\pm)}(s, \xi(s))ds + \varphi(t),$$

and apply Theorem 7.4. Except for the case of $\delta_0^{(-)}$, we can also apply a result of McKean (1960), which claims that the equation (7.95) without the local time $\varphi(t)$ has a unique non-negative solution $\xi(t) \geq 0$, if $\omega(0) > 0$.

Negative eigenvalues:

If $\varepsilon \geq 3$, then $\delta_0^{(+)} \leq -1/2$, and hence the origin is an "exit" boundary point by Feller's test (cf (2.93) in Section 2.11). Therefore, there is no invariant measure and hence the function $\{\psi_0^{(+)}(x)\}^2$ cannot be an invariant measure. Therefore, $\varepsilon \geq 3$, namely, the non-positive eigenvalue $\lambda_0^{(+)} = (2 - \sqrt{1+\varepsilon})/2$ is not allowed.

The ground state:

If $0 \leq \varepsilon < 3$, then $-1/2 < \delta_0^{(+)} \leq 0$, and hence the origin is a "regular" boundary point (*accessible*), cf. (2.93), and we get a unique solution $\{\omega(t), \varphi(t)\}$ of equation (7.95), or we can apply Theorem 2.5 directly on $\mathbf{R}^1$. This example shows that even with the *positive* singular potential at the origin there is the so-called "tunneling", namely the diffusion particle can go through the origin.

If $0 \leq \kappa < 1$, then $0 \leq \delta_0^{(-)} < 1/2$, and hence the origin is also a "regular" (*accessible*) boundary point. Therefore, we get a unique solution $\{\omega(t), \varphi(t)\}$ of the equation (7.95), or we can apply Theorem 2.5 directly on $\mathbf{R}^1$.

The first excited state:

On the contrary, the origin is "entrance" (*inaccessible*), cf. (2.93), for diffusion processes determined by the excited states $\psi_1^{(-)}(x)$ and $\psi_1^{(+)}(x)$ because of Feller's test, since $\delta_1^{(-)}, \delta_1^{(+)} \geq 1/2$. In these cases $\varphi(t) \equiv 0$ for almost all Brownian paths $\omega(t)$ in equation (7.95), since $\xi(t)$ does not hit the origin.

Remark. Instead of the truncation method discussed above, we consider a positive solution of equation (7.80) on $[\alpha, \infty)$ with $\alpha > 0$, and then let $\alpha \downarrow 0$. At the boundary α we impose the mixed boundary condition

$$\beta\psi(\alpha) + (1 - \beta)\frac{d\psi}{dx}(\alpha) = 0, \quad 0 < \beta < 1.$$

(i) If we impose $\lim_{\alpha \downarrow 0} \psi(\alpha) = \infty$, then $\lim_{\alpha \downarrow 0} \frac{d\psi}{dx}(\alpha) = -\infty$ is independent of β, and we get the solution $\psi_0^{(+)}(x)$.

(ii) If we impose $\lim_{\alpha \downarrow 0} \psi(\alpha) = 0$ and choose β properly depending on α so that $\lim_{\alpha \downarrow 0} \frac{d\psi}{dx}(\alpha) = +\infty$, then we get the solution $\psi_0^{(-)}(x)$.

Chapter VIII

The Schrödinger Equation can be a Boltzmann Equation

8.1. Large Deviations[1]

Let us consider the Schrödinger process $\{X_t, Q\}$ determined by the triplet $\{P^c_{(s,x)}, \mu_a, \mu_b\}$ (cf. Chapter 5), namely Q is Csiszar's projection on the subset $\mathbf{A}_{a,b}$ (see (8.4) below) of the renormalized process $\bar{P}$, or a diffusion process constructed by the method of Chapter 6 for a given function $\phi(t, x)$. In the latter case we take the creation and killing $c = -L\phi/\phi$ induced by $\phi(t, x)$, and another function $\hat{\phi}(a, x)$ such that $\mu_a(x) = \hat{\phi}(a, x)\phi(a, x)$, and moreover assume the admissibility condition (5.12) so that $Q = Q_{\mu_a}$ is the Csiszar projection of $\bar{P}$. This case occurs, if a solution $\psi = e^{R+iS}$ of the Schrödinger equation is given beforehand and if we define $\phi = e^{R+S}$ and $\hat{\phi} = e^{R-S}$ in terms of R and S of $\psi = e^{R+iS}$.

Our main theme in this chapter is this:

Find a sequence $\mathbf{Q}^{(n)}$ of systems of interacting n-diffusion processes such that

$$\mathbf{Q}^{(n)} \text{ converges to } Q^\infty, \text{ the infinite product of } Q,$$
$$\text{as } n \text{ tends to infinity,}$$

a precise definition of which will be given in (8.2). The meaning of the convergence is this: If the total number n of particles in the system $\mathbf{Q}^{(n)}$ tends to infinity, then the interacting n-particles in the system become asymptotically independent of each other and indistinguishable, namely every particle will move independently according to the same probability law Q.

---

[1] Cf. Chapter 11, in which a main theorem on large deviations is shown

To be precise we consider a system $\{\Omega^n,\ \mathbf{Q}^{(n,k)}\}$ of interacting n-diffusion particles, where we allow one more parameter k, which will be explained soon in the following.

The drift coefficient $\mathbf{b}_i^{n,k}(t,x) = \mathbf{b}_i^{n,k}(t,x,L_n(x))$, $x = (x_1, \dots, x_n) \in (\mathbf{R}^d)^n$, $i = 1, 2, \dots, n$, of the interacting diffusion processes $(X^1, \dots, X^n)$ must be found so that the propagation of chaos holds. The coefficient $\mathbf{b}_i^{n,k}(t,x)$ represents the interaction of the i-th particle X^i with the other particles: This means, the i-th particle X^i obeys a stochastic differential equation

$$X_t^i = X_a^i + \int_a^t \mathbf{b}_i^{n,k}(r,X_r)dr + \int_a^t \sigma(r,X_r) \cdot dB_r^i \ ,$$

where $X_t = (X_t^1, \dots, X_t^n)$ and B_t^i, $i = 1, 2, \dots, n$, denote mutually independent d-dimensional Brownian motions. The process X_t depends on n and k, which are not indicated to avoid complication of notations.

The **empirical distribution** of the system of interacting diffusion processes $(X_1, \dots, X_n)$ is defined by

$$(8.1) \qquad L_n(\omega) = \frac{1}{n} \sum_{i=1}^n \delta_{\omega_i}, \quad \omega = (\omega_1, \dots, \omega_n) \in \Omega^n.$$

The standard definition of the propagation of chaos is given as follows: For the system $\{\Omega^n,\ \mathbf{Q}^{(n,k)}\}$ of Markov processes, the **propagation of chaos** holds with the limiting distribution Q as $n \to \infty$ and $k \to \infty$, if it has the following two properties

(8.2)

       $\mathbf{Q}^{(n,k)}$ converges weakly to the infinite product of Q,

       $\mathbf{Q}^{(n,k)}[L_n]$ converges weakly to Q,

as $n \to \infty$ and $k \to \infty$, where L_n is the empirical distribution defined in (8.1).

This standard definition will be modified (cf. (8.17), (8.18)) to be suitable with a new formulation of the propagation of chaos in terms of large deviations.

In our context the limiting process Q in (8.2) must be the Schrödinger process (Csiszar's projection of the renormalized measure $\overline{P}$) determined by the triplet $\{P^c_{(s,x)}, \mu_a, \mu_b\}$. Thus, our problem is reduced to constructing a sequence $\{Q^{(n,k)}: n, k = 1, 2, ...\}$ of systems of interacting diffusion processes which satisfy (8.2). This is an inverse problem of the propagation of chaos.

To solve this problem we apply theorems on large deviations due to Csiszar (1984), in particular his asymptotic quasi-independence of a sequence $\{Q^{(n,k)}: n, k = 1, 2, ...\}$ of diffusion processes (probability measures) in connection with the Sanov property.

Although the statement itself of the Sanov property (see, Theorem 10.1) does not contain any topological concept, to prove it we need a topological structure on the space $M_1(\Omega)$ of probability measures on Ω, but we need no topological structure on Ω. In fact, Csiszar (1984) defines an intrinsic topology of $M_1(\Omega)$ which depends only on the measurable structure of Ω.

Csiszar's τ_0-**topology** is given in terms of the basic *neighbourhoods* of $P \in M_1(\Omega)$:

$$(8.3) \quad U(P, \varepsilon, \boldsymbol{P}_k) = \{R \in M_1(\Omega) : |R(\Omega_i) - P(\Omega_i)| < \varepsilon, \text{ for } i = 1, 2, ..., k,$$
$$\text{and } R << P \text{ on } \sigma(\boldsymbol{P}_k), \text{ i.e., } R(\Omega_i) = 0, \text{ if } P(\Omega_i) = 0\},$$

where

$$\boldsymbol{P}_k = \boldsymbol{P}_k(\Omega) = \{\Omega_1, \cdots, \Omega_k\}, k = 1, 2, ..., \text{ are finite measurable}$$
$$\text{partitions of } \Omega,$$

$$\sigma(\boldsymbol{P}_k) \text{ is the } \sigma\text{-field generated by the partition } \boldsymbol{P}_k.$$

Thus the neighbourhood $U(P, \varepsilon, \boldsymbol{P}_k)$ depends only on a finite measurable partition $\boldsymbol{P}_k$ and ε.

Let the subset $\boldsymbol{A}_{a,b}$ of $M_1(\Omega)$ be defined by (5.11), *i.e.*,

$$(8.4) \qquad \boldsymbol{A}_{a,b} = \{P \in M_1(\Omega) : P \circ X_r^{-1} = \mu_r, \text{ for } r = a, b\},$$

where $\mu_r \in M_1(\mathbf{R}^d)$, $r = a, b$, are fixed. We assume that the set $\boldsymbol{A}_{a,b}$ is admissible for the reference process $\overline{P}$ (cf. (5.12)), so that the Csiszar projection Q of $\overline{P}$ on the set $\boldsymbol{A}_{a,b}$ exists.

The set $\mathbf{A}_{a,b}$ has played an important role in discussing variational principle in Chapter 5. However, the lateral conditions defining the subset $\mathbf{A}_{a,b}$ is too strict for an application of Sanov's theorem. One of the troubles with $\mathbf{A}_{a,b}$ is this: the set of inner points of $\mathbf{A}_{a,b}$ in Csiszar's τ_0-topology is an empty set and hence the lower bound in Sanov's theorem[2] becomes useless. Therefore, loosening the lateral conditions at the initial and terminal times $t = a, b$ of the subset $\mathbf{A}_{a,b}$, we approximate it by a family of subsets $\mathbf{A}(\varepsilon, k)$ of $\mathbf{M}_1(\Omega)$ defined by

$$(8.5) \quad \mathbf{A}(\varepsilon, k) = \{P \in \mathbf{M}_1(\Omega) : \left| P[X_r \in B_i] - \mu_r(B_i) \right| \leq \frac{\varepsilon}{2^k}, for \; \forall \; B_i \in \mathcal{P}_k(\mathbf{R}^d),$$

$$and \; P \circ X_r^{-1} << \overline{P} \circ X_r^{-1} \; on \; \sigma(\mathcal{P}_k(\mathbf{R}^d)), r = a, b\},$$

for $\varepsilon > 0$ and $k = 1, 2, \dots$, where $\overline{P}$ is a prescribed reference process, and $\mathcal{P}_k(\mathbf{R}^d) = \{B_1, \dots, B_{2^k}\}$, $k = 1, 2, \dots$, are finite measurable partitions of $\mathbf{R}^d$ such that $\mathcal{P}_{k+1}(\mathbf{R}^d)$ is a refinement of $\mathcal{P}_k(\mathbf{R}^d)$:

$$\sigma(\mathcal{P}_k(\mathbf{R}^d)) \subset \sigma(\mathcal{P}_{k+1}(\mathbf{R}^d)) \quad and \quad \sigma(\mathcal{P}_k(\mathbf{R}^d)) \uparrow \sigma(\mathbf{R}^d), \; as \; k \uparrow \infty.$$

The dependence on $\varepsilon > 0$ of the subset $\mathbf{A}(\varepsilon, k)$ is not substantial but technical. In the proof of Theorem 8.1 we will use it in (8.11).

Let $\overline{\mathbf{P}} = \overline{P}^n$ be the n-fold product of $\overline{P}$. The **conditional probability** $\mathbf{P}^{(n, k)}$ of $\overline{\mathbf{P}}$ is defined, for a fixed $\varepsilon > 0$, by

$$(8.6) \qquad \mathbf{P}^{(n, k)}[\,\cdot\,] = \mathbf{P}^{(n)}_{\mathbf{A}(\varepsilon, k)}[\,\cdot\,] = \overline{\mathbf{P}}[\,\cdot\,|L_n \in \mathbf{A}(\varepsilon, k)],$$

with which the motion of the n-independent particles governed by $\overline{\mathbf{P}}$ is controlled so that the empirical distribution L_n becomes not very different from the prescribed probability distributions μ_a and μ_b at the initial and terminal times $t = a, b$ in the sense of (8.5). Therefore, although it is clear because of the conditioning, it should be emphasized that the process $\{(X_1, \dots, X_n), \mathbf{P}^{(n, k)}\}$ with the conditional probability $\mathbf{P}^{(n, k)}$ does not have the Markov property and $(X_1, \dots, X_n)$ is not independent any more.

Then we will formulate the fundamental assertions of the theory of large deviations in a generalized form as an "approximate Sanov property" and (approximately) asymptotic quasi-independence.

---

[2] Cf. Lemma 11.5 in Chapter 11

Theorem 8.1. (Aebi-Nagasawa (1992)) *Assume the conditions of admissibility (5.12) and integrability (5.19).*

(i) *Let $\mathbf{A}_{a,b}$ be the subset of $\mathbf{M}_1(\Omega)$ defined in (8.4) and $\mathbf{A}(\varepsilon, k)$ be the one defined in (8.5), for a fixed $\varepsilon > 0$. Then*

$$(8.7) \qquad \mathbf{A}_{a,b} = \bigcap_{k \in \mathbf{N}} \mathbf{A}(\varepsilon, k),$$

*and the "**Approximate Sanov Property**" holds for the subset $\mathbf{A}_{a,b}$:*

$$(8.8) \qquad \lim_{k \to \infty} \lim_{n \to \infty} \frac{1}{n} \log \overline{\mathbf{P}}[L_n \in \mathbf{A}(\varepsilon, k)] = - \inf_{P \in \mathbf{A}_{a,b}} \mathrm{H}(P \mid \overline{P})$$

$$= - \mathrm{H}(Q \mid \overline{P}),$$

where the infimum is attained by Csiszar's projection Q (the Schrödinger process determined by the triplet $\{\mathrm{P}^c_{(s,x)}, \mu_a, \mu_b\}$) of the reference process $\overline{P}$ (the renormalized process). In other words, the probability that the rare event $\{L_n \in \mathbf{A}(\varepsilon, k)\}$ occurs is given by

$$\overline{\mathbf{P}}[L_n \in \mathbf{A}(\varepsilon, k)] \approx e^{-n \mathrm{H}(Q \mid \overline{P})}, \quad \text{as } n \to \infty \text{ and } k \to \infty.$$

(ii) *The conditional process $\{(X_1, \ldots, X_n), \mathbf{P}^{(n,k)}\}$ is (approximately) asymptotically quasi-independent with the limiting distribution Q; namely,*

$$(8.9) \qquad \lim_{k \to \infty} \lim_{n \to \infty} \frac{1}{n} \mathrm{H}(\mathbf{P}^{(n,k)} \mid (Q_k)^n) = 0,$$

and Q_k converges to Q in entropy as k tends to infinity, where Q_k denotes the Csiszar projection of the reference process $\overline{P}$ on the set $\mathbf{A}(\varepsilon, k)$.

Proof. A key step of the proof is to show the inclusion

$$(8.10) \qquad \mathbf{A}^\circ(\varepsilon_1, k) \subset \mathbf{A}(\varepsilon_1, k) \subset \mathbf{A}^\circ(\varepsilon_2, k) \subset \mathbf{A}(\varepsilon_2, k),$$

where $0 < \varepsilon_1 < \varepsilon_2$ and $\mathbf{A}^\circ(\varepsilon, k)$ denotes the interior of $\mathbf{A}(\varepsilon, k)$ with respect to Csiszar's τ_0-topology.

Non-trivial inclusion in (8.10) is the second one which follows from

Lemma 8.1. *For any* $\varepsilon_0 > 0$

$$(8.11) \qquad \underset{0 < \varepsilon < \varepsilon_0}{\cup} \mathbf{A}(\varepsilon, k) \subset \mathbf{A}^\circ(\varepsilon_0, k).$$

Proof. Let $P_0 \in \underset{0 < \varepsilon < \varepsilon_0}{\cup} \mathbf{A}(\varepsilon, k)$, i.e., $P_0 \in \mathbf{A}(\varepsilon_1, k)$, for $0 < \varepsilon_1 < \varepsilon_0$. For each $B_i \in \mathcal{P}_k(\mathbf{R}^d)$ define Ω_i^r by

$$\Omega_i^r = \{\omega : X_r(\omega) \in B_i\}, \quad \text{for } r = a, b,$$

and a finite measurable partition $\mathcal{P}$ of Ω by

$$\mathcal{P} = \mathcal{P}(\Omega) = \{\Omega_i^a \cap \Omega_j^b : i, j = 1, 2, \dots, k\}.$$

Let us take $P \in U(P_0, \varepsilon_2, \mathcal{P})$. Then $P \ll P_0$ on $\sigma(\mathcal{P})$, and moreover $P_0 \circ X_r^{-1} \ll \bar{P} \circ X_r^{-1}$ and hence $P \circ X_r^{-1} \ll \bar{P} \circ X_r^{-1}$ on $\sigma(\mathcal{P}_k(\mathbf{R}^d))$, for $r = a, b$. For $\forall B_i \in \mathcal{P}_k(\mathbf{R}^d)$ we have

$$|P[X_r \in B_i] - q_r(B_i)| \le |P[\Omega_i^r] - P_0[\Omega_i^r]| + |P_0[X_r \in B_i] - q_r(B_i)|$$

$$\le k\varepsilon_2 + \frac{\varepsilon_1}{2^k}, \quad \text{for } r = a, b.$$

Therefore, if we choose $\varepsilon_2 > 0$ such that $k\varepsilon_2 + \dfrac{\varepsilon_1}{2^k} \le \dfrac{\varepsilon_0}{2^k}$, then $P \in \mathbf{A}(\varepsilon_0, k)$, i.e., $U(P_0, \varepsilon_2, \mathcal{P}) \subset \mathbf{A}(\varepsilon_0, k)$, which implies (8.11), and completes the proof.

We denote for simplicity

$$\mathrm{H}(\mathbf{A} \mid \bar{P}) = \inf_{P \in \mathbf{A}} \mathrm{H}(P \mid \bar{P}).$$

Then (8.10) combined with (8.7) yields

$$(8.12) \qquad \lim_{k \to \infty} \mathrm{H}(\mathbf{A}^\circ(\varepsilon, k) \mid \bar{P}) = \lim_{k \to \infty} \mathrm{H}(\mathbf{A}(\varepsilon, k) \mid \bar{P}) = \mathrm{H}(Q \mid \bar{P}),$$

where Q is the Csiszar projection on the subset $\mathbf{A}_{a,b}$ of $\bar{P}$.

Since $Q \in \mathbf{A}(\varepsilon, k)$ and Q_k is the Csiszar projection on the set $\mathbf{A}(\varepsilon, k)$ of the reference process $\bar{P}$, Csiszar's inequality (10.12) implies

$$H(Q \mid \bar{P}) - H(Q_k \mid \bar{P}) \geq H(Q \mid Q_k),$$

the right-hand side of which vanishes because of (8.12), and hence we have

(8.13) $$\lim_{k \to \infty} H(Q \mid Q_k) = 0;$$

namely, Q_k converges to Q in entropy and hence in variation (cf. Lemma 10.2).

To apply a theorem of Csiszar on the Sanov property (cf. Chapter 11) to the subset $\mathbf{A}(\varepsilon, k)$ we need

Lemma 8.2. *The subset* $\mathbf{A}(\varepsilon, k) \subset \mathbf{M}_1(\Omega)$ *is completely convex*[3] *and also variation closed.*

Proof. Let $(\Lambda, \mathcal{A}, \mu)$ be an arbitrary probability space and $\eta(\lambda, B)$ be a probability kernel on $\Lambda \times \sigma(\Omega)$ such that $\eta(\lambda, \cdot) \in \mathbf{A}(\varepsilon, k)$, for $\forall \lambda \in \Lambda$. Then, for any $B_i \in \mathbf{P}_k(\mathbf{R}^d)$,

$$\mid \mu\eta[X_r \in B_i] - q_r(B_i) \mid \leq \int \mid \eta(\lambda, \{X_r \in B_i\}) - q_r(B_i) \mid \mu(d\lambda)$$

$$\leq \frac{\varepsilon}{2^k}, \text{ for } r = a, b,$$

where

$$\mu\eta(\cdot) = \int \mu(d\lambda)\eta(\lambda, \cdot).$$

Moreover, it is clear that $\mu\eta \circ X_r^{-1} \ll \bar{P} \circ X_r^{-1}$ on $\sigma(\mathbf{P}_k(\mathbf{R}^d))$, for $r = a, b$. Therefore, $\mu\eta \in \mathbf{A}(\varepsilon, k)$, i.e., the set $\mathbf{A}(\varepsilon, k)$ is completely convex.

To show that the set $\mathbf{A}(\varepsilon, k)$ is variation closed, let $\{P_n\} \subset \mathbf{A}(\varepsilon, k)$ be a sequence which converges to $P \in \mathbf{M}_1(\Omega)$ in variation, i.e.,

$$\| P_n - P \|_{Var} = \int \left| \frac{dP_n}{d\bar{P}} - \frac{dP}{d\bar{P}} \right| d\bar{P} \to 0, \text{ as } n \to \infty,$$

with respect to $\bar{P}$. Then, for any $B_i \in \mathbf{P}_k(\mathbf{R}^d)$ and $r = a, b,$

---

[3] Cf. Section 2 of Chapter 11

$$\left| P[X_r \in B_i] - q_r(B_i) \right| \le \left| P[X_r \in B_i] - P_n[X_r \in B_i] \right| + \left| P_n[X_r \in B_i] - q_r(B_i) \right|,$$

where the second term on the right-hand side is evaluated as

$$| P_n[X_r \in B_i] - q_r(B_i) | \le \frac{\varepsilon}{2^k},$$

and the first term vanishes, namely

$$| P[X_r \in B_i] - P_n[X_r \in B_i] | \le \| P_n - P \|_{Var} \to 0, \ as \ n \to \infty.$$

Combining these inequalities, we have

$$| P[X_r \in B_i] - q_r(B_i) | \le \frac{\varepsilon}{2^k},$$

and hence $P \in \mathbf{A}(\varepsilon, k)$, i.e., the set $\mathbf{A}(\varepsilon, k)$ is variation closed, which completes the proof.

Therefore, applying a theorem of Csiszar (cf. Theorem 11.1) to the set $\mathbf{A}(\varepsilon, k)$, we have the Sanov property of the subset $\mathbf{A}(\varepsilon, k)$:

$$(8.14) \quad - H(\mathbf{A}^{\circ}(\varepsilon, k) \,|\, \overline{P}) \le \liminf_{n \to \infty} \frac{1}{n} \log \overline{P}[L_n \in \mathbf{A}(\varepsilon, k)]$$

$$\le \limsup_{n \to \infty} \frac{1}{n} \log \overline{P}[L_n \in \mathbf{A}(\varepsilon, k)] \le - H(\mathbf{A}(\varepsilon, k) \,|\, \overline{P}),$$

namely

$$\lim_{n \to \infty} \frac{1}{n} \log \overline{P}[L_n \in \mathbf{A}(\varepsilon, k)] = - H(Q_k \,|\, \overline{P}),$$

where $H(\mathbf{A}(\varepsilon, k) \,|\, \overline{P}) = H(Q_k \,|\, \overline{P})$ with the Csiszar projection Q_k on the subset $\mathbf{A}(\varepsilon, k)$ of $\overline{P}$. The inequality (8.14), together with (8.12), yields the approximate Sanov property (8.8).

The second assertion on the asymptotic quasi-independence (8.9) of $\mathbf{P}^{(n, k)}$ follows from Csiszar's inequality (cf. equation (11.10) in Theorem 11.1)

$$(8.15) \quad 0 \le \frac{1}{n} H(\mathbf{P}^{(n, k)} \,|\, (Q_k)^n) \le - H(\mathbf{A}(\varepsilon, k) \,|\, \overline{P}) - \frac{1}{n} \log \overline{P}[L_n \in \mathbf{A}(\varepsilon, k)],$$

together with the approximate Sanov property (8.8); namely the right-hand side of (8.15) vanishes as $n \to \infty$ and $k \to \infty$. This, combined with (8.13), proves the second assertion on the asymptotic quasi-independence of $\mathbf{P}^{(n,k)}$ and completes the proof of the theorem.

8.2 The Propagation of Chaos in Terms of Large Deviations

The conditional process $\{(X_1, \ldots, X_n), \mathbf{P}^{(n,k)}\}$ in Theorem 8.1 is not Markovian as we have remarked already. However, we can find a unique **Markovian modification** $\mathbf{Q}^{(n,k)}$ of $\mathbf{P}^{(n,k)}$, applying Theorem 5.4, the entropy condition of which is satisfied automatically in this case as will be seen.

Let $\mathbf{P}$ and $\mathbf{P}^0$ be probability measures on the space of right-continuous paths on $\mathbf{R}^d$ with $t \in [a, b]$. Then $\mathbf{P}^0$ is called a **Markovian modification** of $\mathbf{P}$, if $\mathbf{P}^0$ is Markovian and the marginal distributions at time t of $\mathbf{P}$ and $\mathbf{P}^0$ coincide for $\forall\, t \in [a, b]$.

Theorem 8.2. *For the conditional process* $\mathbf{P}^{(n,k)}$ *(cf. Definition (8.1)) there exists a unique Markovian modification* $\mathbf{Q}^{(n,k)}$ *such that*

$$(8.16) \qquad H(\mathbf{Q}^{(n,k)} \,|\, (Q_k)^n) \leq H(\mathbf{P}^{(n,k)} \,|\, (Q_k)^n).$$

Proof. We apply Theorem 5.4 for $\mathbf{P} = \mathbf{P}^{(n,k)}$ and $\mathbf{Q} = (Q_k)^n$, where Q_k is the Csiszar projection on the set $\mathbf{A}(\varepsilon, k)$ of $\bar{P}$, setting $\mu_t = \mathbf{P}^{(n,k)} \circ X_t^{-1}$ in the set $\mathbf{\mathit{A}}$ defined at (5.43). Because $H(\mathbf{P}^{(n,k)} \,|\, (Q_k)^n) < \infty$ is clear from (8.9), the admissibility condition of the set $\mathbf{\mathit{A}}$ for $\mathbf{Q} = (Q_k)^n$ is satisfied, and hence we have the Markovian modification $\mathbf{Q}^{(n,k)}$, which is the Csiszar projection of $\mathbf{Q} = (Q_k)^n$ on the set $\mathbf{\mathit{A}}$. This completes the proof.

We define "the propagation of chaos" in a stronger form to be fit for our treatment in terms of large deviations as follows:

For a sequence of systems $\{(X_1, \ldots, X_n), \mathbf{Q}^{(n,k)}\}$ of interacting diffusion processes, the **propagation of chaos** *in entropy* holds with a limiting distribution Q as $n \to \infty$ and $k \to \infty$, if for arbitrary but fixed $m \geq 1$ and for the marginal distribution $\mathbf{Q}_m^{(n,k)}$ of $\mathbf{Q}^{(n,k)}$ on Ω^m

$\mathbf{Q}_m^{(n,k)}$ *converges to the m-fold product* Q^m *of* Q *in entropy,*

(8.17)

$\mathbf{Q}^{(n,k)}[L_n]$ *converges to the* Q *in entropy,*

as $n \to \infty$ and then $k \to \infty$.

Each convergence in (8.17) consists of two steps. In the first property this means that $H(\mathbf{Q}_m^{(n,k)} | (Q_k)^m) \to 0$ as $n \to \infty$ and then $H(Q^m | (Q_k)^m) \to 0$ as $k \to \infty$. The second property should be understood similarly.

Besides the propagation of chaos in entropy (8.17), we can formulate **the propagation of chaos *in variation*** as follows:

$\mathbf{Q}_m^{(n,k)}$ *converges to the m-fold product* Q^m *of* Q *in variation,*

(8.18)

$\mathbf{Q}^{(n,k)}[L_n]$ *converges to the* Q *in variation,*

as $n \to \infty$ and then $k \to \infty$.

Then we have

Lemma 8.3. (i) *The propagation of chaos in entropy* (8.17) *implies the propagation of chaos in variation* (8.18).

(ii) *The propagation of chaos in variation* (8.18) *implies the (weak) propagation of chaos* (8.2).

Proof. Since the second property of (8.18) can be shown similarly, we prove only the first one. If (8.17) holds, then, with the help of Lemma 10.2, we have

$$\{\| \mathbf{Q}_m^{(n,k)} - (Q_k)^m \|_{Var}\}^2 \le 2\,H(\mathbf{Q}_m^{(n,k)} | (Q_k)^m) \to 0,$$

$$\{\| Q^m - (Q_k)^m \|_{Var}\}^2 \le 2\,H(Q^m | (Q_k)^m) \to 0,$$

as $n \to \infty$ and $k \to \infty$, and hence

$$\| \mathbf{Q}_m^{(n,k)} - Q^m \|_{Var} \le \| \mathbf{Q}_m^{(n,k)} - (Q_k)^m \|_{Var} + \| (Q_k)^m - Q^m \|_{Var} \to 0,$$

as $n \to \infty$ and $k \to \infty$. The propagation of chaos (8.2) clearly follows from the propagation of chaos *in variation* (8.18). This completes the proof.

Our main theorem in this section states:

Theorem 8.3. (Aebi-Nagasawa (1992)) *Assume the admissibility condition* (5.14) *and the integrability condition* (5.19).

Let $\{(X_1, \cdots, X_n), \mathbf{Q}^{(n,k)}\}$ *be the system of interacting diffusion process with* $\mathbf{Q}^{(n,k)}$ *which is the Markovian modification of* $\mathbf{P}^{(n,k)}$ *given in Theorem 8.2 such that*

$$\mathrm{H}(\mathbf{Q}^{(n,k)} \mid (Q_k)^n) \leq \mathrm{H}(\mathbf{P}^{(n,k)} \mid (Q_k)^n).$$

Then:

(i) *The Markovian modification* $\{(X_1, \dots, X_n), \mathbf{Q}^{(n,k)}\}$ *is a system of interacting diffusion processes with the Markovian drift coefficient* $\mathbf{b}^{n,k}(t, \mathbf{x})$ (= *interaction*), $\mathbf{x} = (x_1, \dots, x_n) \in (\mathbf{R}^d)^n$, *such that*

(8.19) $\qquad \mathbf{b}^{n,k}(t, \mathbf{x}) = \{\mathbf{b}_i^{n,k}(t, \mathbf{x}, L_n(\mathbf{x})) : i = 1, 2, \dots, n\},$

where $\mathbf{b}_i^{n,k}$ *is the drift vector of* X_i *under* $(Q_k)^n$.

(ii) *The **approximate Sanov property***

$$\lim_{k \to \infty} \lim_{n \to \infty} \frac{1}{n} \log \overline{\mathbf{P}}[L_n \in \mathbf{A}(\varepsilon, k)] = - \inf_{P \in \mathbf{A}_{a,b}} \mathrm{H}(P \mid \overline{P})$$

$$= - \mathrm{H}(Q \mid \overline{P})$$

holds, where the infimum is attained by the Schrödinger process Q *(Csiszar's projection) determined by* $\{\mathbf{P}_{(s,x)}^c, \mu_a, \mu_b\}$. *In other words, the probability that the rare event* $\{L_n \in \mathbf{A}(\varepsilon, k)\}$ *occurs is given by*

$$\overline{\mathbf{P}}[L_n \in \mathbf{A}(\varepsilon, k)] \approx e^{-n\mathrm{H}(Q \mid \overline{P})}, \quad as \ n \to \infty \ and \ k \to \infty.$$

(iii) *The Markovian modification* $\mathbf{Q}^{(n,k)}$ *is asymptotically quasi-independent with the Schrödinger process* Q *as limiting distribution*;

(8.20) $\qquad \lim_{k \to \infty} \lim_{n \to \infty} \frac{1}{n} \mathrm{H}(\mathbf{Q}^{(n,k)} \mid (Q_k)^n) = 0,$

and Q_k *converges to* Q *in entropy as* $k \to \infty$.

(iv) *Moreover, the **propagation of chaos in entropy** holds for the Markovian modification* $\{(X_1, \ldots, X_n), \mathbf{Q}^{(n,k)}\}$ *with the Schrödinger process Q as limiting distribution, when $n \to \infty$ and $k \to \infty$.*

Proof.   Theorem 8.2 implies the existence of the Markovian modification $\mathbf{Q}^{(n,k)}$. To show assertion (i) we first note that the modification $\mathbf{Q}^{(n,k)}$ is absolutely continuous with respect to $(Q_k)^n$ and moreover $Q_k \ll \overline{P}$. There exists, consequently, a measurable (vector) function $\mathbf{b}^{n,k}(t, x)$ and a Brownian motion $\{B_t, (Q_k)^n\}$ such that

$$X_t^i = X_a^i + \int_a^t \mathbf{b}_i^{n,k}(r, X_r)dr + \int_a^t \sigma(r, X_r) \cdot dB_r^i \, ,$$

by the generalized Maruyama-Girsanov theorem (cf. Liptser-Shiriyayev (1977)). Assertion (ii) has been shown in Theorem 8.1. In assertion (iii) the asymptotic quasi-independence of $\mathbf{Q}^{(n,k)}$, namely convergence (8.20) follows from (8.9) with the help of (8.16).

For assertion (iv) we note first that the marginal distributions of the measure $\mathbf{P}^{(n,k)}$ on product spaces $\Omega_{mr+1} \times \cdots \times \Omega_{m(r+1)}$, $r = 0, 1, 2, \ldots$, are identically equal to $\mathbf{P}_m^{(n,k)}$. Therefore, the marginal distribution $\mathbf{Q}_m^{(n,k)}$ of $\mathbf{Q}^{(n,k)}$ on Ω^m converges to $(Q_k)^m$ in entropy as $n \to \infty$ by Lemma 11.3, since

(8.21) $$H(\mathbf{Q}_m^{(n,k)} \,|\, (Q_k)^m) \leq H(\mathbf{P}_m^{(n,k)} \,|\, (Q_k)^m),$$

and then $(Q_k)^m$ converges to Q^m in entropy as $k \to \infty$ by (8.13). Therefore, the first property of (8.17) of the propagation of chaos in entropy holds.

It remains to show the second property of (8.17) of the propagation of chaos in entropy. Since $\mathbf{Q}^{(n,k)}$ is the Markovian modification of $\mathbf{P}^{(n,k)}$, it is sufficient to show that $\mathbf{P}^{(n,k)}[L_n]$ converges in entropy to Q. Let us denote by $P_\Omega^{(n,k)}$ the marginal distribution of $\mathbf{P}^{(n,k)}$ on Ω. Then we have

(8.22) $$\mathbf{P}^{(n,k)}[L_n(B)] = \frac{1}{n} \sum_{i=1}^{n} \mathbf{P}^{(n,k)}[1_B(\omega_i)] = P_\Omega^{(n,k)}[B].$$

Therefore, an application of Lemma 11.3 yields that $\mathbf{P}^{(n,k)}[L_n]$ converges to Q_k in entropy as $n \to \infty$, and then Q_k converges to Q in entropy as $k \to \infty$ by (8.13). This completes the proof.

The assertion (iv) of Theorem 8.3 guarantees the existence of a sequence of systems of interacting diffusion processes which we have been looking for; namely, we can claim that the Schrödinger process $\{X_t, Q\}$ describes the movement of a "typical" particle of a system of interacting diffusion particles, when the size n of the system is sufficiently large.

8.3. Statistical Mechanics for Schrödinger Equations

We summarize the results as a statistical mechanics for a sequence of systems of interacting diffusion processes:[4]

Let $\{\mu_t, t \in [a,b]\}$, be a flow of non-negative distribution densities (in particular $\mu_t = |\psi_t|^2$, where ψ_t, $t \in [a,b]$, a flow of complex valued functions with the normalization $\int |\psi_t(x)|^2 dx = 1$). Instead of the flow μ_t we can begin with a diffusion equation with creation and killing $c(t,x)$ or a Schrödinger equation with a potential $V(t,x)$, which are equivalent with each other as we have shown in Chapter 4.

(First step) There exists, based on the duality relation (time reversal), a diffusion (Schrödinger) process $\{X_t, Q\}$ such that its distribution density coincides with the given flow μ_t (say, $|\psi_t|^2$) and Schrödinger's factorization $\mu_t = \hat{\phi}_t \phi_t$ holds, where $\mu_t dx = Q \circ X_t^{-1}$, namely it is the distribution of the diffusion (Schrödinger) process (say, of a single particle!).

(Second step) Moreover, there exists a sequence $\{(X_t^1, \ldots, X_t^n), Q^{(n,k)}\}$ of systems of interacting diffusion processes, such that for the $Q^{(n,k)}$ the propagation of chaos in entropy holds, especially $Q^{(n,k)}[L_n]$ which coincides with the marginal distribution of $Q^{(n,k)}$ converges in entropy to the Schrödinger process Q as $n \to \infty$ and $k \to \infty$ in this order. Namely, we can regard μ_t (say, $\overline{\psi}\psi$!) as the spatial statistical distribution density of infinitely many interacting diffusion particles.

In our statistical theory the flow, the Schrödinger equation, and the diffusion equations are **macroscopic** descriptions of a system of interacting particles; the diffusion process $\{X_t, Q\}$ gives an **intermediate** description of the movement of a *"typical particle"* in the system; the system of interacting diffusion processes $\{(X_t^1, \ldots, X_t^n), Q^{(n,k)}\}$ is a **microscopic**

---

[4] Cf. Nagasawa (1980, 88, 90)

description. Hence we can complete the diagram in Proposition 4.2 as follows

$$
\begin{array}{ccc}
L^2\text{-}theory & & Transformation \\
\Downarrow & & \Downarrow
\end{array}
$$

Algebraic Theory $\Leftrightarrow$ Wave Theory $\Leftrightarrow$ Diffusion Theory

| Commutation relation | Schrödinger equation | Time reversal (duality) of diffusion equation |

the propagation of chaos $\Rightarrow$ $\Uparrow$

Statistical Mechanics

Systems of interacting diffusion particles

The result we have obtained for the second step connecting the intermediate and microscopic is not quite satisfactory, because assertion (i) of Theorem 8.3 gives no explicit form of interaction between diffusion particles. Nonetheless I would like to claim

Proposition 8.1. *The Schrödinger equation is a "Boltzmann equation" for a system of interacting particles* (= *the system of interacting diffusion processes* $\{(X_1, \ldots, X_n), \mathbf{Q}^{(n,k)}\}$ *in Theorem 8.3*), *as* $n \to \infty$ *and* $k \to \infty$.

I have said "Boltzmann" for the Schrödinger equation here, placing it in the context of the propagation of chaos as formulated in (8.17) and (8.18), and taking into account the equivalence of Schrödinger and diffusion equations established in Chapter 4. However, if we want to keep a close analogy with classical statistical mechanics, it would be relevant to call the distribution $\mathbf{Q}^{(n,k)}$ of the system of n-particles "*the microcanonical ensemble* (or *distribution*)" (cf. (8.18)) of, say, "*Schrödinger gas*", and the distribution Q itself of the diffusion process the "*Gibbs measure*". In fact, in the simplest time-independent (reversible) case, the distribution density μ of $Q \circ X_t^{-1}$ is represented as

$$
\mu(x) = e^{U(x)},
$$

if the drift coefficient is given by a potential function $U(x)$

$$a(x) = \frac{1}{2}\sigma^T\sigma\,\nabla U(x),$$

as Kolmogoroff had already observed in 1937. Moreover, we can keep a further analogy with classical statistical mechanics; Ω = the phase space, $\overline{P}$ = the "normalized" Liouville measure, and so on (see., e.g. Lanford (1973)). Cf. Nagasawa (1990) and also Aebi-Nagasawa (1992).

8.4. Some Comments

In the last chapter entitled "*Une analogie entre la mécanique ondulatoire et quelques problèmes de probabilités en physique classique*" of Schrödinger (1932), he begins the first paragraph with

> "Le sujet que je vais aborder maintenant n'est pas intimement lié aux questions dont il s'est agi dans les chapitres précédents. Tout d'abord vous aurez l'impression de choses qui ne sont pas du tout liées. Il s'agit d'un problème classique: problème de probabilités dans la théorie du mouvement brownien. Mais en fin de compte, il *ressortira* une analogie avec la mécanique ondulatoire, qui fut si frappante pour moi lorsque je l'eus trouvée, qu'il m'est difficile de la croire purement accidentelle."

Then he justifies his formulation of Brownian motions given already in Schrödinger (1931) in two ways. One is based on Eddington's interpretation (1928) of the factorization $\mu_t = \psi_t\overline{\psi}_t$ in quantum theory:

> "The $\psi\psi^*$ is obtained by introducing two symmetrical systems of ψ waves travelling in opposite directions in time; one of these must presumably correspond to probable inference from what is known (or is stated) to have been the condition at a later time."[5]

Schrödinger's factorization $\mu_t = \hat{\phi}_t\phi_t$ of distribution densities is exactly its real-valued counterpart, which is the product of forward and backward predictions $\hat{\phi}_t$ and ϕ_t, respectively. Namely, what we do is the prediction of *intermediate states*, given a pair of data, one at a starting time in the past and another one at a terminal time in the future (but, as we have remarked,

---

[5] Quoted from Schrödinger (1932). Here ψ^* denotes the complex conjugate of ψ

this is the description in the "imaginary or fictitious" evolution). Let us call this *"Eddington-Schrödinger's prediction"*. This point has been fully discussed already in chapters 3, 4, and 5.

Schrödinger's second justification of his formulation of Brownian motions is essentially in the contexts of "large deviations"; namely, if they are placed in ordinary diffusion theory, these processes occur as very rare events, and hence they should (and can) be considered in terms of large deviations. This was realized by Föllmer (1988), in which he applied Sanov's theorem to independent Brownian motions to mathematize Schrödinger's idea. Influenced by Föllmer (1988), Dawson-Gorostitza-Wakolbinger (1990) treated the problem in the case of bounded $c(t, x)$ (i.e., $\phi(t, x)$ has no zeros). Theorems 8.1 and 8.3 generalize their results as "approximate Sanov property (8.8)" for unbounded $c(t, x)$ which is indispensable in quantum mechanics.

As a matter of fact there were two obstacles in treating large deviations for Schrödinger equations. The first one is this; the subset $\mathbf{A}_{a,b}$ defined in (8.4) has no inner point with respect to Csiszar's τ_0-topology, so that the lower bounds of the Sanov property becomes useless. The second one is ; the empirical distribution $L_n(\omega)$ defined in (8.1) does not belong to the subset $\mathbf{A}_{a,b}$, which makes the standard Sanov assertion meaningless. We have seen that a possible formulation to overcome these difficulties is the *approximate Sanov Property* (8.8) given in Theorem 8.1. A similar way of taking the limit doubly in a prearranged order as in (8.8) is well known in statistical mechanics (cf. Lanford (1973)).

We have already remarked in Chapter 6 that a natural way of interpreting Schrödinger equations is to consider them as statistical equations, especially when segregation occurs, namely, as a "Boltzmann equation" of a system of interacting diffusion processes, and a limit theorem of this type has been shown in the framework of the propagation of chaos in some special cases in Theorems 7.1 and 7.3. However, as we have shown, if we revise the assertion on asymptotic quasi-independence in Theorem 8.1 properly, we can prove the existence of systems of interacting diffusion processes in general cases in higher dimensions as stated in Theorem 8.3. Therefore, my attempt to find out a statistical mechanical structure for the Schrödinger equation can be considered as a natural step along the line of Schrödinger's second justification in terms of large deviations.

Summarizing, we have placed quantum mechanics at its natural position in the long historical evolution of statistical mechanics originated by Ludwig Boltzmann.

Chapter IX

Applications of the Statistical Model for Schrödinger Equations

9.1. Segregation of a Monkey Population

Based on the statistical mechanics for Schrödinger equations discussed in the preceding chapters, segregation of monkey populations was discussed in Nagasawa (1980) as a first trial:

"Assume there is a feeding place in a mountain (perhaps in Japan) where we find a population of monkeys (we assume that the movement of a monkey can be approximated as a sample path of a diffusion process, since he is hopping and jumping from one place to another). The feeding place attracts monkeys and most of them move toward the place and will eventually be distributed with the peak density at the centre (one-core distribution). Besides this, there is another equilibrium distribution, where monkeys come closer but do not reach the feeding place, and are distributed on a circle making a doughnut around the feeding place. When the one-core distribution is realized, the population is in the state of the lowest excitation, and in the doughnut distribution the population is an excited state. Moreover, it can be shown that if the population stays as a single group, no other equilibrium distribution exists, that is, in a state of higher excitation the population cannot keep itself as a single party but splits into at least two groups. Once it is divided into two parties, it can take one of three possible equilibrium distributions." We call this *segregation of a population*. If two groups of monkeys are observed around the feeding place, we find three possible distributions, which are distinguished by different eigenvalues (= degrees of excitation of the population).[1]

We consider the problem in the two-dimensional space and set an attractive potential, say,

---

[1] Cf. also Nagasawa-Barth-Wakolbinger (1981)

$$V(x) = \kappa^2(x^2 + y^2),$$

at the origin.

In the p-representation we have a pair of diffusion equations

$$\frac{\partial \phi}{\partial t} + \frac{1}{2}\Delta\phi - V(x)\phi = 0,$$

and

$$-\frac{\partial \widehat{\phi}}{\partial t} + \frac{1}{2}\Delta\widehat{\phi} - V(x)\widehat{\phi} = 0.$$

As shown in Chapter 3, the distribution density of our diffusion process $\{X_t, Q\}$ is given by

$$\mu = \widehat{\phi}\,\phi.$$

Let us assume one of equilibrium (stationary) states is attained, namely

$$\widehat{\phi}(t, x) = e^{-\lambda t}\varphi(x),$$

$$\phi(t, x) = e^{\lambda t}\varphi(x).$$

Substituting them into the diffusion equations, we see immediately that $\varphi(x)$ must be a solution of an eigenvalue problem

(9.1) $-\frac{1}{2}\Delta\varphi + V(x)\varphi = \lambda\varphi.$

Therefore, the distribution density is given in this case by

$$\mu = \widehat{\phi}\phi = \varphi^2.$$

So far we have considered diffusion equations. However, as we have shown in Chapter 4, instead of the p-representation we can adopt another description in terms of the Schrödinger equation

$$i\frac{\partial \psi}{\partial t} + \frac{1}{2}\Delta\psi - V(x)\psi = 0,$$

which is equivalent to our diffusion description. Considering a stationary state, namely, substituting

$$\psi(t, x) = e^{-i\lambda t}\varphi(x),$$

we see easily that the function $\varphi(x)$ must also be a solution of eigenvalue problem (9.1).

Since excited states are degenerated, we allow in general superpositions

$$\varphi = \alpha_1\varphi_1 + \alpha_2\varphi_2,$$

where φ_1 and φ_2 are eigenfunctions associated to an excited eigenvalue.

It should be emphasized here that both real- and complex-valued descriptions explained above are *fictitious*, and hence we don't see either the sample-path behaviour of the process or the effect of the potential function $V(x)$ on the movement of monkeys. In this sense the fictitious description is not appropriate to analyze sample paths.

To investigate the sample-path behaviour of the diffusion process $\{X_t, Q\}$ we need the q-representation. In this *real* description of the q-representation we can analyze the movement of a (typical) monkey in terms of the induced drift coefficient $a = \nabla(\log \varphi)$. Namely, the transition probability q of the diffusion process is governed by a diffusion equation

$$\frac{\partial q}{\partial t} + \frac{1}{2}\Delta q + \nabla(\log \varphi)\cdot\nabla q = 0.$$

We see immediately that if there are two groups of monkeys around the feeding place, there is strong repulsive drift which segregates the two groups besides attractive environmental drift toward the feeding place.

We should notice here that the distribution density $\mu = \varphi^2$ has two different meanings, and has been treated as such; namely, as a distribution density of a single monkey (i.e. the diffusion process $\{X_t, Q\}$)

$$\mu(x)dx = \varphi(x)^2dx = Q\circ X_t^{-1},$$

on the one hand, and as a spatial statistical distribution

$$Q = \lim_{k \to \infty} \lim_{n \to \infty} Q^{(n, k)}[L_n]$$

of the groups of monkeys (cf. (8.18)) on the other hand.

In this statistical model of a monkey population we need a limit theorem: When the population size n becomes large enough, there is, as a limit, a spatial equilibrium distribution of the population $\mu(x) = \varphi^2(x)$ (cf. Chapters 7 and 8).

Eigenvalue problem (9.1) is qualitatively applied in this monkey population model.

9.2. An Eigenvalue Problem

As we have seen in the example of a monkey population, the spatial distribution density of a system of a finite number of interacting diffusion particles converges to $\mu = \varphi^2$, where φ is a solution of a (stationary) Schrödinger equation, when the population size tends to infinity. A detailed analysis of the statistical model has been done in Chapters 7 and 8.

As an example, let us consider a simple Schrödinger equation in one dimension

$$(9.2) \qquad \tfrac{1}{2}\sigma^2 \frac{d^2\varphi}{dx^2} + (\lambda - k|x|)\varphi = 0, \ \ k > 0,$$

which will be applied in the following sections. Substituting

$$x = \alpha y, \mu = \lambda(k\alpha)^{-1} \text{ and } \alpha = (\sigma^2/2k)^{1/3},$$

we get

$$(9.3) \qquad \frac{d^2\varphi}{dy^2} + (\mu - |y|)\varphi = 0.$$

The eigenvalues of (9.3) are the zeros of

$$J_{1/3}(\tfrac{2}{3}\mu^{3/2}) + J_{-1/3}(\tfrac{2}{3}\mu^{3/2}) \ \text{ and } \ J_{2/3}(\tfrac{2}{3}\mu^{3/2}) - J_{-2/3}(\tfrac{2}{3}\mu^{3/2})$$

(cf. Titchmarsh (1962), pp. 90-92). The eigenvalues of (9.3) are given in Tables 2, 3 and 5 (cf. Nagasawa (1981), Nagasawa-Yasue (1982)). The potential

$$(9.4) \qquad V(x) = k|x|, \ \ k > 0,$$

in (9.2) induces a *constant* attractive force toward the origin independent of the distance from the origin. Therefore, equation (9.2) is for a *string* model.

The statistical model for the Schrödinger equation in (9.2) was applied to septation of *Escherichia coli* and to the mass spectrum of mesons.

9.3. Septation of *Escherichia Coli* [2]

"The experiments by M.Yamada and Y. Hirota on a mutant *Escherichia coli* [strain PAT 84(fts A)] show the following: If it is kept at low temperature (30°C), it behaves normally, i.e., it grows up gradually and splits into two cells when the cell length becomes approximately twice as long as the normal size. At high temperature (41°C) it can survive, and grow exponentially without septation, becoming 20-30 (sometimes several hundred) times longer and will eventually die out. A notable peculiarity is that if the temperature is lowered down to 30°C, in 2-4 h (the cell length becomes four to eight times longer), the *E. coli* can and will start septation again after c. 20-30 min. If it is, for example, four times longer, the most frequent septation site is 1 : 3 [cf. Fig 1 and Table 1, which are due to Yamada & Hirota, Nat. Inst. of Genetics, Mishima, Japan (personal communication)]. However, if it is kept at high temperature for more than 3 h, the cell length becomes more than eight times as long as the normal size and it can no longer split, even though the temperature is lowered".

TABLE 1

Position of septa after temperature shift down

Pattern of division	Number of filament	
	Exp. 1	Exp. 2
⌊_____⌋	11	10
⌊__⌊_____⌋	57	34
⌊__⌊_____⌊_____⌋	13	38
⌊_____⌊_____⌋	10	7
⌊____⌊_____⌊_____⌋	1	0
⌊___⌊___⌊___⌊_____⌋	3	4
Others	5	7
Total	100	100
Division sites per filament	1.14	1.33

"We consider an *E. coli* as *a population* (or collection) *of molecules* (perhaps A, G, C, U, etc.) which constitute DNA, RNA, proteins, etc., in the *E. coli*". "We assume that the movement of an ideal molecule can be

[2] Cf. Nagasawa (1981)

described as a sample path of a diffusion process which has the equilibrium distribution density. Moreover, we simplify the model to be a one-dimensional diffusion process, as we are interested in the cell length of *E. coli*. Since DNA, RNA, protein, etc. are very long and tangle together, we assume that an ideal molecule is attracted to the centre, i.e. we regard *E. coli* as a distribution of a diffusion process on a real line under an attractive environment potential (environment here is not of the *E. coli*, but of the ideal molecule *in the E. coli*)". Then we consider an eigenvalue problem (Schrödinger equation) (9.2).

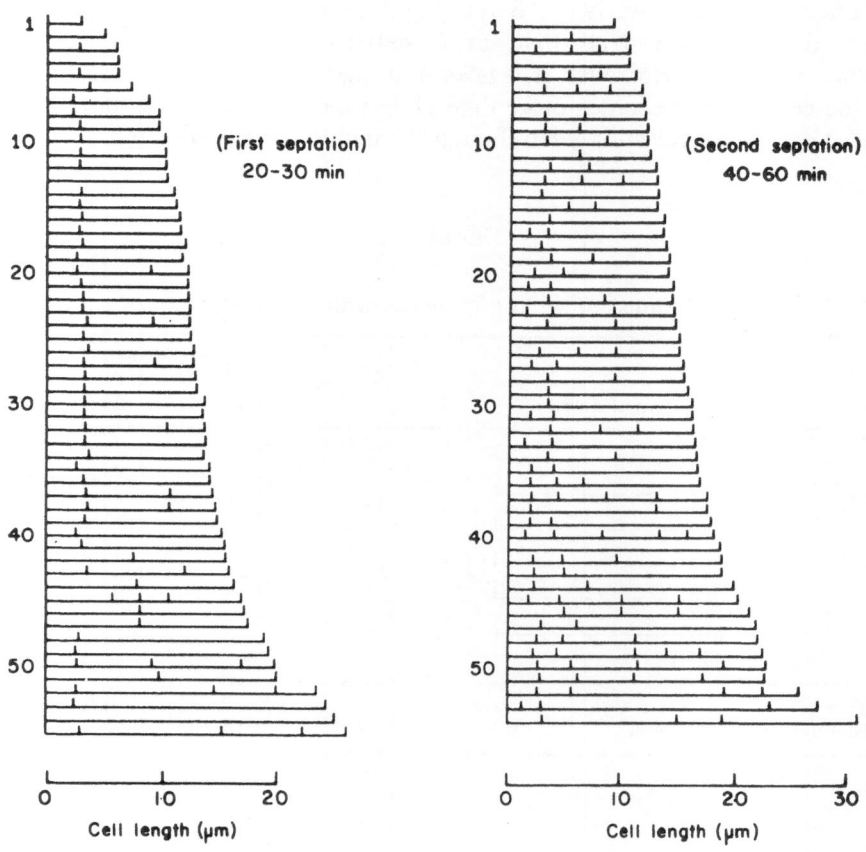

FIG. 1. Position of septa after temperature shift down.

TABLE 2

$J_{2/3}(x) - J_{-2/3}(x) = 0$	$J_{1/3}(x) + J_{-1/3}(x) = 0$
$x_1 = 0.685\,548$	$x_2 = 2.383\,447$
$x_3 = 3.902\,765$	$x_4 = 5.510\,196$
$x_5 = 7.054\,930$	$x_6 = 8.647\,358$
$x_7 = 10.200\,688$	$x_8 = 11.786\,843$
$x_9 = 13.344\,503$	$x_{10} = 14.927\,207$
$x_{11} = 16.487\,475$	$x_{12} = 18.067\,995$
$x_{13} = 19.630\,008$	$x_{14} = 21.209\,021$
$x_{15} = 22.772\,281$	$x_{16} = 24.350\,193$
$x_{17} = 25.914\,390$	$x_{18} = 27.491\,460$

TABLE 3

Eigenvalues of equation (9.3)

$\mu_1 = 1.018\,793\,3$	$\mu_{10} = 7.944\,131\,9$
$\mu_2 = 2.338\,107\,5$	$\mu_{11} = 8.488\,484\,8$
$\mu_3 = 3.248\,197\,2$	$\mu_{12} = 9.022\,648\,6$
$\mu_4 = 4.087\,949\,1$	$\mu_{13} = 9.535\,447\,0$
$\mu_5 = 4.820\,098\,5$	$\mu_{14} = 10.040\,172$
$\mu_6 = 5.520\,559\,0$	$\mu_{15} = 10.527\,658$
$\mu_7 = 6.163\,306\,1$	$\mu_{16} = 11.008\,522$
$\mu_8 = 6.786\,706\,7$	$\mu_{17} = 11.475\,054$
$\mu_9 = 7.372\,175\,8$	$\mu_{18} = 11.936\,013$

COMPARISON WITH EXPERIMENT

We interpret $\ell_k = 2\mu_k$ as the cell length of an *E. coli* when it is in the equilibrium state of energy (excitation) μ_k, $k \geq 1$.

(A) THE NORMAL SEPTATION

The theory implies first of all the *existence of the minimal cell length* ℓ_1 which is realized as the ground state of μ_1. This is what we have called the "normal cell length". An *E. coli* of this minimal cell length grows up gradually. If it had been able to reach the second state of μ_2, it could have become a double-sized *E. coli*. However, this does not occur. Instead, the *E. coli* splits into two cells, because the sum of the energies of the two pieces of the minimal *E. coli* is lower than the energy of the second equilibrium state (cf. Fig. 2):

$$2\mu_1 < \mu_2,$$

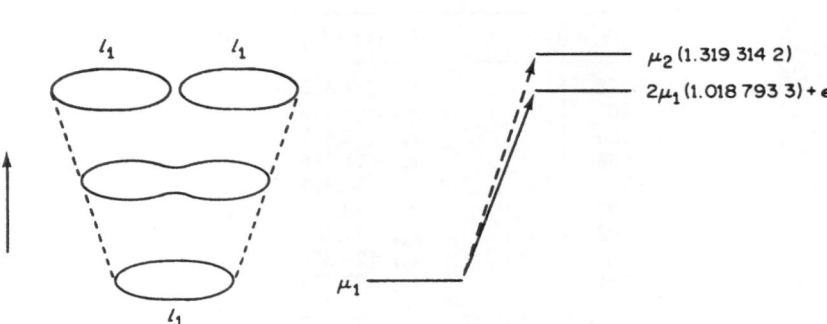

FIG. 2

We can interpret this as the normal septation of *E. coli* at low temperature (30°C). We assume that, in order to realize the septation, some extra energy ($\varepsilon \approx 0.1$) is required for membrane formation at each septation site.

(B) A MUTANT *E. COLI*

At 41°C the septation mechanism is switched off in the mutant *E. coli* which can grow in this high temperature. When the temperature is shifted down to 30°C, the septation mechanism is switched on again and *E. coli* will split after 20-30 min from the switch on time (cf. Fig. 1).

If the mutant *E. coli* is in the state μ_2, i.e., if its cell length has become a little bit longer than double that of the minimal size, it can hardly split, but most probably will grow up toward the state of μ_3, because

$$2\mu_1 + \varepsilon < \mu_2 < \mu_3.$$

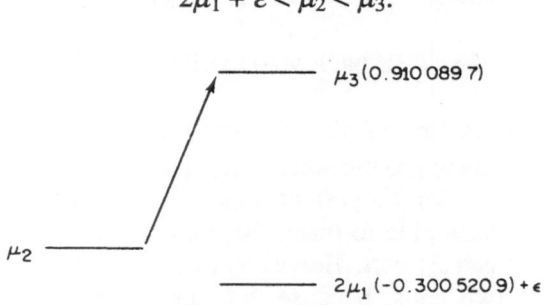

FIG. 3

If the cell length has become three times longer, the *E. coli* will split into two cells of length l_1 and l_2. It can hardly reach the next equilibrium state of μ_4, because

$$\mu_3 < \mu_1 + \mu_2 + \varepsilon < \mu_4.$$

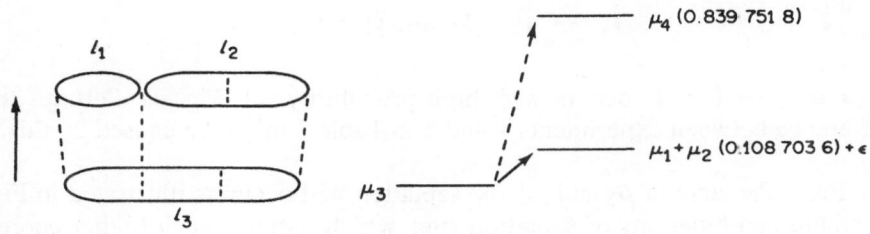

FIG. 4

If the cell length has become four times longer than the minimal size, several combinations of septation sites are possible, namely

$$l_1 + l_3, \quad l_1 + l_2 + l_1, \quad l_2 + l_2, \quad l_1 + l_1 + l_1 + l_1 \,.$$

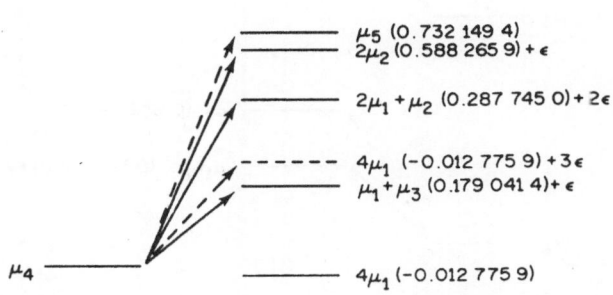

FIG. 5

Even though $4\mu_1 + 3\varepsilon$ is lower than μ_5, one can expect that the occurrence of this septation into four pieces of length l_1 is rare, because μ_4 and $4\mu_1$ are approximately equal and this means that if it could occur the *E. coli* must spend all its energy obtained from outside food sources just for membrane formation at three septation sites without growing. Figure 5 shows that the septation of $l_1 + l_3$ is most probable, $l_1 + l_2 + l_1$ the next,

$l_2 + l_2$ the third, and probably $l_1 + l_1 + l_1 + l_1$ the fourth. This explains well the observed frequency of septation sites (cf. Table 1).

It is not easy to distinguish between the following two patterns:

$$l_4 \rightarrow l_1 + l_2 + l_1,$$

$$l_4 \rightarrow l_1 + l_3 \rightarrow l_1 + (l_2 + l_1),$$

because $l_3 \rightarrow l_2 + l_1$ occurs with high probability (cf. Fig. 4). Perhaps the difference between experiments 1 and 2 in Table 1 might be caused by this.

From the state of μ_5 and μ_6, the septation will occur as illustrated in Fig. 6, where combinations of septation sites which require much higher energy than the next excited state are omitted.

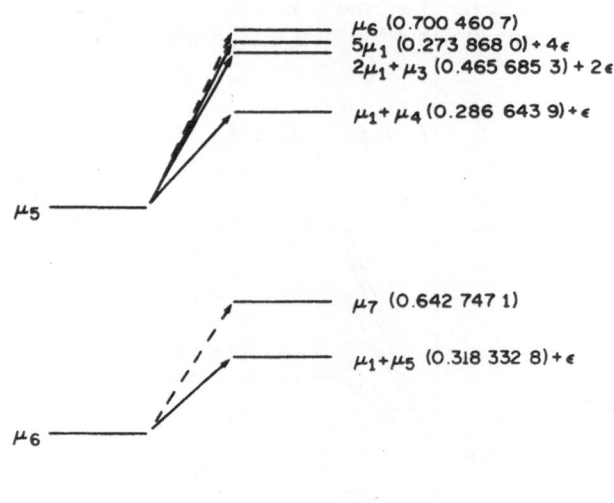

μ_6 (0.700 460 7)
$5\mu_1$ (0.273 868 0) + 4ϵ
$2\mu_1 + \mu_3$ (0.465 685 3) + 2ϵ
$\mu_1 + \mu_4$ (0.286 643 9) + ϵ

μ_5

μ_7 (0.642 747 1)
$\mu_1 + \mu_5$ (0.318 332 8) + ϵ

μ_6

FIG. 6

For the E. coli in the state μ_k ($k = 7, 8, 9$), the septation into $l_1 + l_{k-1}$ can occur, because

$$\mu_k < \mu_1 + \mu_{k-1} + \varepsilon < \mu_{k+1}.$$

There is a threshold number for k; if $k \geq 10$ (or 11, this threshold number depends on ε, and remember that we are assuming $\varepsilon \approx 0.1$), then the order of μ_{k+1} and $\mu_1 + \mu_{k-1} + \varepsilon$ is interchanged:

$$\mu_k < \mu_{k+1} < \mu_1 + \mu_{k-1} + \varepsilon .$$

Therefore, the *E. coli* reaches the next equilibrium state first, if it has arrived at an excited state of $\mu_k > \mu_{10}$ (or μ_{11}). Note that ℓ_{10} (ℓ_{11}) $\approx 8\ell_1$. This means, once an *E. coli* has become more than eight times as long as the minimal size, most probably it will never split any more but just grow up from an equilibrium state to the next higher one and so on (cf. Fig. 7). This is what has been observed in the experiments by Yamada and Hirota.

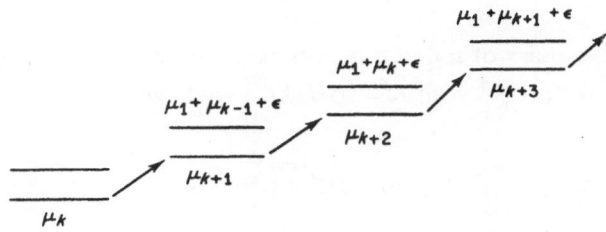

FIG. 7

Remark. The fact that the mutant *E. coli* will eventually die out if it becomes too long indicates that actual environment potential might be $V(x) = k|x| \wedge h$.

9.4. The Mass Spectrum of Mesons

We consider a meson something like a mutant *E. coli* at normal and high temperature. This means we assume equation (9.2). We replace molecules by (infinitely many virtual) glueons which are inside a meson. In the case of *E. coli* we compared eigenvalues not absolutely but relatively, and hence we did not specify the diffusion coefficient. Since we need not only a relative comparison of eigenvalues but also absolute values to determine the mass of mesons, we must find out the *specific diffusion coefficient for glueons*. The mass of a meson in our model is the sum of eigenvalues of (9.2), i.e., the mass of a glueon distribution, a contribution from angular momentum, and the mass of two quarks:

(9.5) $M_{n,j}(q,q') = \lambda_n + j\,m_a + m_{\bar{q}} + m_{q'}$,

where λ_n is an eigenvalue of (9.2) (= the mass of the glueon distribution), j angular momentum of the meson, m_a contribution to the mass per angular momentum, and m_q stands for the mass of a quark q. We denote a meson in our model as (q, ϕ_k, j, q') in terms of two quarks q and q', a glueon distribution ϕ_k, and angular momentum j.

We identify the smallest eigenvalue corresponding to the ground state with the π-meson which is the lightest of all mesons;

$$(u \ or \ d, \phi_1, 0, u \ or \ d) = \pi\text{-meson},$$

and assume the mass of u and d quarks is negligible. Because of the change of variable $\mu = \lambda(k\alpha)^{-1}$ with $\alpha = (\sigma^2/2k)^{1/3}$ we have

(9.6) $\lambda_1 = \mu_1 \left(\dfrac{(\sigma k)^2}{2} \right)^{1/3}$,

$$\mu_1 = 1.018\ 793\ 3$$

(cf Table 5). As the mass of π-meson we take

(9.7) $m(\pi) = 139.566\ 9$ Mev.

From the identification $\lambda_1 = m(\pi)$ we get

(9.8) $\{(\sigma k)^2/2\}^{1/3} = \dfrac{m(\pi)}{\mu_1} = 136.993\ 36$

through which we can determine the specific diffusion coefficient for glueons.[3] . It should be noticed that the value we have obtained above in (9.8) is almost equal to the inverse of the fine structure constant.

There is no meson corresponding to the second eigenvalue λ_2 (see Fig 2), by the same reasoning as for $E.\ coli$.

By comparing the mass of $K^{\pm}$ mesons which contain a single s-quark, we take the mass of s-quark to be 50 Mev;

---

[3] We can not determine σ, but σk

$$(u \text{ or } d, \phi_3, s) = K^{\pm}\text{-meson.}^4$$

In comparison with the mass of $\rho(1)$ $(m = 769)$ and $\omega(1)$ $(m = 783)$ we choose $m_a = \omega(1) - \rho(0) \approx 15$ Mev.

By identifying the mass of D^0 and D mesons which contains a single c-quark, we postulate the mass of c-quark to be 700 Mev;

$$(u \text{ or } d, \phi_{11}, 0, c) = D\text{-meson.}$$

Thus, all parameters in (9.5) are fixed. The mass spectrum of mesons is computed numerically in Nagasawa-Yasue (1982) based on formula (9.5) of the statistical string model (for λ_n see Table 5). Some of them are given in Table 4 .

A comparison with given experimental data shows fairly good agreement except in the case 42 of $F^{\pm}$ (now called $D_s^{\pm}$). For this case, "according to the "*Data Booklet* 1982", the observed mass was *2 021* ±*15* Mev. On the other hand the mass computed by formula (9.5) is $(s, \phi_{12}, 0, c)$ = *1 986* Mev, which is too light to be identified with *2 021* Mev. Therefore, it was not possible to identify $F^{\pm}$ with our composite model of mesons. However, in the "*Data Booklet* 1984" the mass of $F^{\pm}$ was corrected to be *1 971* ±*16* Mev, which agrees with our predicted value *1 986* Mev" (cf. Nagasawa (1985)). According to the "*Stable Particle Summary Table* (1988)", the mass of $D_s^{\pm}$ is *1 969.3* ±*1.1* Mev.

Moreover, we made, based on our statistical string model, a speculative prediction about the mass of $B^{\pm}$ (resp. B^0) which was not known in 1982. The prediction says that it must appear between *5 000* and *5 500* Mev (cf. Nagasawa-Yasue (1982)). In the "*Stable Particle Summary Table* (1988)" the mass of $B^{\pm}$ is given as *5 277.6* ±*1.4* Mev, as predicted.

Therefore, we conclude that the substantial part of our statistical composite model of mesons is compatible with experiments, that is, *a considerable portion of the mass of a meson is contributed by its glueon distribution* (*not only by the mass of quarks*), in contrast to conventional meson models in which it is assumed that the masses of mesons are from quarks, and ignore glueon distributions.

---

[4] We assume heavier quarks need larger eigenvalues, namely more glueons

TABLE 4

Mass (Mev)   *computed*

		Mass (Mev)	computed	
Case 1:	$\pi^{\pm}$	139.5669	139.5669	(postulated)
Case 2:	$K^{\pm}$	493.67	495	(u or $d, \phi_3, 0, s$)
Case 3:	η	549	545	($s, \phi_3, 0, s$)
Case 4:	ρ	769	756	(u or $d, \phi_6, 0, u$ or d))
Case 5:	ω	783	771	(u or $d, \phi_6, 1, u$ or d))
Case 6:	K^*	892	894	(u or $d, \phi_7, 1, s$)
Case 7:	η'	958	944	($s, \phi_7, 0, s$)
Case 8:	S*(0) (975),	δ(0) (980)		*are not well identified*
Case 9:	$\phi(1)$	1 020	1 030	($s, \phi_8, 0, s$)
Case 10:	H(1)	1 190	1 178	(u or $d, \phi_{11}, 1, u$ or d)
Case 11:	B(1)	1 233	1 236	(u or $d, \phi_{12}, 0, u$ or d)
Case 13:	$Q_1(1)$	1 270	1 286	($s, \phi_{12}, 0, u$ or d)
Case 14:	f(2)	1 273	1 266	(u or $d, \phi_{12}, 2, u$ or d)
Case 15:	$A_1(1)$	1 275	1 251	(u or $d, \phi_{12}, 1, u$ or d)
Case 17:	D(1)	1 283	1 278	($s, \phi_{11}, 0, s$)
Case 18:	$\varepsilon(0)$	1 300	1 306	(u or $d, \phi_{13}, 0, u$ or d)
....	....	......	......	
Case 39:	D	1 869	1 863	(u or $d, \phi_{11}, 0, c$)
....	....	......	......	
Case 42:	$F^{\pm}$	2 021	(1 986)	($s, \phi_{12}, 0, c$)
....	....	......	......	
Case 58:	J/ψ	3 097	3 097	($c, \phi_{19}, 0, c$)
Case 59:	$\chi(0)$	3 415	3 390	($c, \phi_{24}, 0, c$)
Case 60:	$\chi(1)$	3 510	3 517	($c, \phi_{26}, 0, c$)
Case 61:	$\chi(2)$	3 556	3 571	($c, \phi_{27}, 1, c$)
Case 62:	η_c'	3 590	3 586	($c, \phi_{27}, 2, c$)
Case 63:	$\psi(1)$	3 686	3 678	($c, \phi_{29}, 1, c$)
Case 64:	$\psi(1)$	3 770	3 768	($c, \phi_{31}, 0, c$)
....	....	......	.....	

Case 12 and Case 16 are not well-established resonances and are omitted.

TABLE 5

n	μ_n	n	λ_n
1	1.018 793 3	1	139.566 90
2	2.338 107 7	2	320.302 89
3	3.248 197 6	3	444.978 25
4	4.087 949 6	4	560.017 87
5	4.820 099 2	5	660.316 77
6	5.520 560 0	6	756.274 54
7	6.163 307 2	7	844.326 00
8	6.786 708 1	8	929.727 16
9	7.372 177 4	9	1009.932 0
10	7.944 133 7	10	1088.285 6
11	8.488 486 7	11	1162.857 8
12	9.022 650 7	12	1236.034 2
13	9.535 449 2	13	1306.283 7
14	10.040 174	14	1375.427 2
15	10.527 660	15	1442.209 0
16	11.008 524	16	1508.083 7
17	11.475 056	17	1571.995 1
18	11.936 016	18	1635.142 9
19	12.384 788	19	1635.142 9
20	12.828 777	20	1757.444 4
21	13.262 219	21	1816.822 7
22	13.691 489	22	1875.629 4
23	14.111 502	23	1933.167 9
24	14.527 830	24	1990.201 7
25	14.935 937	25	2046.109 3
26	15.340 755	26	2101.566 2
27	15.738 201	27	2156.013 4
28	16.132 685	28	2210.054 6
29	16.520 504	29	2263.182 8
30	16.905 634	30	2315.942 7
31	17.284 695	31	2367.871 2
32	17.661 300	32	2419.463 2
33	18.032 345	33	2470.293 5
34	18.401 133	34	2520.814 6
35	18.764 798	35	2570.634 0
36	19.126 381	36	2620.168 0
37	19.483 222	37	2669.052 5
38	19.838 130	38	2717.672 2
39	20.188 631	39	2765.688 3

9.5. Titius-Bode Law

The statistical model is applied to the formation of the orbits of planets, and a statistical interpretation of the Titius-Bode law is given by Albeverio-Blanchard- Høegh-Krohn (1984).

Chapter X

Relative Entropy and Csiszar's Projection

This chapter is devoted to a brief exposition of relative entropy, and to proofs of Csiszar's projection theorem and those on exponential families needed in Chapter 5.

10.1. Relative Entropy

Let $\{\Omega, \mathcal{B}\}$ be a measurable space,

$$\mathbf{M}_1(\Omega) : \textit{the space of probability measures on } \Omega,$$

and the **relative entropy**[1] $H(Q|P)$ of Q with respect to P be defined by

(10.1) $$H(Q|P) = \int (\log \frac{dQ}{dP}) \, dQ, \;\; \textit{if } Q << P, \;\; (= \infty, \; \textit{otherwise}).$$

where $P, Q \in \mathbf{M}_1(\Omega)$. Denoting the Radon-Nikodym derivative of Q with respect to P as

$$q = q_P = \frac{dQ}{dP},$$

the relative entropy $H(Q|P)$ defined by (10.1) can be written in terms of the density function q as

---

[1] It has appeared under various names, cf. Kullback (1959), Csiszar (1975, 84)

$$(10.2) \qquad\qquad H(Q\,|\,P) = \int q \log q \, dP.$$

There is another expression of the relative entropy given as the supremum of the ones defined on finite measurable partitions :

$$(10.3) \qquad\qquad H(Q\,|\,P) = \sup_{\mathcal{P}} H_{\mathcal{P}}(Q\,|\,P),$$

where $\mathcal{P} = \{\Omega_1, \dots, \Omega_k\}$ runs over all finite measurable partitions of Ω and

$$H_{\mathcal{P}}(Q\,|\,P) = \sum_{i=1}^{k} Q(\Omega_i) \log \frac{Q(\Omega_i)}{P(\Omega_i)} \, .$$

Formula (10.3) will be applied in the next chapter but in the following form as an inequality

$$(10.4) \qquad\qquad H_{\mathcal{P}}(Q\,|\,P) \le H(Q\,|\,P).$$

This can be easily shown as follows:

Let us denote $f(x) = x \log x$, which is a convex function of $x \ge 0$. Let $\sigma(\mathcal{P})$ be the σ-algebra generated by $\mathcal{P}$. Then

$$H(Q\,|\,P) = P[f(q)] = P\big[P[f(q)\,|\,\sigma(\mathcal{P})]\big],$$

where $q = \dfrac{dQ}{dP}$. Jensen's inequality yields

$$f(P[q\,|\,\sigma(\mathcal{P})]) \le P[f(q)\,|\,\sigma(\mathcal{P})], \quad P\text{-}a.e.;$$

moreover it is clear that

$$f(P[q\,|\,\sigma(\mathcal{P})]) = \sum_{i=1}^{k} f\Big(\frac{Q(\Omega_i)}{P(\Omega_i)}\Big) 1_{\Omega_i}, \quad P\text{-}a.e.,$$

and hence combining them, we get

$$\sum_{i=1}^{k} Q(\Omega_i) \log \frac{Q(\Omega_i)}{P(\Omega_i)} \le H(Q \mid P),$$

which is nothing but (10.4).

Lemma 10.1. *Let* $H(Q \mid P)$ *be the relative entropy of* Q *with respect to* P *defined in* (10.1). *Then*

(10.5) $H(Q \mid P) \ge 0,$

where the equality holds if and only if $Q = P$.

Proof. Let us define a function $g(x)$ on $[0, \infty)$ by

(10.6) $g(x) = x \log x - x + 1,$

which is non-negative, strictly convex, and takes the minimum $g(1) = 0$. Therefore,

$$0 \le \int g(q) dP = \int q \log q \, dP = H(Q \mid P),$$

and the minimum is attained with $q = 1$, P-a.e., namely

$$H(Q \mid P) = 0,$$

if and only if $Q = P$. This completes the proof.

The **total variation** $\| Q - P \|_{Var}$ of the difference $P - Q$ is defined by

$$\| Q - P \|_{Var} = \inf_{f \in B_1} | Q(f) - P(f) |,$$

and if $P, Q \ll R$, then

(10.7) $\| Q - P \|_{Var} = \| Q - P \|_R = \int | q_R - p_R | dR.$

We omit "*Var*" or "*R*" at the norm, if it is clear, we denote, for example,

$$\| Q - P \| = \int \, | q_P - 1 | \, dP.$$

The relative entropy $H(Q|P)$ is an important tool to estimate the total variation $\| Q - P \|$ as is shown in

Lemma 10.2. *Between the total variation and the relative entropy the following inequality holds:*

(10.8) $\{ \| Q - P \|_{Var} \}^2 \leq 2 H(Q \, | \, P), \quad for \ P, Q \in \mathbf{M}_1(\Omega).$

Proof. We can assume $Q \ll P$. Denote, for $x > 0$,

$$h(x) = 4 + 2x,$$

and

$$g(x) = x \log x - x + 1.$$

Then, $h(x), g(x) \geq 0$, and

(10.9) $3(x - 1)^2 \leq h(x) g(x),$

because $F(x) = h(x) g(x) - 3(x - 1)^2$ is convex and attains the minimum $F(1) = 0.$[2] The inequality (10.9) implies, with $q = q_P$,

$$3 \{ \int \, | q - 1 | \, dP \}^2 \leq \{ \int \, dP \, \sqrt{3 | q - 1 |^2} \ \}^2 \leq \int \, h(q) \, dP \int g(q) \, dP$$

$$\leq 6 \int g(q) \, dP = 6 \int q \log q \, dP,$$

which is (10.8).

Lemma 10.3. (Parallelogram identity)

(10.10) $H(P|R) + H(Q|R) = 2 \, H(\frac{P+Q}{2} | R) + H(P | \frac{P+Q}{2}) + H(Q | \frac{P+Q}{2}).$

Proof. We can assume $P, Q \ll R$. Expressing all terms on the right-hand side of (10.10) with densities, such as

---

[2] Cf. Kemperman (1967)

$$2H(\frac{P+Q}{2}|R) = 2\int \frac{p_R + q_R}{2} \log \frac{p_R + q_R}{2} dR,$$

$$H(P|\frac{P+Q}{2}) = \int p_R \log \frac{2p_R}{p_R + q_R} dR,$$

$$H(Q|\frac{P+Q}{2}) = \int q_R \log \frac{2q_R}{p_R + q_R} dR,$$

and adding them, we get (10.10).

10.2. Csiszar's Projection

In terms of the relative entropy $H(P|\bar{P})$, we define the projection of a probability measure $\bar{P} \in \mathbf{M}_1(\Omega)$ on a variation-closed convex subset $\mathbf{A} \subset \mathbf{M}_1(\Omega)$. The existence of such a projection plays a crucial role in treating the relative entropy like Riesz's projection theorem does in Hilbert spaces.

Theorem 10.1. (Csiszar (1975)) *Let a reference measure $\bar{P} \in \mathbf{M}_1(\Omega)$ be fixed.*

(i) *If a subset $\mathbf{A}$ of $\mathbf{M}_1(\Omega)$ is convex and variation closed, and the subset $\mathbf{A}$ contains at least one element P with $H(P|\bar{P}) < \infty$, then there exists* **Csiszar's projection** $Q \in \mathbf{A}$ (*on the subset $\mathbf{A}$) of $\bar{P}$ such that*

(10.11) $$\inf_{P \in \mathbf{A}} H(P|\bar{P}) = H(Q|\bar{P}),$$

where Q is uniquely determined.

(ii) $Q \in \mathbf{A}$ *is the Csiszar projection (on the subset $\mathbf{A}$) of $\bar{P}$, if and only if $H(Q|\bar{P}) < \infty$, and*

(10.12) $$H(P|\bar{P}) \geq H(P|Q) + H(Q|\bar{P}), \quad for \ \forall P \in \mathbf{A}.$$

Proof. Denote $\rho = \inf_{P \in \mathbf{A}} H(P|\bar{P})$. Let $P_n \in \mathbf{A}$ be a sequence with $H(P_n|\bar{P}) < \infty$ such that

$$\lim_{n \to \infty} H(P_n | \bar{P}) = \rho.$$

Since the subset $\boldsymbol{A}$ is convex,

$$\frac{P_m + P_n}{2} \in \boldsymbol{A}$$

and hence

$$H(\frac{P_m + P_n}{2} | \bar{P}) \geq \rho.$$

Therefore, with the help of the parallelogram identity (10.10), we get

$$H(P_m | \bar{P}) + H(P_n | \bar{P}) - 2\rho$$

$$\geq H(P_m | \bar{P}) + H(P_n | \bar{P}) - 2H(\frac{P_m + P_n}{2} | \bar{P})$$

$$= H(P_m | \frac{P_m + P_n}{2}) + H(P_n | \frac{P_m + P_n}{2}) \geq 0,$$

from which it follows that

(10.13) $H(P_m | \dfrac{P_m + P_n}{2}), \quad H(P_n | \dfrac{P_m + P_n}{2}) \to 0, \quad as \ m, n \to \infty.$

This implies, with the help of Lemma 10.2,

$$\| P_m - P_n \| \leq \| P_m - \frac{P_m + P_n}{2} \| + \| P_n - \frac{P_m + P_n}{2} \|.$$

$$\leq \sqrt{2H(P_m | \frac{P_m + P_n}{2})} + \sqrt{2H(P_n | \frac{P_m + P_n}{2})} \to 0, \quad as \ m, n \to \infty,$$

namely, the sequence $\{P_n\}$ converges in total variation. Since the subset $\boldsymbol{A}$ is variation closed, there exists an element $Q \in \boldsymbol{A}$ such that $\| P_n - Q \| \to 0$, as $n \to \infty$. Then by the Fatou lemma, denoting by q and p_n the densities of Q and P_n with respect to $\bar{P}$,

$$H(Q | \bar{P}) = \int q \log q \, d\bar{P} \leq \liminf_{n \to \infty} \int p_n \log p_n \, d\bar{P}$$

$$= \liminf_{n \to \infty} H(P_n | \bar{P}) = \rho = \inf_{P \in \boldsymbol{A}} H(P | \bar{P}).$$

On the other hand, since $\inf_{P \in \boldsymbol{A}} H(P|\bar{P}) \leq H(Q|\bar{P})$ is clear, we have

$$\inf_{P \in \boldsymbol{A}} H(P|\bar{P}) = H(Q|\bar{P}).$$

The projection Q (on the set $\boldsymbol{A}$) of $\bar{P}$ is uniquely determined, because of Lemmas 10.1 and 10.3.

For the second assertion, we first note a simple identity

$$(10.14) \qquad H(P|\bar{P}) - H(P|Q) = \int (\bar{p}\log\bar{p} - \bar{p}\log\frac{\bar{p}}{\bar{q}})\, d\bar{P}$$

$$= \int \bar{p}\log\bar{q}\, d\bar{P} = \int \log\bar{q}\, dP,$$

where $\bar{p}$ and $\bar{q}$ denote the Radon-Nikodym derivatives of P and Q with respect to $\bar{P}$. Then we prove

Lemma 10.4. $Q \in \boldsymbol{A}$ *is the Csiszar projection* (*on the subset* $\boldsymbol{A}$) *of* $\bar{P}$, *if and only if* $H(Q|\bar{P}) < \infty$ *and*

$$(10.15) \qquad \int \log\bar{q}\, dP \geq H(Q|\bar{P}), \quad for \ , \forall\, P \in \boldsymbol{A},$$

namely (10.12) *holds.*

Proof. Let $p_\alpha = \alpha\bar{p} + (1 - \alpha)\bar{q}$, $0 \leq \alpha \leq 1$ $(dP_\alpha = p_\alpha d\bar{P})$. Then

$$f_\alpha = f(p_\alpha) = p_\alpha \log p_\alpha$$

is a convex function of α, since it is a composition of a convex function $f(x) = x\log x$ and a linear function p_α. Therefore, $(f_\alpha)' \downarrow$ as $\alpha \downarrow 0$, and moreover

$$(10.16) \qquad (f_\alpha)'|_{\alpha = 0} = (\bar{p} - \bar{q})(\log\bar{q} + 1).$$

Therefore, we get

$$(10.17) \qquad \frac{d}{d\alpha} H(P_\alpha|\bar{P})|_{\alpha=0} = \int \log\bar{q}\, dP - H(Q|\bar{P}).$$

In fact, the left-hand side of (10.17) is equal to

$$\lim_{\alpha \downarrow 0} \frac{1}{\alpha} \{ H(P_\alpha | \overline{P}) - H(P_0 | \overline{P}) \} = \lim_{\alpha \downarrow 0} \frac{1}{\alpha} \int (f_\alpha - f_0) \, d\overline{P}$$

$$= \int (f_\alpha)' |_{\alpha = 0} \, d\overline{P}$$

$$= \int (\overline{p} - \overline{q})(\log \overline{q} + 1) \, d\overline{P}$$

$$= \int \log \overline{q} \, dP - H(Q | \overline{P}),$$

where we have substituted (10.16) at the third equality. Suppose (10.15) is not the case. Then the derivative at (10.17) is negative, and hence we can find $\alpha > 0$ such that

$$H(P_\alpha | \overline{P}) < H(P_0 | \overline{P}).$$

Since $P_0 = Q$ and $P_\alpha \in \textbf{\textit{A}}$, Q is not Csiszar's projection on the subset $\textbf{\textit{A}}$ of $\overline{P}$. Conversely, assume (10.15). Then formula (10.14) combined with (10.15) yields (10.12), and hence Q is Csiszar's projection. Therefore, Q is Csiszar's projection of $\overline{P}$ if and only if inequality (10.15) holds. This completes the proof.

Remark 10.1.   If there exists Csiszar's projection Q, then clearly $H(Q|\overline{P}) < \infty$. Therefore, $\overline{q} \log \overline{q} \in L^1(\overline{P})$, and hence $\log \overline{q} \in L^1(Q)$.

10.3. Exponential Families and Marginal Distributions

The proof of Lemma 10.4 shows more than the assertion of the lemma. Let us formulate first of all a segment property.

Let $Q \in \textbf{\textit{A}}$. If there are $P_1, P_2 \in \textbf{\textit{A}}$, $P_1 \neq Q$ such that

(10.18) $Q = \alpha P_1 + (1 - \alpha) P_2, \quad 0 < \alpha < 1,$

then Q has the **segment property** in $\textbf{\textit{A}}$ *with* $P_1, P_2 \in \textbf{\textit{A}}$.

Lemma 10.5. *If Q is Csiszar's projection on $\mathbf{A}$ of $\overline{P}$ and has the segment property in $\mathbf{A}$ with $P_1, P_2 \in \mathbf{A}$, then*

$$(10.19) \qquad \int \log \overline{q} \, dP_i = \mathrm{H}(Q \mid \overline{P}) < \infty, \quad for \ i = 1, 2,$$

where $\overline{q} = dQ/d\overline{P}$, and the equality holds in (10.12) with $P = P_1, P_2$.

Proof. Because of Lemma 10.4 we have (10.15). Suppose

$$\int \log \overline{q} \, dP_i > \mathrm{H}(Q \mid \overline{P}), \quad for \ i = 1 \ or \ 2.$$

Then (10.18) implies

$$\int \log \overline{q} \, dQ > \mathrm{H}(Q \mid \overline{P}),$$

but this is a contradiction. Thus (10.19) holds, and the last assertion is clear from (10.14) and (10.19), which completes the proof.

Condition (10.18) looks rather restrictive, but it is actually not, when we will apply Lemma 10.5 in the proof of the following theorem 10.2 on exponential families.

Let us define a subset $\mathbf{E} \subset \mathbf{M}_1(\Omega)$ by

$$(10.20) \qquad \mathbf{E} = \{ P \in \mathbf{M}_1(\Omega) : \int f_i \, dP = a_i, \ 1 \le i \le k \},$$

for prescribed bounded measurable functions f_i and $a_i \in \mathbf{R}$, $i = 1, 2, \dots, k$.

Then we have a theorem on exponential families of distributions:

Theorem 10.2. *Let $\mathbf{E}$ be defined in (10.20) and assume there exists at least one $P \in \mathbf{E}$ such that $\mathrm{H}(P \mid \overline{P}) < \infty$. Let $Q \in \mathbf{E}$ with $Q \ll \overline{P}$ and denote $\overline{q} = dQ/d\overline{P}$. Then Q is Csiszar's projection on the set $\mathbf{E}$ of $\overline{P}$, if and only if*

$$(10.21) \quad \overline{q}(\omega) = c \exp(\sum_{i=1}^{k} \alpha_i f_i(\omega)), \quad if \ \overline{q}(\omega) > 0, \ (= 0, \ otherwise).$$

Proof. The subset E is clearly convex and variation closed. Let Q be Csiszar's projection on the set E of $\bar{P}$. Instead of the set E, we consider a subset E_Q of E defined by

$$E_Q = \{P \in E : p \le 2, P = pQ\};$$

then, since $Q \in E_Q$, it is clear that

(10.22) $$\inf_{P \in E_Q} H(P \mid \bar{P}) = \inf_{P \in E} H(P \mid \bar{P}) = H(Q \mid \bar{P}).$$

For $P \in E_Q$ ($P \ne Q$), define $P' = 2Q - P \in E_Q$. Then clearly

$$Q = \frac{P + P'}{2},$$

and hence Q has the segment property in the subset E_Q for every $P, P' \in E_Q$ defined above. Therefore, Lemma 10.5 can be applied, and because of (10.19) we have

$$\int \log \bar{q}\, dP = H(Q \mid \bar{P}), \quad for \ \ \forall P \in E_Q,$$

and hence the inequality (10.12) holds with equality, namely

(10.23) $$H(P \mid \bar{P}) = H(P \mid Q) + H(Q \mid \bar{P}), \quad for \ \ \forall P \in E_Q .$$

Moreover

(10.24) $$\int (p - 1) \log \bar{q}\, dQ = 0, \quad for \ \ \forall P \in E_Q,$$

where $\bar{q} = dQ/d\bar{P}$ and $p = dP/dQ$. Let

(10.25) $$\mathcal{H} = \{h: measurable, \int h\, dQ = 0, and \int f_i h\, dQ = 0, for\ 1 \le i \le k\}.$$

Then it is clear that for $h \in \mathcal{H}$, with $\mid h \mid \le 1$, we have

$$P = (1 + h)Q \in E_Q .$$

Therefore, (10.24) yields

(10.26) $$\int h \log \bar{q} \, dQ = 0, \quad for \ \forall \, h \in \mathcal{H} \cap L^{\infty}(Q).$$

This implies that $\log \bar{q}$ belongs to the closed subspace of $L^1(Q)$ spanned by $\{1, f_i : 1 \leq i \leq k\}$, since the dual space of $L^1(Q)$ is $L^{\infty}(Q)$. This proves expression (10.21) for the Csiszar projection Q.

Conversely, if $\bar{q}(\omega)$ is given by (10.21), then

$$\int \log \bar{q} \, dP = \log c + \sum_{i=1}^{k} \int \alpha_i f_i \, dP$$

$$= \log c + \sum_{i=1}^{k} \alpha_i a_i \ (= const. \text{ independent of } P)$$

$$= \mathrm{H}(Q \,|\, \bar{P}), \quad for \ \forall \, P \in \mathbf{E}_Q.$$

Therefore, by Lemma 10.4, the probability measure Q is the Csiszar projection on $\mathbf{E}_Q$ and hence on $\mathbf{E}$. This completes the proof.

Let S_1 and S_2 be measurable spaces, and $P_i \in \mathbf{M}_1(S_i)$, $i = 1, 2$. Denote by $\mathbf{E}_{P_1 P_2}$ the set of all $P \in \mathbf{M}_1(S_1 \times S_2)$ with marginal distributions P_1 and P_2. We take a reference measure $\bar{P} \in \mathbf{M}_1(S_1 \times S_2)$ such that $\mathrm{H}(P \,|\, \bar{P}) < \infty$ with a $P \in \mathbf{E}_{P_1 P_2}$. Since $\mathbf{E}_{P_1 P_2}$ is convex and variation closed, there exists the Csiszar projection Q on $\mathbf{E}_{P_1 P_2}$ of the reference measure $\bar{P}$.

Theorem 10.3. *Let S_1 and S_2 be Polish spaces, and $P_i \in \mathbf{M}_1(S_i)$, $i = 1, 2$. Assume for a fixed $\bar{P} \in \mathbf{M}_1(S_1 \times S_2)$ there exists $P \in \mathbf{E}_{P_1 P_2}$ such that $\mathrm{H}(P \,|\, \bar{P}) < \infty$, and let Q be the Csiszar projection on $\mathbf{E}_{P_1 P_2}$ of $\bar{P}$. Denote*

(10.27) $$\mathbf{L} = \{f_1(x_1) + f_2(x_2) : \forall \, f_i \in L^1(P_i), i = 1, 2\}.$$

If $\mathbf{L}$ is closed in $L^1(Q)$, then there exists $g(x_1)$ and $h(x_2)$ such that

(10.28) $$\frac{dQ}{d\bar{P}} = g(x_1) h(x_2), \quad for \ (x_1, x_2) \in S_1 \times S_2,$$

where $\log g \in L^1(P_1)$ and $\log h \in L^1(P_2)$.

Proof. The subset $\mathcal{E}_{P_1P_2}$ can be represented as

$$(10.29) \quad \mathcal{E}_{P_1P_2} = \{P \in \mathbf{M}_1(S_1 \times S_2) : \int f_{ij}\, dP = a_{ij},\ i = 1, 2;\ j = 1, 2, \ldots\ \},$$

with

$$f_{ij}(x_1, x_2) = f_{ij}(x_i),\quad i = 1, 2;\ j = 1, 2, \ldots\ ,$$

(10.30)

$$a_{ij} = \int f_{ij}(x_i)\, dP_i,\quad i = 1, 2;\ j = 1, 2, \ldots\ ,$$

where $\{f_{1j}\}$ (resp. $\{f_{2j}\}$) denotes a countable subset of $\mathbf{B}(S_1)$ (resp. $\mathbf{B}(S_2)$) which defines P_1 (resp. P_2). Then we can apply the same arguments in the proof of Theorem 10.2, namely we have, for $\bar{q} = dQ/d\bar{P}$,

$$(10.31) \qquad \int h \log \bar{q}\, dQ = 0,\quad for\ \forall\, h \in \mathcal{H} \cap L^\infty(Q),$$

where

$$\mathcal{H} = \{h\colon measurable,\ \int h\, dQ = 0,\ and\ \int f_{ij}\, h\, dQ = 0,\ for\ \forall\, f_{ij}\},$$

with f_{ij} which appeared in $\mathcal{E}_{P_1P_2}$ in (10.29). This implies that $\log \bar{q}$ belongs to the closed subspace of $L^1(Q)$ spanned by $\{1, f_{ij},\ i = 1, 2, j = 1, 2, \ldots\}$, namely the space L defined at (10.27). Therefore, $\log \bar{q} \in L$, since L is closed in $L^1(Q)$ by the assumption, and hence

$$\bar{q}(x_1, x_2) = g(x_1)h(x_2),$$

and moreover $\log g \in L^1(P_1)$ and $\log h \in L^1(P_2)$. This completes the proof.

Remark 10. 2. Cf. Douglas (1964), Lindenstrauss (1965), and Csiszar (1975). The subset L is not always closed in $L^1(Q)$. The closedness of L depends on the given pair $\{P_1, P_2\}$ and the reference measure $\bar{P}$. Cf. Lindenstrauss (1965), Rüschendorf-Thomsen (preprint) for examples that L is not closed.

Schrödinger's problem of a system of non-linear integral equations was settled in Schrödinger (1931) and discussed by Bernstein (1932), Fortet (1940), Beurling (1960), and Jamison (1974,a;75) for positive kernel functions. Then the method of Csiszar's projection was applied in Föllmer (1988) and Nagasawa (1990).

Theorem 10.4. *Let $p(x_1, x_2)$ be a non-negative measurable kernel function on $S_1 \times S_2$, and let $\mu_i(x_i)$, $i = 1, 2$, be probability densities with respect to fixed measures λ_i, $i = 1, 2$, on Polish spaces S_i (if $S_i = \mathbf{R}^d$, λ_i can be the Lebesgue measure on $\mathbf{R}^d$). Then there exists a pair of measurable functions $\hat{\phi}(x_1)$ and $\phi(x_2)$ such that*

(10.32)
$$\hat{\phi}(x_1) \int p(x_1, x_2)\phi(x_2)\lambda_2(dx_2) = \mu_1(x_1),$$

$$\int \hat{\phi}(x_1)\lambda_1(dx_1)p(x_1, x_2)\phi(x_2) = \mu_2(x_2).$$

The theorem can be formulated in terms of the Csiszar projection. Let us define $\bar{P}$ through

$$\bar{P}(B) = \int \lambda_1(dx_1)k(x_1)p(x_1, x_2)1_\mu(x_1, x_2)1_B(x_1, x_2)\lambda_2(dx_2),$$

where $1_\mu(x_1, x_2)$ denotes the indicator function of a subset $\{(x_1, x_2): \mu_1(x_1)\mu_2(x_2) > 0\}$, and $k(x_1) > 0$ should be chosen so that $\bar{P} \in \mathbf{M}_1(S_1 \times S_2)$. Then we can state Schrödinger's problem in terms of the Csiszar projection as follows.

Theorem 10.5. *Let S_1 and S_2 be Polish spaces, and $P_i \in \mathbf{M}_1(S_i)$, $i = 1, 2$. Assume there exists $P \in \pounds_{P_1P_2}$ with $\mathrm{H}(P|\bar{P}) < \infty$ for a fixed reference measure $\bar{P} \in \mathbf{M}_1(S_1 \times S_2)$ such that $\bar{P} << P_1 \otimes P_2$. Let $Q \in \pounds_{P_1P_2}$ be the Csiszar projection (on the subset $\pounds_{P_1P_2}$) of $\bar{P}$. Then there exists a pair of measurable functions $g(x_1)$ and $h(x_2)$ such that*

(10.33)
$$\frac{dQ}{d\bar{P}} = g(x_1)h(x_2), \quad for \ (x_1, x_2) \in S_1 \times S_2,$$

where $\log g \in L^1(P_1)$ and $\log h \in L^1(P_2)$.

Proof. Define a sequence of subsets

$$\pmb{E}_k = \{P \in \mathbf{M}_1(S_1 \times S_2) : \int f_{ij}\, dP = a_{ij}, i = 1, 2; j = 1, 2, \cdots, k\},$$

with f_{ij} and a_{ij} in (10.30) and denote by Q_k the Csiszar projection (on the subset $\pmb{E}_k$) of $\overline{P}$. Then, Theorem 10.2 implies

$$(10.34) \qquad \frac{dQ_k}{d\overline{P}} = \overline{q}_k \quad with \quad \overline{q}_k(x_1, x_2) = g_k(x_1)h_k(x_2).$$

It is clear that $\pmb{E}_k \downarrow \pmb{E}_{P_1 P_2}$. Therefore,

$$\lim_{k \to \infty} H(Q_k | \overline{P}) = H(Q | \overline{P}).$$

Since $Q \in \pmb{E}_k$, applying Csiszar's inequality (10.12), we have

$$H(Q | \overline{P}) - H(Q_k | \overline{P}) \geq H(Q | Q_k),$$

and hence

$$\lim_{k \to \infty} H(Q | Q_k) = 0,$$

i.e., Q_k converges to Q in entropy and hence in variation as k tends to infinity. Therefore, we can assume that the sequence $\overline{q}_k = g_k(x_1)h_k(x_2)$ converges pointwise to $\overline{q} = g(x_1)h(x_2)$ on $D = \{(x_1, x_2): \overline{p}(x_1, x_2) > 0\}$, where we fix a nice version $\overline{p}(x_1, x_2) = d\overline{P}/d(P_1 \otimes P_2)$, and (10.33) holds. Moreover, $\log g \in L^1(P_1)$ and $\log h \in L^1(P_2)$, since $\log \overline{q} \in \pmb{L} \cap L^1(Q)$, where $\pmb{L}$ is defined at (10.27). This completes the proof.

Remark 10.3.   Cf. Aebi (preprint) for a statement of the form of (10.32) and also Rüschendorf-Thomsen (preprint).

Chapter XI
Large Deviations

A fundamental theorem on large deviations of empirical distributions, which is necessary in Chapter 8, will be shown.

11.1. Lemmas

We prepare simple but useful lemmas to prove a fundamental theorem for large deviations of empirical distributions.

Lemma 11.1. *Let* $Q \in \mathbf{M}_1(\Omega)$ *and* $\mathbf{P} \in \mathbf{M}_1(\Omega^n)$, *whose marginal distributions on* Ω *are denoted by* P_i, $i = 1, 2, \ldots, n$. *Then*

$$(11.1) \qquad \mathrm{H}(\mathbf{P}|Q^n) = \mathrm{H}(\mathbf{P}|\prod_{i=1}^{n} P_i) + \sum_{i=1}^{n} \mathrm{H}(P_i|Q).$$

Proof. We can assume $P_i \ll Q$ and $\mathbf{P} \ll \prod_{i=1}^{n} P_i$. Otherwise both sides of (11.1) are infinity. It holds clearly that

$$\frac{d\mathbf{P}}{dQ^n} = \frac{d\mathbf{P}}{\prod_{i=1}^{n} dP_i} \prod_{i=1}^{n} \frac{dP_i}{dQ},$$

which yields (11.1), after taking the logarithm of both sides and integrating by the measure $\mathbf{P}$.

As a simple application of (11.1), we get an inequality which will be generalized in Lemma 11.3.

Lemma 11.2. *If marginal distributions P_i of $\mathbf{P}$ are identically equal to P, then*

$$(11.2) \qquad\qquad H(P \mid Q) \leq \frac{1}{n} H(\mathbf{P} \mid Q^n).$$

Following Csiszar (1984) we say that a sequence of probability measures $\mathbf{P}^{(n)} \in \mathbf{M}_1(\Omega^n)$ is **asymptotically quasi-independent with a limiting distribution** Q, if

$$(11.3) \qquad\qquad \lim_{n \to \infty} \frac{1}{n} H(\mathbf{P}^{(n)} \mid Q^n) = 0.$$

To clarify the meaning of asymptotic quasi-independence, we apply inequality (10.4) of the preceding chapter to the simplest measurable partition $\{B_n, B_n^c\}$ of Ω^n. Then

$$\mathbf{P}^{(n)}(B_n) \log \frac{\mathbf{P}^{(n)}(B_n)}{Q^n(B_n)} + (1 - \mathbf{P}^{(n)}(B_n)) \log \frac{1 - \mathbf{P}^{(n)}(B_n)}{1 - Q^n(B_n)} \leq H(\mathbf{P}^{(n)} \mid Q^n).$$

Therefore, if

$$(11.4) \qquad\qquad Q^n(B_n) \leq e^{-\alpha n}, \quad \alpha > 0,$$

then (11.3) implies

$$(11.5) \qquad\qquad \lim_{n \to \infty} \mathbf{P}^{(n)}(B_n) = 0.$$

The following is a simple example which is not asymptotically quasi-independent: Assume that $H(\mathbf{P}^{(n)} \mid Q^n) < \infty$, and there exists a measurable subset $A \subset \Omega$ such that

$$Q(A) \leq e^{-\alpha}, \quad \alpha > 0,$$

$$\mathbf{P}^{(n)}(A^n) \geq 1 - \frac{1}{n},$$

namely, $\mathbf{P}^{(n)}$ will concentrate gradually on the cube A^n as n tends to infinity. Then (11.4) holds for A^n, but (11.5) does not. Therefore, $\mathbf{P}^{(n)}$ is not asymptotically quasi-independent.

The following lemma plays an important role when we apply large deviations to the propagation of chaos.

Lemma 11.3. *If a sequence of probability measures* $\mathbf{P}^{(n)} \in \mathbf{M}_1(\Omega^n)$ *is asymptotically quasi-independent with limiting distribution Q, and if the marginal distributions of* $\mathbf{P}^{(n)}$ *on* $\Omega_{mr+1} \times \cdots \times \Omega_{m(r+1)}$, $r = 0, 1, 2, \ldots$, *are identically equal to* $\mathbf{P}_m^{(n)}$, *then* $\mathbf{P}_m^{(n)}$ *converges to* Q^m *in entropy as n tends to infinity for any fixed* $m \geq 1$.

Proof.[1] For sufficiently large n, let us denote $n = mk + r$, where $m \geq 1$ is fixed and $0 \leq r < m$. Then we have

$$\mathrm{H}(\mathbf{P}^{(n)} \mid Q^n) = \int \log \frac{d\mathbf{P}^{(n)}}{(d\mathbf{P}_m^{(n)})^k \otimes d\mathbf{P}_r^{(n)}} \, d\mathbf{P}^{(n)} + \int \log \frac{(d\mathbf{P}_m^{(n)})^k \otimes d\mathbf{P}_r^{(n)}}{dQ^n} \, d\mathbf{P}^{(n)}$$

$$\geq \int \log \frac{(d\mathbf{P}_m^{(n)})^k \otimes d\mathbf{P}_r^{(n)}}{dQ^n} \, d\mathbf{P}^{(n)} \geq k \int \log \frac{d\mathbf{P}_m^{(n)}}{dQ^m} \, d\mathbf{P}_m^{(n)}.$$

Therefore, we have a generalized form of the inequality (10.2)

(10.2')
$$\frac{n-r}{n} \frac{1}{m} \mathrm{H}(\mathbf{P}_m^{(n)} \mid Q^m) \leq \frac{1}{n} \mathrm{H}(\mathbf{P}^{(n)} \mid Q^n),$$

which implies, because of the asymptotic quasi-independence of $\mathbf{P}^{(n)}$,

$$\lim_{n \to \infty} \mathrm{H}(\mathbf{P}_m^{(n)} \mid Q^m) \leq m \lim_{n \to \infty} \frac{1}{n} \mathrm{H}(\mathbf{P}^{(n)} \mid Q^n) = 0,$$

and hence $\mathbf{P}_m^{(n)}$ converges to Q^m in entropy as n tends to infinity.

11.2. Large Deviations of Empirical Distributions

Let $(\Omega^n, \overline{P}^n)$ be the n-fold product of $(\Omega, \overline{P})$. Let us denote by L_n the empirical distribution of $(\omega_1, \omega_2, \ldots, \omega_n)$

(11.6)
$$L_n = \frac{1}{n} \sum_{i=1}^{n} \delta_{\omega_i},$$

and define a sequence of **conditional probabilities** $\mathbf{P}^{(n)}$ by

---

[1] The proof is due to E. Bolthausen

(11.7) $$\mathbf{P}^{(n)}[\cdot] = \mathbf{P}_{\boldsymbol{A}}^{(n)}[\cdot] = \overline{P}^n[\cdot \mid L_n \in \boldsymbol{A}],$$

for a subset $\boldsymbol{A} \subset \mathbf{M}_1(\Omega)$, which is well-defined when $\overline{P}^n[L_n \in \boldsymbol{A}] > 0$. Since we apply the preceding lemma to the conditional probabilities, we use the same notation $\mathbf{P}^{(n)}$ for it.

Let $(\Lambda, \mathcal{A}, \mu)$ be any probability space. A subset $\boldsymbol{A} \subset \mathbf{M}_1(\Omega)$ is **completely convex**, if

(11.8) $$\int \mu(d\lambda)\eta(\lambda, \cdot) \in \boldsymbol{A},$$

for any probability kernel $\eta(\lambda, B)$ defined on $\Lambda \times \mathcal{B}$ such that $\eta(\lambda, \cdot) \in \boldsymbol{A}$.[2]

Lemma 11.4. *Assume $\boldsymbol{A} \subset \mathbf{M}_1(\Omega)$ is completely convex and let $P^{(n)}$ be the (identical) marginal distribution of the conditional probability $\mathbf{P}^{(n)}$ which is defined by (11.7). Then $P^{(n)} \in \boldsymbol{A}$.*

Proof. It is clear that

$$\mathbf{P}^{(n)}[L_n(B)] = \frac{1}{n}\mathbf{P}^{(n)}[\sum_{i=1}^{n} 1_B(\omega_i)]$$

$$= P^{(n)}(B),$$

since $\mathbf{P}^{(n)}$ has an identical marginal distribution $P^{(n)}$ on Ω; namely,

$$P^{(n)}(B) = \int_{\Omega_{\boldsymbol{A}}} L_n(\omega, B)\,\mathbf{P}^{(n)}(d\omega),$$

where we regard the measure $\mathbf{P}^{(n)}$ as a probability measure on a subset $\Omega_{\boldsymbol{A}} = \{\omega : L_n(\omega, \cdot) \in \boldsymbol{A}\}$. Therefore, we have $P^{(n)} \in \boldsymbol{A}$, since the subset $\boldsymbol{A}$ is completely convex.

Even though the statement itself of the Sanov property does not contain any topological concept (see (11.13)), to prove it we need a topological structure on the space $\mathbf{M}_1(\Omega)$ of probability measures on Ω (see (11.12)). Csiszar (1984) defines an intrinsic topology of $\mathbf{M}_1(\Omega)$, which depends only

---

[2] This is a generalization of convexity

on the measurable structure of Ω. This is the so-called Csiszar τ_0-**topology**:

The basic *neighbourhoods* of $P \in \mathbf{M}_1(\Omega)$ are given by

$$U(P, \varepsilon, \boldsymbol{P}_k) = \{R \in \mathbf{M}_1(\Omega) : \left| R(\Omega_i) - P(\Omega_i) \right| < \varepsilon, \text{ for } i = 1, 2, \ldots, k,$$
$$\text{and } R \ll P \text{ on } \sigma(\boldsymbol{P}_k), \text{ i.e., } R(\Omega_i) = 0, \text{ if } P(\Omega_i) = 0\},$$

where

$$\boldsymbol{P}_k = \boldsymbol{P}_k(\Omega) = \{\Omega_1, \ldots, \Omega_k\}, \, k = 1, 2, \ldots, \text{ are finite measurable}$$
$$\text{partitions of } \Omega,$$

$\sigma(\boldsymbol{P}_k)$ *is the σ-field generated by the partition* $\boldsymbol{P}_k$.

Theorem 11.1. (Csiszar (1984))

(i) *Let* $\boldsymbol{A} \subset \mathbf{M}_1(\Omega)$ *be completely convex and variation closed, and moreover assume there is* $P \in \boldsymbol{A}$ *such that* $\mathrm{H}(P \mid \overline{P}) < \infty$. *Then there exists the Csiszar projection* Q *on the subset* $\boldsymbol{A}$ *of* $\overline{P} \in \mathbf{M}_1(\Omega)$, *and*

$$(11.9) \qquad \frac{1}{n} \log \overline{P}^n [L_n \in \boldsymbol{A}] \leq -\mathrm{H}(\boldsymbol{A} \mid \overline{P})$$

holds for every n, more precisely

$$(11.10) \qquad \frac{1}{n} \log \overline{P}^n [L_n \in \boldsymbol{A}] \leq -\mathrm{H}(\boldsymbol{A} \mid \overline{P}) - \frac{1}{n} \mathrm{H}(\mathbf{P}^{(n)} \mid Q^n)$$

holds, where $\mathbf{P}^{(n)} \in \mathbf{M}_1(\Omega^n)$ *is the conditional probability defined in* (11.7) *and*

$$(11.11) \qquad \mathrm{H}(\boldsymbol{A} \mid \overline{P}) = \inf_{P \in \boldsymbol{A}} \mathrm{H}(P \mid \overline{P}) = \mathrm{H}(Q \mid \overline{P}).$$

(ii) *Let* $\boldsymbol{A}^\circ$ *denote the interior of the subset* $\boldsymbol{A}$ *with respect to* τ_0-*topology. If*

$$(11.12) \qquad \mathrm{H}(\boldsymbol{A}^\circ \mid \overline{P}) = \mathrm{H}(\boldsymbol{A} \mid \overline{P}),$$

then the subset $\boldsymbol{A}$ *has the Sanov property*

$$(11.13) \qquad \lim_{n \to \infty} \frac{1}{n} \log \overline{P}^n [L_n \in \boldsymbol{A}] = -\mathrm{H}(\boldsymbol{A} \mid \overline{P}).$$

(iii) *Under condition* (11.12) *the sequence* $\mathbf{P}^{(n)} \in \mathbf{M}_1(\Omega^n)$ *of the conditional probabilities defined by* (11.7) *is asymptotically quasi-independent with limiting distribution* Q (= *Csiszar's projection on the subset* $\mathbf{A}$ *of* $\overline{P} \in \mathbf{M}_1(\Omega))$, *that is,*

$$(11.14) \qquad \lim_{n \to \infty} \frac{1}{n} H(\mathbf{P}^{(n)} | Q^n) = 0.$$

Moreover, the marginal distribution $P^{(n)}$ *on* Ω *of* $\mathbf{P}^{(n)}$ *converges to* Q *in entropy*

$$\lim_{n \to \infty} H(P^{(n)} | Q) = 0.$$

Proof. It is enough to prove the refined inequality (11.10) for the first statement. By the definition of the conditional probability $\mathbf{P}^{(n)}$, we have

$$(11.15) \qquad H(\mathbf{P}^{(n)} | \overline{P}^n) = -\log \overline{P}^n[L_n \in \mathbf{A}].$$

On the other hand, applying Lemma 11.1 with $\mathbf{P} = \mathbf{P}^{(n)}$ and $Q = \overline{P}$, we have

$$(11.16) \qquad H(\mathbf{P}^{(n)} | \overline{P}^n) = H(\mathbf{P}^{(n)} | P^n) + n H(P | \overline{P}),$$

where $P = P^{(n)}$ denotes the marginal distribution of $\mathbf{P}^{(n)}$. Now with $Q =$ *the Csiszar projection of* $\overline{P}$ on $\mathbf{A}$, and $\mathbf{P} = \mathbf{P}^{(n)}$, Lemma 11.1 yields

$$(11.17) \qquad H(\mathbf{P}^{(n)} | Q^n) = H(\mathbf{P}^{(n)} | P^n) + n H(P | Q).$$

Hence, combining (11.16) and (11.17), we have

$$(11.18) \qquad H(\mathbf{P}^{(n)} | \overline{P}^n) = H(\mathbf{P}^{(n)} | Q^n) + n\{H(P | \overline{P}) - H(P | Q)\}.$$

On the other hand, Lemma 11.4 implies $P = P^{(n)} \in \mathbf{A}$, and hence

$$(11.19) \qquad H(P | \overline{P}) \geq H(P | Q) + H(Q | \overline{P}),$$

because of inequality (10.12) in Theorem 10.1. Therefore, substituting (11.19) into (11.18), we have finally,

$$(11.20) \quad -\log \overline{P}^n[L_n \in \mathbf{A}] = H(\mathbf{P}^{(n)} | \overline{P}^n) \geq H(\mathbf{P}^{(n)} | Q^n) + n H(Q | \overline{P}),$$

which is inequality (11.10).

The third assertion follows immediately from (11.20), if we assume the Sanov property (11.13): In fact, let us write inequality (11.20) as

$$(11.21) \qquad \frac{1}{n}H(\mathbf{P}^{(n)} | Q^n) \le -\frac{1}{n}\log \overline{P}^n[L_n \in \mathbf{A}] - H(Q | \overline{P}).$$

Then it is clear that (11.14) holds, since the right-hand side of (11.21) vanishes as $n \to \infty$, because of the Sanov property (11.13). The assertion on the marginal distribution $P^{(n)}$ has shown already in Lemma 11.3.

For the second assertion on the Sanov property it is enough to prove

Lemma 11.5. *Let $\mathbf{A} \subset \mathbf{M}_1(\Omega)$ be such that for each $P \in \mathbf{A}$ there is a τ_0-neighborhood $U(P, \varepsilon, \mathcal{P})$ with $U(P, \varepsilon, \mathcal{P}) \cap \mathbf{M}^f(\Omega) \subset \mathbf{A}$, where $\mathbf{M}^f(\Omega)$ denotes the set of all finite atoms. Then*

$$(11.22) \qquad -H(\mathbf{A} | \overline{P}) \le \liminf_{n \to \infty} \frac{1}{n} \log \overline{P}^n[L_n \in \mathbf{A}],$$

which holds also for $\mathbf{A} \subset \mathbf{M}_1(\Omega)$ with $H(\mathbf{A}° | \overline{P}) = H(\mathbf{A} | \overline{P})$.

Proof. We can assume $H(\mathbf{A} | \overline{P}) < \infty$. For a $\delta > 0$, choose any $P \in \mathbf{A}$ such that

$$(11.23) \qquad H(P | \overline{P}) < H(\mathbf{A} | \overline{P}) + \delta,$$

and take a finite measurable partition $\mathcal{P} = \{\Omega_1, \dots, \Omega_k\}$ of Ω and $\varepsilon > 0$ such that $U(P, \varepsilon, \mathcal{P}) \cap \mathbf{M}^f(\Omega) \subset \mathbf{A}$. Choose $0 < \varepsilon' < \varepsilon$ small enough so that

$$(11.24) \qquad \left| r_i \log \frac{r_i}{\overline{P}(\Omega_i)} - P(\Omega_i) \log \frac{P(\Omega_i)}{\overline{P}(\Omega_i)} \right| < \frac{\delta}{k}, \quad i = 1, \dots, k,$$

for r_i with

$$(11.25) \qquad |r_i - P(\Omega_i)| < \varepsilon', \quad (r_i = 0, \text{ if } P(\Omega_i) = 0).$$

For sufficiently large n there exist $m_1, \dots, m_k$ such that $\sum_{i=1}^{k} m_i = n$ and $r_i = m_i/n$ $(i = 1, 2, \dots, k)$ satisfy (11.25). Then

(11.26) $\bar{P}^n[L_n \in \boldsymbol{A}] \geq \bar{P}^n[L_n \in U(P, \varepsilon, \boldsymbol{P}) \cap \mathbf{M}^f(\Omega)]$

$$\geq \bar{P}^n[L_n(\Omega_i) = r_i, 1 \leq i \leq k] = \frac{n!}{m_1! \dots m_k!} \prod_{i=1}^{k} (\bar{P}[\Omega_i])^{m_i}$$

$$\geq (n+1)^{-k} \exp(-n \sum_{i=1}^{k} r_i \log \frac{r_i}{\bar{P}(\Omega_i)}),$$

where we have applied Stirling's formula

$$\frac{n!}{m_1! \dots m_k!} \geq (n+1)^{-k} \exp(-n \sum_{i=1}^{k} r_i \log r_i), \quad r_i = m_i/n.$$

On the other hand, because of (11.24) we have

(11.27) $$\sum_{i=1}^{k} r_i \log \frac{r_i}{\bar{P}(\Omega_i)} \leq \sum_{i=1}^{k} P(\Omega_i) \log \frac{P(\Omega_i)}{\bar{P}(\Omega_i)} + \delta$$

$$\leq \mathrm{H}(P \mid \bar{P}) + \delta$$

$$\leq \mathrm{H}(\boldsymbol{A} \mid \bar{P}) + 2\delta,$$

where inequality (10.4) of the relative entropy is employed in the second inequality, and (11.23) is applied to the third inequality. Therefore, from (11.26) together with (11.27), we get

$$- 2\delta - \mathrm{H}(\boldsymbol{A} \mid \bar{P}) \leq \liminf_{n \to \infty} \frac{1}{n} \log \bar{P}^n[L_n \in \boldsymbol{A}],$$

from which (11.22) follows.

Let $\mathrm{H}(\boldsymbol{A}^\circ \mid \bar{P}) = \mathrm{H}(\boldsymbol{A} \mid \bar{P})$. Then we can apply the above arguments to the set $\boldsymbol{A}^\circ$, since $\boldsymbol{A}^\circ$ is open and not empty, and there exists a τ_0-neighborhood $U(P, \varepsilon, \boldsymbol{P})$ with $U(P, \varepsilon, \boldsymbol{P}) \cap \mathbf{M}^f(\Omega) \subset \boldsymbol{A}^\circ$. This completes the proof.

Remark. One of important contributions of Csiszar (1984) is inequality (11.10), with which we can estimate the speed of convergence (11.21), namely the asymptotic quasi-independence (11.14). For further information on large deviations we refer to Azencott (1980), Varadhan (1984), Deuschel-Stroock (1989).

Chapter XII

Non-Linearity Induced by the Branching Property

The purpose of this chapter is to show how the non-linearity of a special kind appears from the branching property. The usual branching property induces a rather restrictive non-linearity. To extend it, the branching property will be generalized. Therefore, this chapter does not belong to the main context of the monograph. It is for the reader's convenience in referring to, and is not necessary in reading the main text.

12.1. Branching Property

Branching Markov processes are mathematical models for random evolution which contains creation of particles. Consider a system of particles and divide the system arbitrarily into two groups indexed by a and b. Particles in each group can move and create descendants without interference between the two groups. This is a special kind of independence characterizing branching processes, which can be formulated in terms of the branching property (or multiplicativity) of the transition probability of a strong Markov process on a state space endowed with multiplication.

A measurable space S is *multiplicative* if

(12.1)   *for $a, b \in S$ a multiplication $a \cdot b \in S$ is defined and measurable,*

(12.2)   *there exists an element $\delta \in S$ such that*

$$\delta \cdot a = a \cdot \delta = a, \text{ for all } a \in S.$$

The point $\delta \in S$ stands for the state of no existing particle, and $a \in S$, $a \neq \delta$, denotes a group of particles.

Examples of multiplicative spaces are:

(i) $N = \{0, 1, \dots \}$ is multiplicative with $a \cdot b = a + b$, and $\delta = 0$.

When one is interested in just the number of particles, one can take this state space.

(ii) Let E be a measurable space (later we take $E = \mathbf{R}^d$), E^n the n-fold product of E, and $E^0 = \{\delta\}$, where δ is an extra point. Taking two points $a = (x_1, \dots, x_n) \in E^n$ and $b = (y_1, \dots, y_m) \in E^m$, let us define a product $a \cdot b = (a, b) = (x_1, \dots, x_n, y_1, \dots, y_m) \in E^{n+m}$. Then $S = \cup_{n=0}^\infty E^n$ is a multiplicative space, if we define $\delta \cdot a = a \cdot \delta = a$, for $a \in S$.

When we shall consider branching diffusion processes, this state space will be adopted.

(iii) Let $\mathbf{M}$ be a subset of the space of σ-finite measures on a measurable space and assume $a \cdot b = a + b \in \mathbf{M}$, for $a, b \in \mathbf{M}$, and $\delta = 0 \in \mathbf{M}$. Then the space $\mathbf{M}$ is multiplicative with an appropriate σ-field on $\mathbf{M}$.

Branching processes on this state space are often called "continuous state (or measure-valued) branching processes". One can interpret a measure $a \in \mathbf{M}$ to be a "cloud of (infinitely many) particles".

(iv) Let $\widetilde{E}^n$ be the quotient space of E^n with the equivalence relation of the permutations of coordinates. The multiplication of $a \in \widetilde{E}^n$ and $b \in \widetilde{E}^m$ is defined to be the element in $\widetilde{E}^{n+m}$ which contains (a, b). Then $\widetilde{S} = \cup_{n=0}^\infty \widetilde{E}^n$ is a multiplicative space.

We can identify $a = (x_1, \dots, x_n) \in \widetilde{E}^n$ with $\mu = \sum_{i=1}^n \delta_{x_i} \in \mathbf{M}^f = \{all\ finite\ atoms\}$ and embed $\widetilde{S}$ into the space $\mathbf{M}^f$.

One can adopt this state space when particles in a system are indistinguishable.

For finite (or σ-finite) measures μ and ν on a multiplicative space S the convolution is defined by

(12.3) $$\int \mu * \nu(db)\, g(b) = \int \mu(db_1)\nu(db_2)\, g(b_1 \cdot b_2),$$

for any non-negative measurable function g on S.

A transition probability $\mathbf{P}_t(a, db)$ on a multiplicative space S with $\mathbf{P}_t(\delta, \{\delta\}) = 1$ has the *branching* (or *multiplicative*) *property* if it satisfies

$$(12.4) \qquad \mathbf{P}_t(a \cdot b, \cdot) = \mathbf{P}_t(a, \cdot) * \mathbf{P}_t(b, \cdot), \quad for \quad \forall a, b \in S,$$

where $*$ denotes the convolution of measures defined in (12.3).

Formula (12.4) represents the independence of evolution of two groups indexed by a and b at the initial time, as mentioned already.

If the transition probability of a strong Markov process on a multiplicative space S has the branching property (12.4), we call it a *branching Markov process*.

It is easy to see that the branching property (12.4) is equivalent to

$$(12.5) \qquad \mathbf{P}_t(\mu * \nu, \cdot) = \mathbf{P}_t(\mu, \cdot) * \mathbf{P}_t(\nu, \cdot),$$

for any finite measures μ and ν on S, where $\mathbf{P}_t(\mu, \cdot) = \int \mu(da) \mathbf{P}_t(a, \cdot)$.

If a function g on a multiplicative space S satisfies

$$(12.6) \qquad g(a \cdot b) = g(a) g(b), \quad for \quad a, b \in S,$$

then g is called *multiplicative*.

Theorem 12.1. *Let a transition probability $\mathbf{P}_t$ on a multiplicative space S have the branching property (12.4). If a bounded measurable function g on S is multiplicative, then $\mathbf{P}_t g$ is also multiplicative, i.e.,*

$$(12.7) \qquad \mathbf{P}_t g(a \cdot b) = \mathbf{P}_t g(a) \mathbf{P}_t g(b) \quad and \quad \mathbf{P}_t g(\delta) = 1.$$

Proof. If g is multiplicative, then $g(\delta) = 1$. Therefore, $\mathbf{P}_t g(\delta) = 1$, because $\mathbf{P}_t(\delta, \{\delta\}) = 1$. Formulae (12.4) and (12.6) yield

$$\mathbf{P}_t g(a \cdot b) = \int \mathbf{P}_t(a, dc_1) \mathbf{P}_t(b, dc_2) g(c_1) g(c_2) = \mathbf{P}_t g(a) \mathbf{P}_t g(b),$$

which completes the proof.

We consider the multiplicative space $S = \cup_{n=0}^{\infty} E^n$ (or the symmetric $\widetilde{S} = \cup_{n=0}^{\infty} \widetilde{E}^n$). If a function g on S (or $\widetilde{S}$) is multiplicative, then it is clear that g is determined by its restriction $g|_E$ on E. Let us denote $f(x) = g|_E(x)$, then

$$(12.8) \qquad g(x_1, \dots, x_n) = \prod_{i=1}^{n} f(x_i), \quad and \quad g(\delta) = 1.$$

We introduce a special notation $g = \hat{f}$ for functions with the property (12.8), and denote $\hat{B}_1(S) = \{\hat{f} : measurable\ function\ f\ on\ E\ with\ |f| \leq 1\}$.

Then Theorem 12.1 implies

Theorem 12.2. *Let P_t be a branching (multiplicative) transition probability (or semi-group) on $S = \cup_{n=0}^{\infty} E^n$ (or $\widetilde{S} = \cup_{n=0}^{\infty} \widetilde{E}^n$).*

Then the semi-group P_t maps $\hat{B}_1(S)$ into $\hat{B}_1(S)$, i.e.,

$$(12.9) \qquad\qquad P_t \hat{f} = \widehat{(P_t \hat{f})|_E},$$

for $f \in B_1(E) = \{f : measurable\ function\ on\ E\ with\ |f| \leq 1\}$.

Property (12.9) is often called simply the "branching property" (cf. Ikeda-Nagasawa-Watanabe (1968, 69)).

Remark. When E consists of a single point, identifying E^n with n, we have $S = N = \{0, 1, 2, \dots\}$. In this case the branching property (12.9) is nothing but the multiplication rule of the generating function:

$$(12.10) \qquad F_t(m + n, f) = F_t(m, f) F_t(n, f),$$

where f is a constant with $|f| \leq 1$ and

$$(12.11) \qquad F_t(m, f) = P_t \hat{f}(m) = \sum_{k=0}^{\infty} P_t(m, k) f^k.$$

The function $F_t(m, f)$ is a non-linear function of f. The non-linearity of branching processes will appear in this way, as will be explained in the next section.

12.2. Non-Linear Equations of Branching Processes

In this section we consider a strong Markov process $\{X_t, P_a, a \in S\}$ on the multiplicative space $S = \cup_{n=0}^{\infty} E^n$ (or $\widetilde{S} = \cup_{n=0}^{\infty} \widetilde{E}^n$), with $E = \mathbf{R}^d$.

If its semi-group

$$(12.12) \qquad \mathbf{P}_t g(a) = P_a[g(X_t)], \quad for \quad a \in S,$$

has the branching property (12.9), it is called simply a *branching Markov process*.

Let $T(\omega)$ be the first branching time of a single particle, *i.e.*,

$$(12.13) \qquad T(\omega) = \inf \{t : X_0(\omega) \in E, \text{ and } X_t(\omega) \notin E\},$$

and define a semi-group of a *single particle* by

$$(12.14) \qquad P_t^0 f(x) = P_x[f(X_t): t < T],$$

$$= \int P_t^0(x, dy) f(y), \quad for \quad x \in E,$$

for bounded measurable f on E.

We consider from now on branching diffusion processes. This means that a single particle up to the first branching time $\{X_t, t < T, P_x, x \in \mathbf{R}^d\}$ is a diffusion process determined by an elliptic operator A given in (2.1) which is killed at T.

Furthermore, let us define the joint distribution of the branching time and the location at which the branching occurs :

$$(12.15) \qquad K(x, ds, dy) = P_x[T \in ds, X_{T-} \in dy],$$

and the "branching law"

$$(12.16) \qquad F(y, u) = q_0(y) + \sum_{n=2}^{\infty} q_n(y) \int_{E^n} \pi_n(y, da) \widehat{u}(a),$$

where $u \in B_1(\mathbf{R}^d)$, $q_n(y) = P_y[X_T \in E^n]$, and $\pi_n(y, B)$ is a probability kernel on $E \times E^n$, which will be called a *branching distribution*. Therefore, $q_n(y) \geq 0$ and

(12.17)
$$q_0(y) + \sum_{n=2}^{\infty} q_n(y) = 1.$$

The $q_n(y)$ is the probability that a single particle splits into n particles at y. The n descendants are distributed according to the distribution $\pi_n(y, \cdot)$.

If descendants start from the point where the mother particle dies, then the branching distribution is a Dirac measure and we have

(12.18)
$$\int_{E^n} \pi_n(y, da)\hat{u}(a) = u^n(y),$$

and hence

(12.19)
$$F(y, u) = q_0(y) + \sum_{n=2}^{\infty} q_n(y)u^n(y).$$

This is a typical case of the branching law of branching diffusion processes.

Theorem 12.3. *Let*

(12.20)
$$u(t, x) = \mathbf{P}_t \hat{f}(x), \quad x \in \mathbf{R}^d, \quad f \in B_1(\mathbf{R}^d).$$

Then it satisfies a non-linear integral equation

(12.21)
$$u(t, x) = \mathbf{P}_t^0 f(x) + \int_{[0, t] \times \mathbf{R}^d} K(x, ds, dy)F(y, u_{t-s}).$$

Proof. By the strong Markov property we have

(12.22) $$\mathbf{P}_x[\hat{f}(X_t)] = \mathbf{P}_x[\hat{f}(X_t); t < T] + \mathbf{P}_x[\mathbf{P}_{X_T}[\hat{f}(X_{t-s})]|_{s=T}; t \geq T],$$

where

(12.23)
$$\mathbf{P}_{X_T}[\hat{f}(X_{t-s})]|_{s=T} = \widehat{u_{t-s}}(X_T)|_{s=T},$$

because of the branching property

$$P_a[\hat{f}(X_{t-s})] = \mathbf{P}_{t-s}\hat{f}(a) = \widehat{u_{t-s}}(a), \quad a \in E^n.$$

Hence, the second term on the right-hand side of (12.22) turns out to be

$$\int_{[0,t]\times\mathbf{R}^d} \mathbf{P}_x[T \in ds, X_{T_-} \in dy\,;\, \mathbf{P}_{X_{T_-}}[\,\widehat{u_{t-T}}(X_T)]]$$

$$= \int_{[0,t]\times\mathbf{R}^d} K(x, ds, dy)F(y, u_{t-s}),$$

because $\mathbf{P}_{X_{T_-}}[\widehat{u_{t-s}}(X_T)] = F(X_{T_-}, u_{t-s})$. The first term on the right-hand side of (12.22) coincides with $P_t^o f(x)$. Therefore, $u(t,x)$ satisfies the integral equation (12.21). This completes the proof.

Lemma 12.1. *Let the semi-group P_t^o defined in (12.14) be given by the Kac semi-group*

(12.24) $$P_t^o f(x) = \mathbf{P}_x[m_t f(X_t)],$$

where

$$m_t = \exp\left(-\int_0^t c(X_s)ds\right)$$

with a non-negative measurable function $c(x)$, and $\{X_t, \mathbf{P}_x\}$ denotes a diffusion process determined by the elliptic operator A of (2.1). Then

(12.25) $$K(x, ds, dy) = P_t^o(x, dy)c(y)ds.$$

Proof. Definition (12.15) combined with (12.24) yields

$$K(x, ds, dy) = \mathbf{P}_x[T \le s + ds, X_{T_-} \in dy] - \mathbf{P}_x[T \le s, X_{T_-} \in dy]$$

$$= -\mathbf{P}_x[dm_s 1_{dy}(X_s)]$$

$$= \mathbf{P}_x[m_s c(X_s)ds 1_{dy}(X_s)]$$

$$= P_s^o(x, dy)c(y)ds.$$

Therefore, we have

Theorem 12.4. *Let* $u(t,x) = \mathbf{P}_t \hat{f}(x)$, $x \in \mathbf{R}^d$, $f \in B_1(\mathbf{R}^d)$. *Under condition* (12.24), *the function* $u(t,x)$ *defined in* (12.20) *satisfies an integral equation*

$$(12.26) \qquad u(t,x) = \mathbf{P}_t^0 f(x) + \int_{[0,t]\times\mathbf{R}^d} \mathbf{P}_s^0(x,dy)c(y)dsF(y, u_{t-s}).$$

If $F(y,u)$ *is given by* (12.19), *then the function* $u(t,x)$ *is a weak solution of the non-linear parabolic equation*

$$(12.27) \qquad \frac{\partial u}{\partial t} = \tfrac{1}{2}\Delta u + b(x)\cdot\nabla u - c(x)u + c(x)\{q_0(x) + \sum_{n=2}^{\infty} q_n(x)u^n\}.$$

Remark. If $f \geq 0$, $u(t,x) = \mathbf{P}_t\hat{f}(x)$ gives the minimal solution of the non-linear equation (12.27) in terms of branching processes. For the minimal and maximal solutions of (12.27) cf., e.g. Nagasawa (1977).

Equation (12.27) which is induced by a branching diffusion process is rather restrictive as a quasi-linear diffusion equation, because of the killing term $-c(x)u$. Therefore, for example, a quasi-linear equation

$$\frac{\partial u}{\partial t} = \tfrac{1}{2}\Delta u + b(x)\cdot\nabla u + c(x)u^2$$

cannot be treated in the framework of ordinary branching diffusion processes. In order to treat such an equation in the framework of branching diffusion processes, we must have a way of getting rid of the killing term $-c(x)u$ from the equation in (12.27). In the next section we will discuss a generalized branching property, which enables us to handle such equations.

12.3. Quasi-Linear Parabolic Equations

As the first step we generalize the branching property so that we get probabilistic solutions of the quasi-linear parabolic differential equation

$$(12.28) \qquad \frac{\partial u}{\partial t} = \tfrac{1}{2}\Delta u + b(x)\cdot\nabla u + c(x)\sum_{n=0}^{\infty} q_n(x)u^n,$$

where $q_n(x)$, $n \geq 0$, may take positive and negative values in general under the condition

(12.29)
$$\sum_{n=0}^{\infty} |q_n(x)| = 1.$$

First of all we define a *generalized multiplicative space*:

(12.30)
$$\mathbf{S} = S \times N \times J$$

where $S = \cup_{n=0}^{\infty} E^n$ (or $\widetilde{S} = \cup_{n=0}^{\infty} \widetilde{E}^n$), $N = \{0, 1, 2, \dots\}$, and $J = \{0, 1\}$.

A measurable function g on $\mathbf{S}$ is called *multiplicative on* $\mathbf{S}$ (in a generalized sense) if

(12.31)
$$g(a \cdot b, k, j) = (-1)^j \lambda^k g(a) g(b),$$

$$g(\delta, 0, 0) = 1,$$

where $g(a) = g(a, 0, 0)$ and λ is a fixed positive constant. We will set $\lambda = 2$ later.

For simplicity we assume (12.24) holds, and hence (12.25). Let us denote a strong Markov process on the extended multiplicative space $\mathbf{S}$ by

(12.32)
$$\mathbf{X}_t = (X_t, k_t, j_t),$$

where the meanings of the second and third terms will be explained in the following.

Let $a_t(\omega) = \int_0^t c(X_s(\omega)) ds$ be the Kac additive functional of a single particle and set

$$a_t(\widetilde{\omega}) = a_t(\omega_1) + \cdots + a_t(\omega_n), \quad for \ \ \widetilde{\omega} = (\omega_1, \dots, \omega_n),$$

if there are n-particles. Moreover, we assume

$$k_t(\widetilde{\omega}, \omega') = p_{a_t(\widetilde{\omega})}(\omega'),$$

where $p_t(\omega')$ is a Poisson process which is independent of X_t. We can interpret the process k_t as the sum of (random) ages of existing particles.

Concerning the third variable j_t, we assume that it remains unchanged until the branching time, and when a single particle x_t splits into n particles at t, the process j_t changes its value if $q_n(x_t) < 0$, but remains unchanged if $q_n(x_t) > 0$.

The process X_t evolves as the branching Markov process with $\{ |q_n(x)| \}$ discussed in the preceding sections.

Let us define for $f \in B_1(E)$

$$(12.33) \qquad \tilde{f}(a, k, j) = (-1)^j \lambda^k \hat{f}(a), \quad a \in S.$$

Then $\tilde{f}$ is clearly multiplicative in the sense of (12.31).

Furthermore we assume:

(12.34)   *At the branching time the process splits into n particles with the*
          probability $|q_n(y)|$. The n-particles start from $((y, \dots, y), k, 0)$
          if $q_n(y) > 0$, but they start from $((y, \dots, y), k, 1)$ if $q_n(y) < 0$.

Under the assumptions on k_t and j_t, and the branching rule (12.34), a strong Markov process $\{X_t = (X_t, k_t, j_t), P_{(a,k,j)}\}$ on S is called a *branching diffusion process with age and sign*, if it satisfies the extended branching property:

$$(12.35) \qquad\qquad P_t \tilde{f} = \overline{(P_t f)}|_E.$$

Then we have

Theorem 12.5.   Let $\{X_t = (X_t, k_t, j_t), P_{(a,k,j)}, (a, k, j) \in S\}$ be   a *branching diffusion process with age and sign, and set $\lambda = 2$ in (12.33). If the function $u(t, x) = P_t \tilde{f}(x, 0, 0)$ is well-defined,[1] then it satisfies a non-linear integral equation*

$$(12.36) \quad u(t, x) = P_t f(x) + \int_{[0, t] \times E} P_t(x, dy) c(y) ds \sum_{n=0}^{\infty} q_n(y)(u_{t-s}(y))^n,$$

where

$$P_t f(x) = P_x[f(X_t)]$$

---

is the semi-group of the diffusion process determined by the elliptic operator A given in (2.1).[2] Therefore, $u(t, x)$ is a weak solution of the quasi-linear diffusion equation (12.28).

Proof. We observe a single particle until the first branching time. Then we have, because of the property of the process k_t ,

$$P_{(x,0,0)}[f(X_t)2^{k_t} ; t < T] = P_x[m_t f(X_t) \sum_{k=0}^{\infty} e^{-a_t} \frac{(2a_t)^k}{k!}]$$

$$= P_x[e^{-a_t} f(X_t) \sum_{k=0}^{\infty} e^{-a_t} \frac{(2a_t)^k}{k!}]$$

$$= P_x[f(X_t)] = P_t f(x),$$

which gives the first term on the right-hand side of (12.36). The same manipulation as above shows that

$$K((x,0,0), ds, dy) = P_s(x, dy)c(y)ds.$$

Therefore, we get the second term on the right-hand side of (12.36) because of (12.34) and of the extended branching property (12.35). This completes the proof.

Remark. For the construction of branching processes with age and sign see Sirao (1968), Nagasawa (1968, 72). For an application of the generalized branching property to a probabilistic treatment of the blowing up of solutions of non-linear integral equations, cf. Nagasawa-Sirao (1969).[3]

Remark. If we choose $|q_1(x)| \equiv 1$ in (12.28), then the equation reduces to a linear diffusion equation

$$\frac{\partial u}{\partial t} = \frac{1}{2} \Delta u + b(x) \cdot \nabla u + c(x)q_1(x)u.$$

In this manner we can handle diffusion equations with creation and killing (cf. Nagasawa (1969)), in other words, in terms of diffusion processes with age ($\lambda = 2$), since there is no branching. This is what is remarked in Section 2.7 as a model of diffusion processes with creation and killing.

---

[2] Notice that there is no killing

[3] This contains a generalization of Fujita (1966)

12.4. Branching Markov Processes with Non-Linear Drift

Introducing further an additional structure to the multiplicative state space, one can get non-linear drift through an extended branching property.

Let us denote

$$S^{(0)} = \cup_{n=0}^{\infty} E^n,$$

and $S^{(1)}$ is defined to be the collection of all elements of the form

$$(a_0, D(a_1), \dots, D(a_m)), \quad m = 1, 2, \dots,$$

where $a_i \in S^{(0)}$, at least one of $a_i \in S^{(0)}$ is not equal to δ. All elements obtained through permutation must be included in $S^{(1)}$. We apply a convention such as

$$(\delta, D(a_1), D\delta, D(a_3)) \;\Rightarrow\; (D(a_1), D(a_3)),$$

through which we delete δ which represents the state of non-existence of particles. The meaning of the notation D will be seen in (12.39). The space $S^{(n)}$, $n \geq 2$, is the collection of all elements of the form

$$(a_0, D(a_1), \dots, D(a_m)), \quad m = 1, 2, \dots,$$

where $a_0 \in S^{(0)} \cup S^{(1)} \cup \cdots \cup S^{(n-1)}$, $a_j \in S^{(n-1)}, j \geq 1$, and of all elements obtained through permutation. Finally an enlarged state space is defined by

(12.37) $$S = \bigcup_{n=0}^{\infty} S^{(n)}.$$

We define a multiplication in S through

(12.38) $$a \cdot b = (a, b), \quad \text{for} \quad a, b \in S,$$

$$\delta a = a \cdot \delta = a.$$

Then the space S is multiplicative.

For $a = (x_0, \mathrm{D}(x_1), \dots, \mathrm{D}(x_m)) \in S^{(1)}$ we set

(12.39) $$\hat{f}(a) = \hat{f}(x_0)\mathrm{D}\hat{f}(x_1) \cdots \mathrm{D}\hat{f}(x_m),$$

where

$$\mathrm{D}\hat{f}(x) = \sum_{i=1}^{n} \mathrm{D}_i \hat{f}(x_1, \dots, x_d), \quad x = (x_1, \dots, x_d),$$

and D_i denotes the first order differential operator applied to the i-th coordinate x_i.

We can construct strong Markov processes with the extended branching property on the enlarged state space S, with the help of the revival theorem which will be explained in the next section. In terms of the extended branching Markov process we can get a solution $u(t, x)$ for a non-linear integral equation

(12.40) $$u(t,x) = P_t f(x) + \int_0^t c\,ds \int_E P_s(x, dy) \sum_{p,q} c_{p,q}(u(t-s,y))^p (\mathrm{D}u(t-s,y))^q,$$

in other words, a weak solution to

$$\frac{\partial u}{\partial t} = \frac{1}{2} \Delta u + c \sum_{p,q} c_{p,q} u^p (\mathrm{D}u)^q.$$

For a construction of the process, we refer to Nagasawa (1968, 72).

The above statement claims that one can represent a solution of the equation (12.40) in terms of an extended branching diffusion process on S theoretically. Practically, however, the enlarged state space is too complicated to compute a stochastic solution $u(t, x)$ in concrete cases.

12.5. Revival of a Markov Process

We will apply the "revival theorem" of Markov processes to a probabilistic construction of branching processes.

When we discuss Markov processes with finite life time, we adopt a standard convention that we attach an extra point Δ to our state space S, define the life time

$$\zeta(\omega) = \sup\{t : X_t(\omega) \in S\},$$

and then set

$$X_t(\omega) = \Delta, \quad for \quad \forall\, t \ge \zeta(\omega).$$

Moreover, we always set $f(\Delta) = 0$ for any function f defined on S.

Let $\{X_t, \zeta, P_a, a \in S\}$ be a Markov process with finite life time. "Revival" of the process means that we let the process start again from the point $X_{\zeta-(\omega)}(\omega)$ with a "revival" distribution.

A probability kernel $N(\omega, db)$ on $\Omega \times \mathcal{B}(S)$ is called a *revival* (renewal) *kernel*, if

$$(12.41) \qquad N(\theta_t\omega, \cdot) = N(\omega, \cdot), \quad if \; t < \zeta(\omega),$$

$$(12.42) \qquad N(\omega, \cdot) = \delta_\Delta, \quad if \; \zeta(\omega) = 0.$$

A typical example of a revival kernel is

$$(12.43) \qquad N(\omega, \cdot) = p(X_{\zeta-}(\omega), \cdot)$$

with a probability kernel $p(a, db)$ on $S \times \mathcal{B}(S)$, and in addition (12.42).

With a given revival kernel $N(\omega, \cdot)$, we define a probability kernel on $\Omega \times \mathcal{B}(\Omega)$ by

$$(12.44) \qquad Q(\omega, B) = \int_S N(\omega, db)P_b[B].$$

Applying Ionescu-Tulcea's theorem (cf. e.g. Doob (1953), Neveu (1965)), we get a Markov chain on the "state space" Ω, namely a system of probability measures Π_ω defined on $\{\Omega^\infty, \mathcal{B}(\Omega)^\infty\}$ such that

$$(12.45) \quad \Pi_{\omega_1}[f] = \int Q(\omega_1, d\omega_2) \cdots Q(\omega_{n-1}, d\omega_n)f(\omega_1, \dots, \omega_n),$$

for any bounded measurable function f on Ω^n.

Then we define a system of probability measures on $\{\Omega^\infty, \mathcal{B}(\Omega)^\infty\}$ by

$$(12.46) \qquad \mathbf{P}_a[f] = \int_\Omega P_a(d\omega)\Pi_\omega[f] \,,$$

for bounded measurable function f on Ω^∞.

Let us define a sequence of *revival times* $r_k(\widetilde{\omega})$ by

$$(12.47) \qquad r_k(\widetilde{\omega}) = \sum_{i=1}^{k} \zeta(\omega_i), \quad for \;\; \widetilde{\omega} = (\omega_1, \omega_2, \dots),$$

and a process $X_t(\widetilde{\omega})$ by

$$(12.48) \qquad X_t(\widetilde{\omega}) = X_t(\omega_1), \qquad if \;\; t < r_1(\widetilde{\omega})$$

$$= X_{t-r_{k-1}}(\omega_k), \quad if \;\; r_{k-1}(\widetilde{\omega}) \le t < r_k(\widetilde{\omega})$$

$$= \Delta, \qquad\qquad if \;\; t \ge r_\infty(\widetilde{\omega}),$$

for $\widetilde{\omega} = (\omega_1, \omega_2, \dots)$, where $r_\infty(\widetilde{\omega}) = \lim_{k \to \infty} r_k(\widetilde{\omega})$.

Theorem 12.6. *Let* $\{X_t, \zeta, \mathbf{P}_a, a \in S\}$ *be a Markov process with finite life time. Then*:

(i) *The revival process* $\{X_t, r_\infty, \mathbf{P}_a, a \in S\}$ *is a Markov process satisfying*

(12.49) $\{X_t : t < r_1, \mathbf{P}_a\}$ *is equivalent to the given process* $\{X_t : t < \zeta, \mathbf{P}_a\}$,

and

$$(12.50) \qquad \mathbf{P}_a[F(\omega_1)g(X_{r_1})] = \mathbf{P}_a[F(\omega_1)N(\omega_1, g)],$$

where F is a bounded measurable function on Ω and g a bounded measurable function on S.

(ii) *If the given process is strong Markov (right-continuous), so is the revival process.*

Proof is immediate because of the construction. For details see Ikeda-Nagasawa-Watanabe (1968,69), Nagasawa (1977), and Meyer (1975).

Theorem 12.7. *Denote*

(12.51) $\phi(a, dt, db) = \mathbf{P}_a[r_1 \in dt, X_{r_1} \in db].$

Then $u(t, a) = \mathbf{P}_t f(a) = \mathbf{P}_a[f(X_t)]$, with a non-negative measurable function f on S, is a minimal solution of an integral equation

(12.52) $u(t, a) = \mathbf{P}_t f(a) + \displaystyle\int_{[0, t] \times S} \phi(a, dr, db) u(t - r, b),$

where $\mathbf{P}_t f$ is the semi-group of the given process.

Proof. Since $\mathbf{P}_t f(a) = \displaystyle\sum_{k=0}^{\infty} \mathbf{P}_t^{(k)} f(a)$, where $\mathbf{P}_t^{(0)} f = \mathbf{P}_t f$ and

$$\mathbf{P}_t^{(k)} f(a) = \mathbf{P}_a[f(X_t): r_k \leq t < r_{k+1}],$$

the assertion follows immediately.

12.6. Construction of Branching Markov Processes

For a given Markov process $\{X_t, \zeta, \mathbf{P}_x, x \in E\}$ on a state space E we define the direct product $\{X_t, \zeta, \mathbf{P}_a, a = (x_1, \ldots, x_n) \in E^n\}$ on Ω^n, for each $n \geq 1$:

$$\zeta(\widetilde{\omega}) = \min\{\zeta(\omega_k): k = 1, 2, \ldots, n\},$$

$$X_t(\widetilde{\omega}) = (X_t(\omega_1), \ldots, X_t(\omega_n)), \quad if \ \ t < \zeta(\widetilde{\omega}),$$

$$= \Delta, \qquad\qquad\qquad if \ \ t \geq \zeta(\widetilde{\omega}),$$

$$\mathbf{P}_a = \mathbf{P}_{x_1} \otimes \cdots \otimes \mathbf{P}_{x_n}, \quad for \ \ a = (x_1, \ldots, x_n) \in E^n.$$

Moreover, we consider a Markov process on $S = \cup_{n=0}^{\infty} E^n$, $E^0 = \{\delta\}$, where δ is an extra point,

(12.53) $\{X_t, \ \zeta, \ \mathbf{P}_a, a \in S = \cup_{n=0}^{\infty} E^n\}$

defined on $\Omega = \cup_{n=0}^{\infty} \Omega^n$, $\Omega^{(0)} = \{\omega_\delta\}$, where ω_δ is an extra point such that $X_t(\omega_\delta) = \delta$, for all $t \geq 0$.

Let $\pi_m(x, db)$ be a probability kernel on $E \times \mathcal{B}(E^m)$ and define a probability kernel on $E \times \mathcal{B}(S)$ by

$$(12.54) \qquad \pi(x, db) = \sum_{m=0}^{\infty} q_m(x) \pi_m(x, db),$$

where $q_m(x)$ are non-negative measurable functions such that

$$\sum_{m=0}^{\infty} q_m(x) = 1,$$

and π_0 is a point measure at $\{\delta\}$. We will call $\{q_m(x),\ \pi_m(x, db),\ m \geq 0\}$ a *branching law*. The kernel $\pi(x, db)$ governs the branching of a single particle.

For $a = (x_1, \dots, x_n)$ we define a kernel $\pi^{(i)}$, $i \leq n$, by

$$(12.55) \quad \int_S \pi^{(i)}(a, db) f(b) = q_0(x_i) f(x_1, \dots, x_{i-1}, x_{i+1}, \dots, x_n)$$

$$+ \sum_{m=1}^{\infty} q_m(x_i) \int_{E^m} \pi_m(x_i, db) f(x_1, \dots, x_{i-1}, b, x_{i-1}, \dots, x_n),$$

namely $\pi^{(i)}$ gives the distribution of offspring just after the i-th particle among the n-particles branches.

In terms of the $\pi^{(i)}$ we define a revival kernel $N(\widetilde{\omega}, db)$ on $\Omega \times \mathcal{B}(S)$: When $\widetilde{\omega} \in \Omega^n$, $\widetilde{\omega} = (\omega_1, \dots, \omega_n)$,

$$N(\widetilde{\omega}, db) = \sum_{i=1}^{n} 1_{\{\zeta(\widetilde{\omega}) = \zeta(\omega_i)\}}(\widetilde{\omega})\ \pi^{(i)}(X_{\zeta-}(\widetilde{\omega}), db),$$

(12.56)

$$N(\omega_\delta,\ db) = a \text{ point measure on } \delta.$$

Then we have

Theorem 12.8. *Given a Markov process* $\{X_t, \zeta, P_x, x \in E\}$ *on* E *with finite life time such that*

$$P_x[\exists X_{\zeta-} \in E] = 1,\ and\ P_x[\zeta = t] = 0,\ for\ \forall\ t \geq 0,$$

and a branching law $\{q_m(x),\ \pi_m(x, db),\ m \geq 0\}$. *Then there exists a branching Markov process* $\{X_t, r_\infty, \mathbf{P}_a, S = \cup_{n=0}^\infty E^n\}$ *such that*

(12.57) $\qquad \mathbf{P}_x[f(X_t); t < r_1] = \mathbf{P}_x[f(X_t); t < \zeta], \quad for \ x \in E,$

(12.58) $\qquad \mathbf{P}_a[r_1 \in ds, X_{r_1} \in db] = \mathbf{P}_a[r_1 \in ds, N(\cdot, db)], \quad for \ a \in S,$

where $N(\cdot, db)$ *is defined in* (12.56).

Proof. We apply Theorem 12.6 to the revival kernel $N(\widetilde{\omega}, db)$ defined in (12.56) and the process $\{X_t, \zeta, \mathbf{P}_a, a \in S = \cup_{n=0}^\infty E^n\}$ given in (12.53). Then we get a revival process $\{X_t, r_\infty, \mathbf{P}_a, S = \cup_{n=0}^\infty E^n\}$. It is clear that (12.57) and (12.58) hold because of the construction of the process. Therefore, it is enough to prove the branching property of the revival process.

Let us denote

(12.59)
$$P_t^0 = P_t,$$
$$P_t^k f(a) = \mathbf{P}_a[f(X_t); r_k \leq t < r_{k+1}], \quad for \ k \geq 1.$$

Then, because of (12.60) which will be shown below we have

$$\mathbf{P}_t(a \cdot b, \cdot) = \sum_{k=0}^\infty P_t^k(a \cdot b, \cdot) = \sum_{k=0}^\infty \sum_{i=0}^k P_t^i(a, \cdot) * P_t^{k-i}(b, \cdot)$$

$$= \sum_{k=0}^\infty P_t^k(a, \cdot) * \sum_{k=0}^\infty P_t^k(b, \cdot) = \mathbf{P}_t(a, \cdot) * \mathbf{P}_t(b, \cdot).$$

This completes the proof of the theorem.

Lemma 12.2. *Let* $P_t^k f(a)$ *be defined at* (12.59). *Then*

(12.60) $\qquad P_t^k(a \cdot b, \cdot) = \sum_{i=0}^k P_t^i(a, \cdot) * P_t^{k-i}(b, \cdot), \quad k = 0, 1, 2, \dots .$

Proof. We prove (12.60) by induction. For $k = 0$ it is evident because of the construction. Let

$$\phi(a, ds, db) = \mathbf{P}_a[r_1 \in ds, X_{r_1} \in db].$$

Then the formulae (12.55) and (12.56) yield

(12.61) $\qquad \phi(a \cdot b, ds, dc) = \phi(a, ds, dc_1) P_s(b, dc_2) + P_s(a, dc_1) \phi(b, ds, dc_2),$

where $c = c_1 \cdot c_2$. The strong Markov property of X_t at r_k implies

(12.62) $\qquad P_t^k f(a) = \int_{[0,t] \times S} \phi(a, ds, db) P_{t-s}^{k-1} f(b).$

Therefore, applying (12.61), and (12.62), we have

$$P_t^{k+1} f(a \cdot b) = \int_{[0,t] \times S} \phi(a \cdot b, ds, dc) P_{t-s}^k f(c)$$

$$= \int_{[0,t] \times S \times S} \{ \phi(a, ds, dc_1) P_s(b, dc_2) + P_s(a, dc_1) \phi(b, ds, dc_2) \}$$

$$\times \sum_{i=0}^{k} \int_{S \times S} P_{t-s}^i(c_1, de_1) P_{t-s}^{k-i}(c_2, de_2) f(e_1 \cdot e_2).$$

Moreover, since

(12.63) $\qquad P_s P_{t-s}^k f(a) = \int_{[s,t] \times S} \phi(a, dr, db) P_{t-r}^{k-1} f(b),$

denoting $F^k(a, s, B) = P_s P_{t-s}^k(a, B)$, we can rewrite as follows:

$$P_t^{k+1} f(a \cdot b) = - \sum_{i=0}^{k} \int_{[0,t] \times S \times S} \{ dF^{i+1}(a, s, de_1) F^{k-i}(b, s, de_2)$$

$$+ F^i(a, s, de_1) dF^{k+1-i}(b, s, de_2) \} f(e_1 \cdot e_2)$$

$$= - \sum_{i=0}^{k+1} \int_{[0,t] \times S \times S} \{ dF^i(a, s, de_1) F^{k+1-i}(b, s, de_2)$$

$$+ F^i(a, s, de_1) dF^{k+1-i}(b, s, de_2) \} f(e_1 \cdot e_2),$$

where we have added $(-dF^0)F^{k+1}$ and $-F^{k+1}(dF^0)$ $(dF^0 = 0 \ !)$.

Therefore

$$P_t^{k+1} f(a \cdot b) = \sum_{i=0}^{k+1} \int_{S \times S} F^i(a, 0, de_1) F^{k+1-i}(b, 0, de_2) f(e_1 \cdot e_2)$$

$$= \sum_{i=0}^{k+1} \int_{S \times S} P_t^i(a, de_1) P_t^{k+1-i}(b, de_2) f(e_1 \cdot e_2).$$

This completes the proof of Lemma 12.2.

Remark. It is immediate to extend the construction which has been discussed above to the more general case of branching Markov processes with age and sign (cf. Nagasawa (1968)).

Appendix

a.1. Fényes' "Equation of Motion" of Probability Densities

We consider diffusion equations in duality (diffusion processes with time reversal), which implies the duality relation (3.34') and (3.35'); namely

(A.1) $$\mathbf{u} = \frac{a + \hat{a}}{2} = \nabla R, \qquad \mathbf{v} = \frac{a - \hat{a}}{2} = \nabla S,$$

(A.2) $$\frac{\partial R}{\partial t} + \frac{1}{2} \Delta S + \nabla S \cdot \nabla R + \boldsymbol{b} \cdot \nabla R = 0,$$

where $(\sigma^T \sigma)^{ij} = \delta^{ij}$ is assumed. Moreover, we set $\boldsymbol{b} \equiv 0$ in this section for simplicity.

Influenced by Schrödinger (1931), Fényes (1952) considered a diffusion process for quantum mechanics.[1] Fényes defines a Lagrangian

(A.3) $$L(t) = \int \{ \frac{\partial S}{\partial t} + \frac{1}{2} (\nabla S)^2 + V + \frac{1}{2} (\frac{1}{2} \frac{\nabla \mu}{\mu})^2 \} \, \mu \, dx,$$

where $\mu_t(x) = e^{2R(t,x)}$ denotes the distribution density of the diffusion process and V is a potential function. He calls the term $\Pi(\mu) = \frac{1}{2} (\frac{1}{2} \frac{\nabla \mu}{\mu})^2 \mu$ in (A.3) "Diffusionsdruck". Since

(A.4) $$\frac{1}{2} \frac{\nabla \mu}{\mu} = \nabla R,$$

---

[1] For further literatures on the subject cf. chapter 9 of Jammer (1974)

his Lagrangian (A.3) can be represented in terms of $R(t,x)$ and $S(t,x)$ as

$$L(t) = \int \{ \frac{\partial S}{\partial t} + \frac{1}{2}(\nabla S)^2 + \frac{1}{2}(\nabla R)^2 + V \} \mu \, dx.$$

Applying the variational principle

$$\delta \int_a^b L(t) \, dt = 0,$$

Fényes obtains

$$(A.5) \quad \frac{\partial S}{\partial t} + \frac{1}{2}\{\nabla(R+S)\}^2 - (\nabla(R+S)) \cdot (\frac{1}{2}\frac{\nabla\mu}{\mu}) + (\frac{1}{2}\frac{\nabla\mu}{\mu})^2 - \frac{1}{4}\frac{\Delta\mu}{\mu} + V = 0,$$

and calls it "Bewegungsgleichung" of probability densities.

Postulating equation (A.5), he demonstrates that the function $\psi = e^{R+iS}$ satisfies the Schrödinger equation

$$i \frac{\partial \psi}{\partial t} + \frac{1}{2}\Delta\psi - V(t,x)\psi = 0.$$

We can show this as follows: substituting (A.4) and

$$(A.6) \qquad\qquad \frac{1}{4}\frac{\Delta\mu}{\mu} = \frac{1}{2}\Delta R + (\nabla R)^2$$

into (A.5), we have

$$(A.7) \qquad\qquad \frac{\partial S}{\partial t} + \frac{1}{2}(\nabla S)^2 - \frac{1}{2}(\nabla R)^2 - \frac{1}{2}\Delta R + V = 0.$$

Thus equation (4.8) holds. Moreover (4.9) is nothing but (A.2). Therefore, $\psi = e^{R+iS}$ satisfies the Schrödinger equation by Lemma 4.1.

However, as will be shown, equation (A.5) holds automatically in diffusion theory with $V = -c - 2 \, \partial S/\partial t - (\nabla S)^2$,[2] where c is the creation and killing induced by $\phi = e^{R+S}$, and hence (A.5) is not something to be postulated additionally.

---

[2] Cf. Section 4.3 for the non-linear dependence of V and c

In fact:

Lemma A.1. *Assume the duality relation, i.e., (A.1) and (A.2). Let* $\phi = e^{R+S}$ *and* $c = -L\phi/\phi$ *be the creation and killing induced by* ϕ.[3] *Then equation (A.5) holds with* $V = -c - 2\,\partial S/\partial t - (\nabla S)^2$.

Proof. Lemma 4.2 claims

$$-c = \{-\frac{\partial S}{\partial t} + \frac{1}{2}\Delta R + \frac{1}{2}(\nabla R)^2 - \frac{1}{2}(\nabla S)^2\}$$

$$+\{\frac{\partial R}{\partial t} + \frac{1}{2}\Delta S + (\nabla S)\cdot(\nabla R)\}$$

$$+\{2\frac{\partial S}{\partial t} + (\nabla S)^2\},$$

where the second line of the right-hand side vanishes because of (A.2). Therefore, with

(A.8) $$V = -c - 2\frac{\partial S}{\partial t} - (\nabla S)^2,$$

equation (A.7) holds and hence (A.5).

Since we have shown in Chapter 4 that diffusion and Schrödinger equations are equivalent, it is clear that equation (A.5) is not necessary to be put in between. Nonetheless, one might argue that equation (A.5) is "the equation of motion" and it is physically meaningful or at least helpful. However, as we have seen in Chapter 4, the diffusion equation itself is the equation of motion, from which the Schrödinger equation follows, and hence the variational principle of Fényes with the Lagrangian $L(t)$ should better be regarded as a way of characterizing the Schrödinger process as we have done in Chapters 5 and 6.

a.2. Stochastic Mechanics

Nelson (1966, 67) played an important role in bringing people's attention to the subject, and many publications followed.

---

[3] Cf. (3.40), where L is defined at (3.39)

He deduced in his paper the Schrödinger equation from his stochastic mechanics (= stochastic Newtonian equation) as follows:

Let $B(t)$ and $\widehat{B}(t)$ be defined at (3.32), i.e.,

$$B(t) = \frac{\partial}{\partial t} + \frac{1}{2}\Delta + b(t,x)\cdot\nabla + a(t,x)\cdot\nabla,$$

$$\widehat{B}(t) = -\frac{\partial}{\partial t} + \frac{1}{2}\Delta - b(t,x)\cdot\nabla + \widehat{a}(t,x)\cdot\nabla.$$

Nelson (1966) introduces "the mean acceleration"

(A.9) $\alpha(t,x) = -\frac{1}{2}\{B(t)\widehat{B}(t)x + \widehat{B}(t)B(t)x\}.$

Actually he defines forward and backward derivatives D and $-D_*$ applied to X_t instead of $B(t)$ and $\widehat{B}(t)$ applied to x, respectively, which are the same thing for diffusion processes (for non-Markovian cases we need to look at the operators D and D_* more carefully, for this cf. Föllmer (1986)).

In the following we assume $(\sigma^T\sigma)^{ij} = \delta^{ij}$ (cf. Dankel (1971), Dohrn-Guerra (1978), Meyer (1980/81) for the case of manifolds).

Assuming the duality relation (A.1) and (A.2) he gets, through formal manipulation of vector calculus,

(A.10) $\alpha(t,x) = -\frac{1}{2}\{B(t)(-b + \widehat{a}) + \widehat{B}(t)(b + a)\}$

$$= -\{-\frac{\partial v}{\partial t} + \frac{1}{2}\Delta u + \frac{1}{2}(\widehat{a}\cdot\nabla)a + \frac{1}{2}(a\cdot\nabla)\widehat{a} - (b\cdot\nabla)v - (v\cdot\nabla)b - v\times\mathrm{curl}\,b\}$$

$$+ \{\frac{\partial b}{\partial t} + \frac{1}{2}\nabla(b^2) - (b + v)\times\mathrm{curl}\,b\}.$$

Then it is shown that the Schrödinger equation can be deduced from his "stochastic Newtonian equation"

(A.11) $\alpha(t,x) = -\nabla V + \frac{\partial b}{\partial t} + \frac{1}{2}\nabla(b^2) - (b + v)\times\mathrm{curl}\,b,$[4]

---

[4] The right-hand side is the force on a charged particle in an electromagnetic field

which is called "stochastic mechanics" ("stochastic quantization").

In fact, it is easy to see, because of (A.1) and with formal manipulation of vector calculus,

$$(A.12) \quad \nabla\left\{-\frac{\partial S}{\partial t} + \frac{1}{2}\Delta R + \frac{1}{2}(\nabla R)^2 - \frac{1}{2}(\nabla S)^2 - b\cdot\nabla S\right\}$$

$$= -\frac{\partial \mathbf{v}}{\partial t} + \frac{1}{2}\Delta\mathbf{u} + \frac{1}{2}(\hat{a}\cdot\nabla)a + \frac{1}{2}(a\cdot\nabla)\hat{a} - (b\cdot\nabla)\mathbf{v} - (\mathbf{v}\cdot\nabla)b - \mathbf{v}\times\text{curl }b.$$

Therefore, substituting (A.12) into the first line of the rightmost equation of (A.10), one gets

$$(A.13) \quad \alpha(t,x) = -\nabla\left\{-\frac{\partial S}{\partial t} + \frac{1}{2}\Delta R + \frac{1}{2}(\nabla R)^2 - \frac{1}{2}(\nabla S)^2 - b\cdot\nabla S\right\}$$

$$+ \left\{\frac{\partial b}{\partial t} + \frac{1}{2}\nabla(b^2) - (b + \mathbf{v})\times\text{curl }b\right\}.$$

Substituting this into the left-hand side of (A.11), one gets

$$(A.14) \quad \nabla\left\{-\frac{\partial S}{\partial t} + \frac{1}{2}\Delta R + \frac{1}{2}(\nabla R)^2 - \frac{1}{2}(\nabla S)^2 - b\cdot\nabla S\right\} = \nabla V,$$

namely, except for an ambiguity with a function K such that $\nabla K = 0$ (we can set $K \equiv 0$), equality (4.8) holds. In addition, equation (4.9) holds, since it is nothing but (A.2). Therefore, $\psi = e^{R + iS}$ satisfies the Schrödinger equation (4.1) by Lemma 4.1.

His arguments explained above caused some kinds of irritation among conventional (mathematical) physicists. His "Newtonian equation (A.11)" (especially "acceleration" (A.9)) was controversial (they considered it artificial). Moreover, there has been criticism that equation (A.11) is not at all practical and one must solve the Schrödinger equation in any case.

As a matter of fact it can be shown that equation (A.11) is a simple consequence of time reversal (duality) of diffusion processes:

Lemma A.2. (Nagasaawa (1991)) *Assume the duality relation, i.e., (A.1) and (A.2). Set $\phi = e^{R+S}$ and $c = -L\phi/\phi$ be the creation and killing induced by ϕ. Define $\alpha(t,x)$ by (A.9). Then equation (A.11) holds with*

$$V = -c - 2\,\partial S/\partial t - (\nabla S)^2 - 2\,\boldsymbol{b}\cdot\nabla S.^{[5]}$$

Proof. By Lemma 4.2 equation (4.11) holds. The second line of the right-hand side of (4.11) vanishes by (A.2); we have consequently

$$-c = \{-\frac{\partial S}{\partial t} + \frac{1}{2}\,\Delta R + \frac{1}{2}(\nabla R)^2 - \frac{1}{2}(\nabla S)^2 - \boldsymbol{b}\cdot\nabla S\}$$

$$+ \{2\frac{\partial S}{\partial t} + (\nabla S)^2 + 2\,\boldsymbol{b}\cdot\nabla S\},$$

and hence

(A.15) $$-\nabla c = \nabla\{-\frac{\partial S}{\partial t} + \frac{1}{2}\,\Delta R + \frac{1}{2}(\nabla R)^2 - \frac{1}{2}(\nabla S)^2 - \boldsymbol{b}\cdot\nabla S\}$$

$$+ \nabla(2\frac{\partial S}{\partial t} + (\nabla S)^2 + 2\,\boldsymbol{b}\cdot\nabla S).$$

After substituting (A.13) into (A.15), we get

$$-\nabla c = -\alpha(t,x) + \{\frac{\partial \boldsymbol{b}}{\partial t} + \frac{1}{2}\,\nabla(\boldsymbol{b}^2) - (\boldsymbol{b} + \mathbf{v})\times\mathrm{curl}\,\boldsymbol{b}\}$$

$$+ \nabla(2\frac{\partial S}{\partial t} + (\nabla S)^2 + 2\,\boldsymbol{b}\cdot\nabla S),$$

which is nothing but equation (A.11) with

$$V = -c - 2\frac{\partial S}{\partial t} - (\nabla S)^2 - 2\,\boldsymbol{b}\cdot\nabla S.$$

This completes the proof.

Remark. If we begin with $c(t,x,\phi)$ defined by (4.16), then we get exactly the equation (A.11) with a given $V(t, x)$.

It has been considered that equation (A.11) is a physically meaningful postulate as the equation of motion or as a sort of quantization, and called often "stochastic quantization", since it deduces the Schrödinger equation. Lemma A.2 shows that equation (A.11) is correct with the V specified there. However, since the Schrödinger equation follows directly from the

---

[5] Cf. Section 4.3 on the non-linear dependence of V and c

pair of diffusion equations in duality without equation (A.11), we can avoid it as being the equation of motion. As already remarked, the diffusion equation itself is the equation of motion, which deduces (actually is equivalent to) the Schrödinger equation.

a.3. Segregation of a Population

Consider a system of interacting diffusion processes (cf. Chapters 7 and 8) and assume the typical particle moves according to a diffusion process with drift $a(x)$ (also with $\hat{a}(x)$ for the time reversed process). We assume that $b(t, x) \equiv 0$ and that $a(x)$ and $\hat{a}(x)$ do not depend on t. Let us define fields of "kinetic energy $K(x)$" and "population pressure $Q(x)$" by

$$K(x) = \frac{1}{2} \{ \frac{1}{2} a(x)^2 + \frac{1}{2} \hat{a}(x)^2 \},$$

$$Q(x) = \frac{1}{2} \frac{1}{\mu(x)} (-\frac{1}{2} \Delta \mu(x)),$$

respectively, where $\mu(x) = \hat{\phi}(x)\phi(x)$. In Nagasawa (1980) it is shown that the law of equilibrium of energy

(A.16) $\qquad K(x) + Q(x) + V(x) = \lambda, \ \ for\ x \in \{x; \mu(x) \neq 0\}$

is equivalent to the stationary Schrödinger equation

$$\frac{1}{2} \Delta \psi(x) + (\lambda - V(x)) \psi(x) = 0,$$

where $\psi(x) = e^{R(x) + i S(x)}$. This follows immediately from a formula

(A.17) $\qquad\qquad K(x) + Q(x) = -\frac{1}{2} \frac{1}{\psi(x)} \Delta \psi(x).$

This formula can be verified as follows: notice first of all

$$-\frac{1}{2} \frac{1}{\psi} \Delta \psi = \frac{1}{2} (\nabla S)^2 - \frac{1}{2} (\nabla R)^2 - \frac{1}{2} \Delta R - i(\frac{1}{2} \Delta S + \nabla R \cdot \nabla S),$$

where the imaginary part vanishes because of (A.2). Secondly (A.6) holds, i.e.,

$$Q(x) = -\frac{1}{4} \frac{1}{\mu} \Delta \mu = - (\nabla R)^2 - \frac{1}{2} \Delta R.$$

Finally, because of the duality relation (A.1),

$$K(x) = \tfrac{1}{2}(\nabla S)^2 + \tfrac{1}{2}(\nabla R)^2.$$

Combining them we have the formula (A.17).

Based on this fact, segregation of a population was discussed (cf. Chapters 7, 8, and 9).

Equation (A.16) was postulated in Nagasawa (1980), but it should not have been done, since it holds automatically with $V = -c - (\nabla S)^2$. In fact (A.15) implies

$$\nabla\{c + (\nabla S)^2\} = \nabla\{\tfrac{1}{2}(\nabla S)^2 - \tfrac{1}{2}(\nabla R)^2 - \tfrac{1}{2}\Delta R\},$$

from which (A.16) follows.

a.4. Euclidean Quantum Mechanics

Analogous to (A.9), Zambrini (1987) defines

(A.18) $\tilde{\alpha}(t, x) = \tfrac{1}{2}\{B(t)B(t)x + \hat{B}(t)\hat{B}(t)x\},$

$((\sigma\sigma^T)^{ij} = \delta^{ij}$ *is also required*). It is easy to show, with the same manipulation as with (A.10),

(A.19) $\tilde{\alpha}(t, x) = \tfrac{1}{2}\{B(t)(b + a) + \hat{B}(t)(-b + \hat{a})\}$

$$= \{-\frac{\partial v}{\partial t} + \tfrac{1}{2}\Delta u + \tfrac{1}{2}(\hat{a}\cdot\nabla)a + \tfrac{1}{2}(a\cdot\nabla)\hat{a} - (b\cdot\nabla)v - (v\cdot\nabla)b - v\times\mathrm{curl}\,b\}$$

$$+ \{\frac{\partial b}{\partial t} + \tfrac{1}{2}\nabla(b^2) - (b + v)\times\mathrm{curl}\,b\} + \{2\frac{\partial v}{\partial t} + \nabla(v^2) + 2\nabla(b\cdot v)\},$$

where **u** and **v** are defined in (A.1). Substituting (A.12) into (A.19), we have

(A.20) $\tilde{\alpha}(t, x) = \nabla\{-\frac{\partial S}{\partial t} + \tfrac{1}{2}\Delta R + \tfrac{1}{2}(\nabla R)^2 - \tfrac{1}{2}(\nabla S)^2 - b\cdot\nabla S\}$

$$+ \{\frac{\partial b}{\partial t} + \tfrac{1}{2}\nabla(b^2) - (b + v)\times\mathrm{curl}\,b\} + \nabla\{2\frac{\partial S}{\partial t} + (\nabla S)^2 + 2\,b\cdot\nabla S\}.$$

He postulates

(A.21) $$\tilde{\alpha}(t,x) = -\nabla c + \frac{\partial \boldsymbol{b}}{\partial t} + \frac{1}{2}\nabla(\boldsymbol{b}^2) - (\boldsymbol{b} + \mathbf{v})\times\text{curl } \boldsymbol{b},^6$$

and calls this "Euclidean quantum mechanics".

After substituting (A.20), equation (A.21) reads

(A.22) $$\nabla\{-\frac{\partial S}{\partial t} + \frac{1}{2}\Delta R + \frac{1}{2}(\nabla R)^2 - \frac{1}{2}(\nabla S)^2 - \boldsymbol{b}\cdot\nabla S\}$$

$$+ \nabla\{2\frac{\partial S}{\partial t} + (\nabla S)^2 + 2\boldsymbol{b}\cdot\nabla S\} = -\nabla c.$$

Therefore, except for an ambiguity with a function K such that $\nabla K = 0$ (we can set $K \equiv 0$), equation (4.8) holds with

(A.23) $$V = -c - 2\frac{\partial S}{\partial t} - (\nabla S)^2 - 2\boldsymbol{b}\cdot\nabla S.$$

Moreover, equation (4.9) is nothing but (A.2). Therefore, as in the case of Nelson, an application of Lemma 4.1 implies that the function $\psi = e^{R+iS}$ satisfies the Schrödinger equation (4.1) with the V defined in (A.23).

As a matter of fact it is easy to see that equation (A.21) holds automatically as a simple consequence of the time reversal of diffusion processes (the duality relation of diffusion processes):

Lemma A.3. (Nagasawa (1991)) *Assume the duality relation, i.e., (A.1) and (A.2). Set $\phi = e^{R+S}$ and let $c = -L\phi/\phi$ be the creation and killing induced by ϕ. Define $\tilde{\alpha}(t,x)$ by (A.18). Then, equation (A.21) holds.*

Proof. As is shown in the proof of Lemma A.2, we have (A.15), into which we substitute (A.20). Then we get

$$-\nabla c = \tilde{\alpha}(t,x) - \{\frac{\partial \boldsymbol{b}}{\partial t} + \frac{1}{2}\nabla(\boldsymbol{b}^2) - (\boldsymbol{b} + \mathbf{v})\times\text{curl } \boldsymbol{b}\},$$

which is nothing but equation (A.21). This completes the proof.

---

[6] He treated the case of $\boldsymbol{b} \equiv 0$, and a and $\hat{a}$ are bounded

Therefore, the remarks which have been given to Nelson's equation (A.11) also applies to the equation (A. 21).

a.5. Remarks

(i) Things are now transparent. We can summarize what we have seen in Sections a.1, 2, 3 and 4 as follows:

Let $\psi = e^{R + iS}$. Then

(A.24) $i\dfrac{\partial \psi}{\partial t} + \dfrac{1}{2}\Delta \psi + i\,b(t,x)\cdot\nabla \psi - V(t,x)\psi$

$$= -\{\frac{\partial S}{\partial t} - \frac{1}{2}\Delta R - \frac{1}{2}(\sigma\nabla R)^2 + \frac{1}{2}(\sigma\nabla S)^2 + b\cdot\nabla S + V\}\psi$$

$$+ i\,\{\frac{\partial R}{\partial t} + \frac{1}{2}\Delta S + (\sigma\nabla S)\cdot(\sigma\nabla R) + b\cdot\nabla R\}\psi.$$

Consider a diffusion process in Schrödinger's representation. Then we have a pair of functions

$$\phi = e^{R + S} \quad and \quad \widehat{\phi} = e^{R - S}$$

such that $\mu = \widehat{\phi}\,\phi$ is the distribution density of the diffusion process. We identify R and S in $\psi = e^{R + iS}$ and $\phi = e^{R + S}$.

Consider now Kolmogoroff's representation. Then we see that the imaginary part in (A.24) vanishes, since $\mu = \widehat{\phi}\phi = \overline{\psi}\psi = e^{-2R}$ satisfies a pair of Fokker-Planck's equations

$$-\frac{\partial \mu}{\partial t} + \frac{1}{2}\Delta \mu - \frac{1}{\sqrt{\sigma_2}}\nabla\{\sqrt{\sigma_2}(b(t,x) + a(t,x))\,\mu\} = 0,$$

and

$$\frac{\partial \mu}{\partial t} + \frac{1}{2}\Delta \mu - \frac{1}{\sqrt{\sigma_2}}\nabla\{\sqrt{\sigma_2}(-b(t,x) + \widehat{a}(t,x))\,\mu\} = 0,$$

at the same time because of the duality (time reversal); namely (3.35) holds and hence (3.35'), which is (A.2).

Therefore, for the function $\psi = e^{R + iS}$ to satisfy the Schrödinger equation, it is necessary and sufficient that the real part of (A.24) also vanishes:

(A.25) $\qquad \dfrac{\partial S}{\partial t} - \dfrac{1}{2}\Delta R - \dfrac{1}{2}(\sigma\nabla R)^2 + \dfrac{1}{2}(\sigma\nabla S)^2 + b\cdot\nabla S + V = 0.$

We have shown in Theorem 4.1 that equation (A.25) holds automatically with $V = -c\ -2\dfrac{\partial S}{\partial t} - (\sigma\nabla S)^2 - 2b\cdot\nabla S.$

Since this fact was not clearly recognized, equation (A.25) was *postulated* in various different forms: as "Bewegungsgleichung" (A.5) (or "Lagrangian" (A.3)) in Fényes (1952); as "stochastic mechanics (or stochastic quantization)" (A.11) in Nelson (1966); as "equilibrium of energy" (A.16) in Nagasawa (1980); as "the least action principle (with his Lagrangian (5.46))" of Yasue (1981); as "Euclidean quantum mechanics" (A.21) in Zambrini (1987).

(ii) The potential function $V(t, x)$ in equation (A.25) and the creation and killing $c(t, x)$ satisfy

(A.26) $\qquad\qquad c + V + 2\dfrac{\partial S}{\partial t} + (\sigma\nabla S)^2 + 2b\cdot\nabla S = 0,$

where $c(t, x) = -L\phi(t, x)/\phi(t, x)$ is induced by the function $\phi = e^{R + S}$ of the corresponding diffusion process in Schrödinger's representation. For the non-linear dependence (A.26), see Section 4.3. Notice that the formula (A.26) is *not* a requirement but a *consequence* of the transformation

$$\psi = e^{R + iS} \qquad\qquad\qquad \phi = e^{R + S}$$
$$\Leftrightarrow (R, S) \Leftrightarrow$$
$$\overline{\psi} = e^{R - iS} \qquad\qquad\qquad \hat{\phi} = e^{R - S}$$

between the pairs $\{\overline{\psi}, \psi\}$ and $\{\hat{\phi}, \phi\}$.

a.6. Bohmian Mechanics

The drift field (coefficient) $a + b$, where

(A.27) $\qquad\qquad\qquad\qquad a = \nabla S + \nabla R$

(cf. (A.1)) cannot be regarded as a classical velocity field. Although this is clear, let us consider, to be concrete, the first excited state $\varphi(x) = xe^{-x^2/2}$ of the one-dimensional harmonic oscillator. In this case $b \equiv 0$, $S \equiv 0$, and hence

$$a = \frac{1}{x} - x .$$

Suppose it were a classical velocity field, namely assume

(A.28) $$\frac{dx}{dt} = \frac{1}{x} - x ,$$

which implies

(A.29) $$|x^2 - 1| = \kappa e^{-2t}, \quad x \neq \pm 1,$$

where κ is an integration constant. Therefore, x converges to $+1$ or -1 asymptotically as t tends to infinity ($x = \pm 1$ are stable points). The conclusion is absurd. It contradicts the distribution density $|\varphi(x)|^2 = x^2 e^{-x^2}$.

Let us assume in the following that $b \equiv 0$ for simplicity. In the so-called "Bohmian Mechanics" (cf. e.g., Dürr-Goldstein-Zanghi (1992)) it is postulated that

(i) *the Schrödinger equation holds*;

(ii) *a part of the drift coefficient*

(A.30) $$\mathbf{v} = \nabla S$$

must be a classical velocity field associated with the Schrödinger equation (or a wave function $\psi = e^{R + iS}$), namely

(A.31) $$\frac{dx}{dt} = \nabla S.$$

Let us call (A.31) a "velocity field postulate".

As we have shown (cf. Theorem 4.1) the Schrödinger equation is *equivalent* to a diffusion process $\{X_t, Q\}$ (Schrödinger process), or in other words to a pair of diffusion equations in duality. Accordingly, the duality relation (3.34) and (3.35), namely, (A.1) and (A.2) hold. (A.2) is the continuity equation

(A.32) $\frac{\partial \mu}{\partial t} + \text{div}\,(\mathbf{v}\mu) = 0 \,.$

In DGZ (1992) the continuity equation (A.32) seems to be regarded as a support for their velocity field postulate (A.31), because equation (A.32) contains only $\mathbf{v} = \nabla S$ explicitly.

However, equation (A.2), which is the same as (A.32), is nothing else but an equation which relates the pair of quantities S and R. Therefore, their argument in DGZ (1992) is rather weak regarding (A.32) as a justification of taking just ∇S and abandoning the term ∇R from the drift field. As a matter of fact, the continuity equation (A.32) is equivalent to the pair of the Fokker-Planck equations

$$-\frac{\partial \mu}{\partial t} + \frac{1}{2} \Delta \mu - \text{div}\,(a\mu) = 0,$$

(A.33)

$$\frac{\partial \mu}{\partial t} + \frac{1}{2} \Delta \mu - \text{div}\,(\widehat{a}\mu) = 0,$$

where

$$a = \nabla S + \nabla R,$$

$$\widehat{a} = -\nabla S + \nabla R.$$

The most important fact which should be emphasized is this: We have already the motion of a particle(s) which is deduced from (actually equivalent to) the Schrödinger equation, namely the Schrödinger process $\{X_t, Q\}$ (a diffusion process). Therefore, since the Schrödinger equation is postulated in Bohmian mechanics, it is highly artificial to introduce "the velocity field postulate (A.31)" additionally in order to only consider a deterministic motion.[7]

Moreover, the velocity field postulate (A.31) provides us with a peculiar picture of stationary states. Let us consider a stationary state - to be concrete the first excited state of the one-dimensional harmonic oscillator. Then,

$$\nabla S \equiv 0.$$

Therefore, postulate (A.31) implies

---

[7] It should be emphasized that the existence of the Schrödinger process is not something to be postulated, but a conclusion deduced from the Schrödinger equation itself

(A.34) $$\frac{dx}{dt} = 0,$$

and hence the point (particle) x stands still, which contradicts the Schrödinger equation.

Perhaps "Bohmian mechanics" can be regarded as a sort of "semi-classical limit" which I have mentioned in Chapters 1 and 4 : Namely, if the drift coefficient $a = \nabla S + \nabla R$ is large enough compared to the diffusion coefficient so that the Brownian noise may be neglected, one can consider the classical motion defined through

(A.35) $$\frac{dx}{dt} \cong \nabla S + \nabla R$$

as a "semi-classical limit". An important point in (A.35) is this : the right hand side is the drift coefficient $a = \nabla S + \nabla R$ and hence the term ∇R should not be neglected.

The example mentioned above tells us that stationary states in quantum mechanics are not at all "semi-classical". In connection with this, it is an interesting question when one can neglect ∇R in equation (A.35), so that equation (A.31) will turn out to be reasonable.

References

Aebi, R. (1989): MN-transformed α-diffusion with singular drift. Doctoral Dissertation at the University of Zürich.

Aebi, R. (1992): Itô's formula for non-smooth functions. Publ. RIMS Kyoto Univ. **28**, 595-602.

Aebi, R. (1993): Diffusions with singular drift related to wave functions. To appear in Probab. Th. Rel. Fields.

Aebi, R. (preprint): A solution to Schrödinger's problem of non-linear integral equations.

Aebi, R., Nagasawa, M. (1992): Large deviations and the propagation of chaos for Schrödinger processes. Probab. Th. Rel. Fields, **94**, 53-68.

Aizenman, M., Simon, B. (1982): Brownian motion and Harnack inequality for Schrödinger operators. Comm. Pure Appl. Math. **35**, 209-273.

Albeverio, S., Høegh-Krohn, R. (1974): A remark on the connection between stochastic mechanics and the heat equation. J. Math. Phys. **15**, 1745-1748.

Albeverio, S., Høegh-Krohn, R. (1976): Mathematical theory of Feynman path integrals. Lecture Notes in Math. **523**, 1-139.

Albeverio, S., Blanchard, Ph., Høegh-Krohn, R. (1984): A stochastic model for the orbits of planets and satellites: An interpretation of Titius-Boode law. Expositiones Mathematicae, **4**, 365-373.

Amano, K. (1943): *The origin of the theory of heat-radiation and quantum theory*, Selected papers of W. Wien and M. Planck. (Japanese) Dai-nihon Shuppan. Tokyo.

Amano, K. (1948): *History of Quantum Mechanics*. (Japanese) Nihon-Kagaku-Sha. Tokyo.

Arnold, L. (1973): *Stochastische Differentialgleichungen*. R. Oldenbourg Verlag. München Wien.

Azema, J. (1973): Theorie générale des processus et retournement du temps, Ann. Scient. Ec. Norm. Sup. **6**, 459-519.

Azencott, R. (1980): Grandes deviations et applications, In Ecole d'Eté de probabilités de Saint-Flour VIII, Springer Lect. Notes in Math. **774**, 1-176.

Baras, P., Goldstein, J. A. (1984): Remarks on the inverse square potential in quantum mechanics. 31-35 in *"Differential Equations"* ed. Knowles, I., Lewis, R. T. Elsevier Science Publishers B.V. (North Holland).

Bauer, H. (1981): *Probability Theory and Elements of Measure Theory.* Academic Press, London New York.

Bauer, H. (1990): *Mass- und Integrationstheorie.* Walter de Gruyter, Berlin New York.

Bauer, H. (1991): *Wahrscheinlichkeitstheorie.* Walter de Gruyter, Berlin New York.

Bernstein, S. (1932): Sur les liaisons entre les grandeurs aléatoires. Verhand. Internat. Math. Kongr. Zürich, Bd. **1**, 288-309.

Beurling, A. (1960): An automorphism of product measures. Ann. Math. **72**, 189-200.

Billingsley, P. (1968): *Convergence of Probability Measures.* John Wiley & Sons Inc., New York.

Blanchard, Ph., Golin, S. (1987): Diffusion processes with singular drift fields. Comm. Math. Phys. **109**, 421-435.

Blanchard, Ph., Combe, Ph. & Zheng, W. (1987): Mathematical and Physical Aspects of Stochastic Mechanics. Lecture Notes in Phys. **281**, Springer.

Blaquière, A. (1991): Sufficiency conditions for existence of an optimal feedback control in stochastic mechanics. Dynamics and Control **1**, 7-24.

Blaquière, A. (1992): Controllability of a Fokker-Planck equation, the Schrödinger system, and a related stochastic optimal control. Dynamics and Control **2**, 235-253.

Blumenthal, R.M., Getoor, R.K. (1968) : *Markov processes and potential theory.* Academic Press, New York and London.

Born, M. (1926): Zur Quantenmechanik der Stossvorgänge. Z. Phys. **37**, 863-867.

Born, M., Jordan, W. (1925): Zur Quantenmechanik. Z. Phys. **34**, 858-888.

Born, M., Heisenberg, W., Jordan, W. (1926) : Zur Quantenmechanik II. Z. Phys. **35**, 557-615.

Braun, W., Hepp, K. (1977): The Vlasov dynamics and its fluctuations in 1/N limit of interacting classical particles. Comm. Math. Phys. **56**, 101-113.

Brox, T. (1986): A one-dimensional diffusion process in a Wiener medium. Ann. Probab. **14**, 1206-1218.

Carlen, E.A. (1984): Conservative diffusions. Comm. Math. Phys. **94**, 293-315.

Carmona, R. (1979): Processus de diffusion gouverné par la form de Dirichlet de l'opérateur de Schrödinger. Sém. de Probabilité XIII. Lect. Notes Math. **721**, 557-569, Springer-Verlag, Berlin Heidelberg.

Carmona, R. (1985): Probabilistic construction of Nelson processes. Taniguchi Symp. PMMP Katata, 55-81.

Chung, K.L., Walsh, J.B. (1969): To reverse a Markov process. Acta Math. **123**, 225-251.

Chung, K.L., Rao, K.M. (1981): Feynman-Kac formula and Schrödinger equation, Sem. Stoch. Proc. Ed. Cinlar, Chung, Getoor, 1-29, Birkhäuser. Boston-Basel-Stuttgart.

Chung, K.L., Williams, R.J. (1983): *Introduction to Stochastic Integration.* Birkhäuser, Boston-Basel-Stuttgart. 2nd ed. 1988.

Clark, J.M.C. (1970): The representation of functionals of Brownian motion by stochastic integrals. Ann. Math. Stat. **41**, 1281-1295.

Csiszar, I. (1975): I-divergence geometry of probability distribution and minimization problems. The Ann. Probability **3**, 146-158.

Csiszar, I. (1984): Sanov property, generalized I-projection and a conditional limit theorem. The Annals of Probability **12**, 768-793.

Dankel, Th. G. (1971): Mechanics on manifolds and the incorporation of spin into Nelson's stochastic mechanics. Arch. Rat. Mech. Anal. **37**, 192-221.

Dawson, D.A. (1983): Critical dynamics and fluctuation for a mean-field model of cooperative behaviour. J. Stat. Phys. **31**, 29-85.

Dawson, D.A., Gärtner, J. (1984): Long-time fluctuations of weakly interacting diffusions. Technical Report Series of the Laboratory for Research in Statistics and Probability, Carleton University.

Dawson, D.A., Gärtner, J. (1989): Large deviations, free energy functional and quasi-potential for a mean field model of interacting diffusions. Memoirs of the AMS Vol. 78, Number 398.

Dawson, D., Gorostiza, L., Wakolbinger, A. (1990): Schrödinger processes and large deviations. J. Math. Phys. **31**, 2385-2388.

Dellacherie, C., Meyer, P.A. (1978): *Probabilities and potential*. North-Holland Publ. Co.

Deuschel, J.-D., Stroock, D.W. (1989): *Large Deviations*. Academic Press, INC. Boston San Diego New York Berkeley London Sydney Tokyo.

Dirac, P.A.M. (1930, 58): *The principle of Quantum Mechanics*, 1st, 4th ed., Oxford, New York

Dohrn, D., Guerra, F. (1978): Nelson's stochastic mechanics on Riemanian manifolds. Lett. al Nuovo Cimento, **22**, 121-127.

Donsker, M.D., Varadhan, S.R.S. (1975): Asymptotic evaluation of certain Wiener integrals for large time. In Funct. Integ. its Appls. (Ed. A.M. Arthurs), 15-33. Clarendon Press.

Doob, J.L. (1953): *Stochastic Processes*, John Wiley & Sons, Inc., New York.

Doob, J.L. (1984): *Classical Potential Theory and Its Probabilistic Counterpart*. Springer-Verlag, New York Berlin Heidelberg Tokyo.

Dorling, J. (1987): Schrödinger's original interpretation of the Schrödinger equation: rescue attempt. 16-40 in "*Schrödinger, centenary celebration of a polymath*" Ed. Kilmister, C. W., Cambridge University press Cambridge.

Doss, H. (1980): Sur une résolution stochastique de l'équation de Schrödinger à coefficients analytiques. Comm. Math. Phys. **73**, 247-264.

Douglas, R.G. (1964): On extremal measures and subspace density. Michigan Math. J. **11**, 243-246.

Dürr, D., Goldstein, S., Zanghi, N. (1992): Quantum equilibrium and origin of absolute uncertainty. J. Stat. Phys. **67**, 843-905.

Dynkin, E.B. (1965): *Markov Processes*, Vol. I, II. Springer-Verlag, Berlin Göttingen Heidelberg.

Dynkin, E.B. (1985): An application of flows to time shift and time reversal in stochastic processes. Trans. Amer. Math. Soc. **287**, 613-619.

Eddington, A.S. (1928): Gifford lecture. Cf. Schrödinger (1932)

Einstein, A. (1905, a): Über einen die Erzeugung und Verwandlung des Lichtes betreffenden heuristischen Gesichtspunkt. Annalen der Physik, **17**, 132-148.

Einstein, A. (1905, b): Über die von der molekularkinematischen Theorie der Wärme geforderte Bewegung von ruhenden Flüssigieiten suspendierten Teilchen. Annalen der Physik, **17**, 549-560.

Einstein, A. (1905, c): Zur Electrodynamik bewegter Körper. Annalen der Physik, **17**, 891-921.

Einstein, A. (1906): Zur Theorie der Lichterzeugung und Lichtabsorption. Annalen der Physik, **20**, 199-206.

Einstein, A., Podolsky, B., Rosen, N. (1935): Can quantum-mechanical description of physical reality be considered complete ? Physical Rev. **47**, 777-780.

Ezawa, H., Klauder, J.R., Shepp, L.A. (1975): Vestigial effects of singular potentials in diffusion theory and quantum mechanics. J. Math. Phys. **16**, 783-799.

Faris, W. Simon, B. (1975): Degenerate and nondegenerate ground states for Schrödinger operators. Duke Math. J. **42**, 559-567.

Feller, W. (1954): Diffusion processes in one dimension. Trans. Am. Math. Soc. **77**, 1-31.

Feller, W. (1957): Generalized second order differential operators and their lateral conditions. Illinoi J. Math. **1**, 495-504.

Feller, W. (1968): *An introduction to probability theory and its applications* Vol. 1, Third Edition. John Wiley, New York.

Fényes, I. (1952): Eine wahrscheinlichkeitstheoretische Begründung und Interpretation der Quantenmechanik. Zeitschrift der Physik 132, 81-106.

Feynman, R.P. (1948): Space-time approach to non-relativistic quantum mechanics. Reviews of Modern Phys. **22**, 367-387.

Föllmer, H. (1985): An entropy approach to the time reversal of diffusion processes. Proc. 4th IFIP workshop on stochastic differential equations Eds. M. Métivier & E. Pardoux, Lecture Notes in Control and Information, Springer-Verlag.

Föllmer, H. (1986): Time reversal on Wiener space. Bibos-Symposium "Stochastic Processes in Math. Phys." Springer Lecture Notes Math. **1158**, 119-129.

Föllmer, H. (1988): Random fields and diffusion processes. Ecole d'été de Saint Flour XV-XVII (1985-87), Springer Lect. Notes in Math. **1362**.

Föllmer, H., Wakolbinger, A. (1986): Time reversal of infinite dimensional diffusions. Stoch. Processes Appl., **22**, 59-77.

Fortet, R. (1940): Résolution d'un système d'équation de M. Schrödinger. J. Math. Pures Appl. IX, 83-95.

Friedman, A. (1964): *Partial differential equations of parabolic type.* Prentice-Hall, Inc., Englewood Cliffs.

Friedman, A. (1975/76): *Stochastic Differential Equations and Applications*, Vol. 1, 2. Academic Press, New York San Francisco London.

Fujita, H. (1966): On the blowing up of solutions of Cauchy problem for $u_t = \Delta u + u^{1+\alpha}$. J. Fac. Sci. Univ. Tokyo **13**, 109-124.

Fukushima, M. (1980): *Dirichlet forms and Markov processes.* Kodamsha LTD, Tokyo, North-Holland Publ. Co. Amsterdam, Oxford, NewYork.

Fukushima, M., Takeda, M. (1984): A transformation of symmetric Markov processes and Donsker-Varadhan theory. Osaka J. Math. 21, 311-326.

Getoor, R.K., Glover, J. (1984): Riesz decomposition in Markov process theory. Trans. Amer. Math. Soc. **285**, 107-132.

Getoor, R.K., Sharpe, M.J. (1984): Naturality, standardness, and weak duality for Markov processes. Z. Wahrsch. Verw. Gebiete **67**, 1-62.

Gihman, I.I., Skorohod, A.V. (1969): *Introduction to the Theory of Random Processes.* W.B. Saunders Co., Philadelphia London Tronto.

Gihman, I.I., Skorohod, A.V. (1974, 75): *The Theory of Stochastic Processes* I, II. Springer-Verlag, Berlin Heidelberg New York.

Girsanov, I.V. (1960): On transforming a certain class of stochastic processes by absolutely continuous substitution of measures. Theor. Probab. Appl. **5**, 285-301.

Glimm, J., Jaffe, A. (1987): *Quantum Physics, A functional Integral point of view.* Springer-Verlag, New York Berlin Heidelberg London Tokyo.

Grad, H. (1961): The many faces of entropy. Comm. pure and appl. Math. **14**, 323-354.

Guerra, F., Morato, L.M. (1983): Quantization of dynamical system and stochastic control theory. Phys. Rev. D, 1774-1786.

Guerra, F., Pavon, M. (1988): Stochastic variational principles and free energy for dissipative processes. In *Analysis and Control of nonlinear Systems*, C.I. Byrnes, C.F.Martin, & R.E.Seaks Eds. Elsevier Science Publ. B.V., North Holland.

Harris, T.E. (1965): Diffusion with "collisions" between particles. J. Appl. Probab. 2, 323-338.

Hasiminsky, R. Z. (1959): On positive solutions of the equation Au+Vu = 0. Theory Probab. its Appl. 4, 309-318 (English translation).

Heisenberg, W. (1925): Über quanten-theoretische Umdeutung kinetischer und mechanischer Beziehungen. Zs. f. Phys. 33, 879-893.

Heisenberg, W. (1955) : *Das Naturbild der Heutigen Physik*. Rowohlt Taschenbuch Verlag, GmbH, Hamburg.

Hunt, G.A., (1957, 58): Markov processes and potentials I, II, & III. Ill. J. Math. 1, 44-93, 316-369; 2, 151-213.

Hunt, G.A., (1960): Markov chains and Martin boundaries. Ill. J. Math. 4, 316-340.

Ikeda, N., Nagasawa, M., Sato, K. (1964): A time reversal of Markov Processes with killing. Kodai Mth. Sem, Rep, 16, 88-97.

Ikeda, N., Nagasawa, M., Watanabe, S. (1968,69): Branching Markov Processes I, II, III. Journal of Math. Kyoto Univ. 8, 233-278, 365-410; 9, 95-160.

Ikeda, N., Watanabe, S. (1981, 89): *Stochastic differential equations and diffusion processes*. Kodansha Ltd, North-Holland Publ. Co.

Itô, K. (1961): *Lectures on Stochastic Processes*. Tata Institute, Bombay.

Itô, K. (1961): Wiener integrals and Feynman integral. Proc. Fourth Berkeley Symp. om Math. Stat. and Prob. 2, 227-238. Univ. California Press, Berkeley.

Itô, K. (1967): Generalized uniform complex measure in the Hilbertian metric space with their application to the Feynman path integral. Proc. Fifth Berkeley Symp. on Math. Stat. and Prob. II, 145-146. Univ. California Press, Berkeley.

Itô, K., McKean, H.P. (1965): *Diffusion Processes and their sample paths*, Springer-Verlag.

Itô, K., Nisio, M. (1968): On the convergence of sums of independent Banach space valued random variables, Osaka J. Math. 5, 35-48.

Ito, S. (1957): Fundamental solutions of parabolic differential equations and boundary value problems. Japanese Journal of Math. 27, 55-102.

Jamison, B. (1974,a): Reciprocal processes. Z. Wahrsch. Verw. Geb. **30**, 65-86.

Jamison, B. (1974,b): A Martin boundary interpretation of the maximum entropy argument. Z. Warhr. Verw. Geb. **30**, 265-272.

Jamison, B. (1975): The Markov processes of Schrödinger. Z. Wahrsch. Verw. Geb. **32**, 323-331.

Jammer, M. (1974): *The philosophy of quantum mechanics. The interpretations of quantum mechanics in historical perspective.* John Wiley & Sons, Inc. New York.

Kac, M. (1949): On distributions of certain Wiener functionals, Trans. Amer. Math. Soc., **65**, 1-13.

Kac, M. (1951): On some connections between probability theory and differential and integral equations. Proc. Second Berkeley Symp. on Math. Stat. Probab. 189-215. Univ. Calif. Press.

Kac, M. (1980): *Integration in Function Spaces and Some of its Applications.* Pisa.

Karlin, S. Taylor, H.M. (1981): *Second course in Stochastic Processes.* Academic Press.

Kawazu, K., Tamura, Y., Tanaka, H. (1992): Localization of diffusion processes in one-dimensional random environment. J. Math. Soc. Japan **44**, 515-550.

Kemperman, J.H.B. (1967): On the optimum rate of transmitting information. Springer Lec. Notes in Math. **89**, 126-169.

Kilmister, C. W. (1987): *Schrödinger, Centenary celebration of a polymath.* Cambridge Univ. Press, Cambridge, New York, New Rochelle, Melbourne, Sydney.

Kolmogoroff, A. (1931): Analytischen Methoden in Wahrscheinlichkeitsrechnung. Math. Ann. **104**, 415-458.

Kolmogoroff, A. (1933): *Grundbegriffe der Wahrscheinlichkeitsrechnung.* Ergbn. d. Math. 2, Heft 3. Springer-Verlag.

Kolmogoroff, A. (1936): Zur Theorie der Markoffschen Ketten. Math. Ann. **112**, 155-160.

Kolmogoroff, A. (1937): Zur Umkehrbarkeit der statistischen Naturgesetze. Math. Ann. **113**, 766-772.

Kullback, S. (1959): *Information Theory and Statistics*. John Wiley & Sons, Inc. New York.

Kunita, H., Watanabe, T. (1966): On certain reversed processes and their applications to potential theory and boundary theory. J. Math. Mech. **15**, 398-434

Kusuoka, S., Tamura, Y. (1984): Gibbs measures for mean field potentials. J. Fac. Sci. Univ. Tokyo Sect. IA Math. **31**, 223-245.

Kuznezov, S.E. (1973): Construction of Markov processes with random birth and death times. Theor. Veryat. i Prim. **18**, 596-601. (English transl. Theory Probab. Appl.)

Lanford, O.E. (1973): Entropy and equilibrium states in classical statistical mechanics. In "*Statistical Mechanics and Mathematical problems*". Ed. by A. Lenard. Lect. Notes in Physics **20**, 1-113. Springer-Verlag.

Lindenstrauss, J. (1965): A remark on extreme doubly stochastic measure. Amer. Math. Monthly **72**, 379-382.

Lions, P.L., Sznitman, A.S. (1984): Stochastic differential equations with reflecting boundary conditions. Comm. Pure Appl. Math. **37**, 511-537.

Liptser, R.S., Shiryayev, A.N. (1977): *Statistics of random processes* I. General theory. Springer-Verlag, Berlin Göttingen Heidelberg.

Lorentz, H.A. (1915): *The theory of Electrons*. Second Ed. (Dover Publ. Inc. 1952)

Mackey, G.W. (1968): *Induced representations of groups and quantum mechanics*. W.A. Benjamin, INC. New York Amsterdam.

Mackey, G.W. (1978): *Unitary Group Representations in Physics, Probability, and Number theory*. The Benjamin / Cummings Pub. Co. Inc. London Amsterdam.

Maruyama, G. (1954): On the transition probability functions of the Markov processes. Nat. Sci. Rep. Ochanomizu Univ. **5**, 10-20.

McKean, H.P. (1960): The Bessel motion and a singular integral equation. Memoires of Coll. of Sci., Univ. Kyoto, Ser. A, **33**, 317-322.

McKean, H.P. (1966): A class of Markov processes associated with non-linear parabolic equations. Proc. Natl. Acad. Sci. **56**, 1907-1911.

McKean, H.P. (1967): Propagation of chaos for a class of nonlinear parabolic equations. Lecture Series in Differential Equations, Catholic Univ., 41-57.

McKean, H.P. (1969): *Stochastic Integrals*. Academic Press, New York and London

Meyer, P.A. (1962): Fonctionnelles multiplicatives et additives de Markov. Ann. Inst. Fourier, **12**, 125-230.

Meyer, P.A. (1963): La propriété de Markov forte des fonctionnelles multiplicatives. Theory of Prob. its Appl. **8**, 328-334.

Meyer, P.A. (1971): Le retournement du temps, d'après Chung et Walsh. Springer Lect. Notes in Math. **191**, 213-245.

Meyer, P.A. (1975): Renaissance, Recollement, Mélanges, Ralentissement de processus de markov. Ann. Inst. Fourier, Grenoble, **25**, 465-497.

Meyer, P.A. (1980/81): Géométrie différentielle stochastique (bis), Sem. de Probab. XVI, Springer Lect. Notes in Math. **921**, 165-207.

Meyer, P.A. (1992): *Quantum Probability* for Probabilist. Publ. l'Institut de Recherche Mathématique Avancé, Univ. Strasbourg et C.N.R.S..

Meyer, P.A., Zheng, W.A. (1983/84): Construction de processus de Nelson reversibles. Springer Lect. Notes in Math. **1123**, 12-26.

Mitro, J.B. (1979): Dual Markov processes: Construction of a useful auxiliary processes. Z. Wahrscheinlich. verw. Gebiete, **47**, 139-156.

Nagasawa, M. (1961): The adjoint process of a diffusion process with reflecting barrier. Kodai Math. Sem. Rep. **13**, 235-248.

Nagasawa, M. (1964): Time reversal of Markov processes. Nagoya Math. Jour. **24**, 177-204.

Nagasawa, M. (1968): Construction of branching Markov processes with age and sign. Kodai Math. Sem. Rep. 20, 469-508.

Nagasawa, M. (1969): Markov processes with creation and annihilation. Z. Wahrsch. Verw. Geb. **14**, 49-60.

Nagasawa, M. (1970/71): Lecture Notes on Markov Processes. Aarhus University (mimeographed).

Nagasawa, M. (1972): Branching property of Markov processes. Semi. Probab. Strasbourg VI, Springer-Verlag, 177-197.

Nagasawa, M. (1977): Basic Models of Branching Processes. Bull. Int. Stat. Inst. XLVII (2), 423-445.

Nagasawa, M. (1980): Segregation of a population in an environment. J. Math. Biol. **9**, 213-235.

Nagasawa, M. (1981): An application of segregation model for septation of *Escherichia Coli*. J. Theoret. Biol. **90**, 445-455.

Nagasawa, M. (1985): Macroscopic, intermediate, microscopic and mesons. Lect. Notes Phys. **262**, Springer, 427-437.

Nagasawa, M. (1988): A statistical model of segregation of a population. Stochastic modelling in Biology. ed. P. Tautu, World Scientific.

Nagasawa, M. (1989,a): Transformations of diffusion and Schrödinger processes. Probab. Th. Rel. Fields, **82**, 109-136.

Nagasawa, M. (1989,b): Stochastic variational principle of Schrödinger processes. Seminar on stochastic processes. Ed. Cinlar, Chung, Getoor. Birkhäuser.

Nagasawa, M. (1989,c): Remarks to "Stochastic variational principle of Schrödinger processes" (preprint, unpublished).

Nagasawa, M. (1990): Can the Schrödinger equation be a Boltzmann equation ? Evanston 1989, In *"Diffusion processes and Related problems in Analysis"*. Ed. by M. Pinsky, Birkhäuser.

Nagasawa, M. (1991): The equivalence of diffusion and Schrödinger equations: A solution to Schrödinger's conjecture. Proceedings of Locarno Conference 1991, Ed. S. Albeverio, World Scientific, Singapore.

Nagasawa, M. (1992): A Principle of superposition and interference of diffusion processes. Third European Symposium on Analysis and Probability, Jan. 1992 at the Institute Henri Poincaré, to appear in Sém. de Probabilités.

Nagasawa, M., Sato, K. (1963): Some theorems on time change and killing of Markov processes. Kodai Math. Sem. Rep. **15**, 195-219.

Nagasawa, M., Sirao, T. (1969): Probabilistic treatment of the blowing up of solutions for a nonlinear integral equation. Trans. Amer. Math. Soc. **139**, 301-310.

Nagasawa, M., Maruyama, T. (1979): An application of time reversal of Markov processes to a problem of population genetics. Advances in Appl. Probab. **11**, 457-478.

Nagasawa, M., Barth, Th., Wakolbinger, A. (1981): Mathematisches Modellieren: Erregungszustände von Affen am Futterplatz oder von Molekülen in einer Zelle, Univ. Zürich Mitteilungsblatt des Rektorats, 7-9, (Bibos Conferance, Univ. Bielefeld).

306   References

Nagasawa, M., Yasue, K. (1982): A statistical model of mesons. Publ. de l'Inst. Rech. Math. Ava. (CNRS) **33**,1-48, Univ. Strasbourg.

Nagasawa, M., Tanaka, H. (1985): A diffusion process in a singular meandrift-field. Z. Wahrsch. verw. Gebiete **68**, 247-269.

Nagasawa, M., Tanaka, H. (1986): Propagation of chaos for diffusing particles of two types with singular mean field interaction. Probab. Th. Rel. Fields, **71**, 69-83.

Nagasawa, M., Tanaka, H. (1987,a): Diffusion with interactions and collisions between coloured particles and the propagation of chaos. Probab. Th. Rel. Fields, **74**, 161-198.

Nagasawa, M., Tanaka, H. (1987,b): A proof of the propagation of chaos for diffusion processes with drift coefficients not of average form. Tokyo J. Math. **10**, 403-418.

Nelson, E. (1958): The adjoint Markov process. Duke Math. J. **25**, 671-690.

Nelson, E. (1966): Derivation of Schrödinger equation from Newtonian mechanics. Phys. Rev. **150**, 1076-1085.

Nelson, E. (1967): *Dynamical Theories of Brownian Motion*. Princeton University Press.

Nelson, E. (1985): Critical diffusion. Sem. de Probab. XIX. Springer Lect. Notes in Math. **1123**, 1-11.

Neveu, J. (1965): *Mathematical Foundations of the Calculus of Probability*. Holden-Day, Inc., San Fransisco, London, Amsterdam.

Norris, J.R. (1988): Constructions of diffusions with a given density. In: *Stochastic Calculus in Application*, (ed. Norris, J.R). 69-77. Longman.

Ölschläger, K. (1984): A martingale approach to the law of large numbers for weakly interacting stochastic processes. Ann. Probab. **12**, 458-479.

Ölschläger, K. (1985): A law of large numbers for moderately interacting diffusion processes. Z. Wahrsch. Verw. Geb. **69**, 279-322.

Ölschläger, K. (1989): Many-particle systems and the continuum description of their dynamics. Univ. Heidelberg.

Oshima, Y. (1992): Some properties of Markov processes associated with time dependent Dirichlet forms. Osaka J. Math. **29**, 103-127.

Parthasarathy, K.R. (1967): *Probability Measure on Metric Spaces*. Academic Press, New York and London.

Parthasarathy, K.R. (1992): *An Introduction to Quantum Stochastic Calculus*. Birkhäuser Verlag, Basel Boston Berlin.

Pitman, J., Yor, M. (1982): A decomposition of Bessel bridges. Z. Wahr. verw. Gebiete 59, 425-457.

Planck, M. (1900): Über eine Verbesserung der Wien'schen Spectralgleichung. Verh. der D. Phys. Ges. **2**, 202-204

Planck, M. (1900): Zur Theorie des Gesetzes der Energieverteilung im Normalspectrum. Verh. der D. Phys. Ges. **2**, 237-245.

Planck, M. (1901): Über das Gesetz der Energieverteilung im Normalspectrum. Annalen. der Phys. **4**, 553-563.

Port, S.C., Stone, C.J. (1978): *Brownian Motion and Classical Potential Theory*. Academic Press, New York.

Roelly, S. Zessin, H. (preprint): The equivalence of equilibrium principles in statistical mechanics and some applications to large particle systems.

Rüschendorf, L., Thomas, W. (preprint): Note on the Schrödinger equation and *I*-projection.

Schrödinger, E. (1922): Was ist ein Naturgesetz ? (Antrittsrede an der Universität Zürich, 9. Dezember 1922), Sonderdruck aus Die Naturwissenschaften **17**, 1929 9-11, Springer-Verlag.

Schrödinger, E. (1926, I): Quantisierung als Eigenwertproblem (1. Mitteilung). Ann. der Physik, **79**, 336-376.

Schrödinger, E. (1926): Über das Verhältnis der Heisenberg-Born-Jordanschen Quantenmechanik zu der meinen. Ann. der Physik, **79**, 734-756.

Schrödinger, E. (1926, IV): Quantisierung als Eigenwertproblem (4. Mitteilung). Ann. der Physik, **81**, 109-139.

Schrödinger, E. (1930): Zum Heisenbergschen Unschärfeprinzip. Sitzungsberichte der preussischen Akad. der Wissenschaften Physicalisch Mathematische Klasse, 296-303.

Schrödinger, E. (1931): Über die Umkehrung der Naturgesetze. Sitzungsberichte der preussischen Akad. der Wissenschaften Physicalisch Mathematische Klasse, 144-153.

Schrödinger, E. (1932): Sur la théorie relativiste de l'électron et l'interprétation de la mécanique quantique. Ann. Inst. H. Poincaré 2, 269-310.

Schrödinger, E. (1935): Die gegenwärtige Situation in der Quantenmechanik. Die Naturwissenschaften **23**, 807-812, 824-828, 844-849.

Schwinger, J. (1958): *Quantum Electrodynamics* (Ed). Dover Publ., Inc. New York.

Sharpe, M. (1988): *General Theory of Markov Processes*. Academic Press, Inc., Boston San Diego New York.

Shiga, T., Tanaka, H. (1985): Central limit theorem for a system of Markovian particles with mean field interaction. Z. Wahr. verw. Gebiete **69**, 439-459.

Sirao, T. (1968): On signed branching Markov processes with age. Nagoya Math. J. **32,** 155-225.

Skorokhod, A.V. (1961): Stochastic equations for diffusion processes in a bounded region. Th. Probab. Its Appl. **VI**, 264-274.

Spohn, H. (1991): Fixed point of a functional renormalization group for critical wetting. Europhys. Lett., **14**, 689-692.

Stroock, D.W., Varadhan, S.R.S. (1970): *Multidimensional Diffusion Processes*. Springer-Verlag, Berlin Göttingen Heidelberg.

Stummer, W. (1990): The Novikov and Entropy conditions of diffusion processes with singular drift. Doctoral Dissertation at the University of Zürich.

Sturm, Th. (1989): Störung von Hunt-Prozessen durch signierte additive Functionale. Doctoral Dissertation at the University of Erlangen.

Sturm, Th. (1992): Schrödinger operators with arbitrary nonnegative potentials. Operator Theory: Advance and Applications, **57**, 291-306.

Sturm, Th. (preprint): Schrödinger operators with highly singular, oscillating potentials.

Sznitman, A.S. (1984): Non-linear reflecting diffusion processes and propagation of chaos and fluctuations associated. J. Funct. Anal. **56**, 311-336.

Sznitman, A.S. (1984): A propagation of chaos result for Burgers' equation. Probab. Th. Rel. Fields **71**, 581-613.

Sznitman, A.S. (1989): Topics in propagation of chaos. Ecole d'été de probabilités de Saint Flour. Lect. Notes Math. **1464**, 165-251.

Tanaka, H. (1979): Stochastic differential equations with reflecting boundary condition in convex regions. Hiroshima Math. J. **9**, 163-177.

Tanaka, H. (1984): Limit theorems for certain diffusion processes with interaction. In *Stochastic Analysis*, Ed. Itô, K, 469-488, Kinokuniya, North-Holland.

Tanaka, H. (1987): Limit distributions for one-dimensional diffusion processes in self-similar random environments. IMA hydrodynamic behavior and interacting particle systems, Vol 9, 189-210, ed. by G. Papanicolaou, Springer-Verlag.

Takesaki, M. (1979): *Theory of operator algebras* I. Springer-Verlag, New York, Heidelberg, Berlin

Titchmarsh, E.C., (1962): *Eigenfunction expansion associated with second order differential equations*, Part 1, 2nd ed. Oxford; Clarendon Press.

Tomonaga, S. (1946): On a relativistically invariant formulation of the quantum theory of wave fields. Progress of Theoretical Physics **1**, 27-39.

Umegaki, H. (1954): Conditional expectation in an operator algebra, I. Tôhoku Math. J. **6**, 177-181.

Van der Waerden, B.L. (1967): *Sources of Quantum Mechanics.* Dover Publ. Inc., New York

Varadhan, S.R.S. (1984): *Large deviations and applications.* Soc. Ind. Appl. Math. Philadelphia.

Von Neumann, J. (1932): *Die Mathematische Grundlagen der Quantenmechanik.* Springer-Verlag.

Wakolbinger, A., (1989): A simplified variational characterization of Schrödinger processes. J. M. Phys. **27**, 2943-2946.

Wakolbinger, A. (preprint): Schrödinger bridges from 1931 to 1991.

Wakolbinger, A., Stummer, W. (preprint): Remarks on stochastic Newtonian equations, Schrödinger processes, and gauge invariance.

Williams, D. (1979): *Diffusions, Markov processes, and Martingales.* John Wiley & Sons.

Yasue, K. (1981): Stochastic calculus of variations. J. Funct. Anal. **41**, 327-340.

Yasue, K. (1986): The least action principle in quantum theory. Soryusiron Kenkyu (Japanese).

Yosida, K. (1965): *Functional Analysis*. Springer-Verlag, Berlin Göttingen Heidelberg.

Yukawa, H. (1935): On the interaction of elementary particles, I. Proc. Phys.-Math. Soc. Japan **17**, 48-57.

Zhao, Z. (1986): Schrödinger conditional Brownian motion and stochastic calculus of variations. Stochastics **18**, 1-15.

Zambrini, J.C. (1985): Variational processes. Lect. Notes Phys. **262**, 517-529. Springer.

Zambrini, J.C. (1986,a): Stochastic mechanics according to Schrödinger. Phys. Rev. A. **33**, 1532-1548.

Zambrini, J.C. (1986,b): Variational processes and stochastic versions of mechanics. J. Math. Phys. **27**, 2307-2330.

Zambrini, J.C. (1987): Euclidean quantum mechanics. Phys. Rev., A **35**, 3631-3649.

Zheng, W.A., Meyer, P.A. (1984/85): Sur la construction de certaines diffusions. Sem. de Probab. XX. Springer Lect. Notes in Math. **1294**, 334-337.

Zheng, W.A., (1985): Tightness results for laws of diffusion processes. Application to stochastic mechanics. Ann. Inst. H. Poincaré **B21**. 103-124.

Index

The foundations of this outstanding book series were laid in 1944. Until the end of the 1970s, a total of 77 volumes appeared, including works of such distinguished mathematicians as Carathéodory, Nevanlinna, and Shafarevich, to name a few. The series came to its present name and appearance in the 1980s. According to its well-established tradition, only monographs of excellent quality will be published in this collection. Comprehensive, in-depth treatments of areas of current interest are presented to a readership ranging from graduate students to professional mathematicians. Concrete examples and applications both within and beyond the immediate domain of mathematics illustrate the import and consequences of the theory under discussion.

We encourage preparation of manuscripts in TeX for delivery in camera-ready copy which leads to rapid publication, or in electronic form for interfacing with laser printers or typesetters. Proposals should be sent directly to the editors or to: Birkhäuser Verlag, P.O. Box 133, CH-4010 Basel, Switzerland.

Published in the series Monographs in Mathematics since 1983

Volume 78: **Hans Triebel, Theory of Function Spaces.**
1983, 284 pages, hardcover, ISBN 3-7643-1381-1.

Volume 79: **Gennadi M. Henkin/Jürgen Leiterer, Theory of Functions on Complex Manifolds.**
1984, 228 pages, hardcover, ISBN 3-7643-1477-X.

Volume 80: **Enrico Giusti, Minimal Surfaces and Functions of Bounded Variation.** 1984, 240 pages, hardcover, ISBN 3-7643-3153-4.

Volume 81: **Robert J. Zimmer, Ergodic Theory.**
1984, 210 pages, hardcover, ISBN 3-7643-3184-4.

Volume 82: **V. I. Arnold/S. M. Gusein-Zade/A. N. Varchenko, Singularities of Differentiable Maps - Volume I.**
1985, 392 pages, hardcover, ISBN 3-7643-3187-9.

Volume 83: **V. I. Arnold/S. M. Gusein-Zade/A. N. Varchenko, Singularities of Differentiable Maps - Volume II.**
1988, 500 pages, hardcover, ISBN 3-7643-3185-2.

Volume 84: **Hans Triebel, Theory of Function Spaces II.**
1992, 380 pages, hardcover, ISBN 3-7643-2639-5.

Volume 85: **K.R. Parthasarathy, An Introduction to Quantum Stochastic Calculus.**
1992, 300 pages, hardcover, ISBN 3-7643-2697-2.

Volume 86: **Masao Nagasawa, Schrödinger Equations and Diffusion Theory.**
1993, 332 pages, hardcover, ISBN 3-7643-2875-4.

Volume 87: **Jan Prüss, Evolutionary Integral Equations and Applications.**
1993, 392 pages, hardcover, ISBN 3-7643-2876-2.

Birkhäuser Advanced Texts - Basler Lehrbücher

Managing Editors:

H. Amann (Universität Zürich)
H. Kraft (Universität Basel)

This series presents, at an advanced level, introductions to some of the fields of current interest in mathematics. Starting with basic concepts, fundamental results and techniques are covered, and important applications and new developments discussed. The textbooks are suitable as an introduction for students and non-specialists, and they can also be used as background material for advanced courses and seminars.

We encourage preparation of manuscripts in TeX for delivery in camera-ready copy which leads to rapid publication, or in electronic form for interfacing with laser printers or typesetters. Proposals should be sent directly to the editors or to: Birkhäuser Verlag, P.O. Box 133, CH-4010 Basel, Switzerland.

Published in the series **Birkhäuser Advanced Texts - Basler Lehrbücher:**

Lectures in Mathematics - ETH Zürich

*Each year the Eidgenössische Technische Hochschule (ETH) at Zürich invites a selected group of mathematicians to give postgraduate seminars in various areas of pure and applied mathematics. These seminars are directed to an audience of many levels and backgrounds. Now some of the most successful lectures are being published for a wider audience through the **Lectures in Mathematics - ETH Zürich** series. Lively and informal in style, moderate in size and price, these books will appeal to professionals and students alike, bringing a quick understanding of some important areas of current research.*

Previously published:

Randall J. LeVeque, Numerical Methods for Conservation Laws.
Second edition 1992, 214 pages, softcover, ISBN 3-7643-2723-5.

J. Donald Monk, Cardinal Functions on Boolean Algebras.
1990, 152 pages, softcover, ISBN 3-7643-2495-3.

Carl de Boor, Splinefunktionen (german).
1990, 184 pages, softcover, ISBN 3-7643-2514-3.

Daniel Bättig/Horst Knörrer, Singularitäten (german).
1991, 140 pages, softcover, ISBN 3-7643-2616-6.

Anthony J. Tromba, Teichmüller Theory in Riemannian Geometry.
1992, 224 pages, softcover, ISBN 3-7643-2735-9.

Raghavan Narasimhan, Compact Riemann Surfaces.
1992, 128 pages, softcover, ISBN 3-7643-2742-1.

Marc Yor, Some Aspects of Brownian Motion. Part I: Some Special Functionals.
1992, 144 pages, softcover, ISBN 3-7643-2807-X.

Olavi Nevanlinna, Convergence of Iterations for Linear Equations.
1993, 184 pages, softcover, ISBN 3-7643-2865-7.